S0-AAJ-133

GENETICS

A Human Concern

H. Eldon Sutton

Robert P. Wagner

The Genetics Institute and
Department of Zoology
The University of Texas at Austin

GENETICS

A Human Concern

MACMILLAN PUBLISHING COMPANY

NEW YORK

Collier Macmillan Publishers

London

To our silent majority
Beverly and Peggy

Copyright © 1985, Macmillan Publishing Company, a division of Macmillan, Inc.
Printed in the United States of America

All rights reserved. No part of this book may be reproduced or
transmitted in any form or by any means, electronic or mechanical,
including photocopying, recording, or any information storage and
retrieval system, without permission in writing from the Publisher.

Macmillan Publishing Company
866 Third Avenue, New York, New York 10022

Collier Macmillan Canada, Inc.

Library of Congress Cataloging in Publication Data

Sutton, H. Eldon (Harry Eldon)
   Genetics, a human concern.

   Includes bibliographical references and index.
   1. Genetics. 2. Genetics—Social aspects.
I. Wagner, Robert P.   II. Title.
QH430.S93   1985   575.1   83–16282
ISBN 0–02–418320–2

Printing: 1 2 3 4 5 6 7 8   Year: 5 6 7 8 9 0 1 2

ISBN 0-02-418320-2

This book is addressed to those who want to learn genetics of immediate importance to humans, individually and in society. This includes not only the genetics of humans but also the genetics of other organisms that we depend on for food, shelter, medicines, and the innumerable other items we use every day.

This textbook is meant for college and university students who are not pursuing a major in one of the sciences but who desire to learn enough about genetics to be able to make judgments, or at least have informed opinions, about many of the societal issues that have a major element of genetic biology associated with them.

Human society is currently being confronted with problems arising from genetic knowledge and the application of this knowledge to human reproduction and its consequences. Such innovations as artificial fertilization, genetic counseling, prenatal diagnosis and medical treatment, and extreme measures to save the lives of defective infants raise all sorts of ethical and religious questions. Additionally, the problems raised by sterilization of adults, genetic engineering, and induced abortion again rouse strong feelings.

All of these intrusions into the biological functioning of humans have genetic factors not just strongly involved but at the heart of the matter. The problems generated need the attention not only of ethicists and theologians but of lawyers as well. Many cases taken to court to be adjudicated require the participants to have genetic knowledge to be able to arrive at a fair settlement. Furthermore, legislative bodies are more and more inclined to pass laws dealing with genetic problems. In the past, these laws have been concerned mainly with sterilization, but now they

Preface

deal with the whole range of genetic matters. As a consequence, it has become necessary for legislators to become knowledgeable enough about genetics to fashion intelligent laws and for the public to become knowledgeable enough to vote intelligently when these matters come to the ballot box. To adapt a theological phrase, "genetics is the ground of our being;" we can no longer ignore the importance of genetics, as has been done frequently in the past.

In preparing this book, we have assumed that the students using it will not have a strong background in biological science. Some may have minimal knowledge with no formal background of study; others may have had a good course of instruction in high school. To help those with some deficiency in the chemistry and cell biology necessary to understand the basics of present-day genetics, we have included background material in these subjects in Chapters 2 through 5. All students are urged to peruse the material in these chapters, those with deficiencies carefully, and those to whom the material is somewhat familiar as a review.

Most of the illustrations were researched by Philip C. Wagner, and original drawings made by him were used by Macmillan artists to prepare the final product. We are much indebted to PCW for the talent, time, and energy as well as creativity he has given to this endeavor. Many persons contributed photographs and helped us locate illustrations. In addition to those identified in the figure credits, Gerald P. Hodge of the University of Michigan was very helpful in providing information on the Habsburgs. We express our appreciation to all these upon whom we imposed and who responded so willingly.

We wish to thank those who have given (initially anonymously) of their time to read the manuscript in its various stages of preparation and offered their criticisms. The comments have been invaluable to us, but we have written about things as we see them, and all errors of commission are ours, as well as all opinions, though we have tried to keep opinions to a minimum. In a subject such as this, it is impossible to write without some bias, and whatever is written will always find some disagreement among those who are just as knowledgeable of the field as we think we are. We offer our thanks to Joseph Grossfield, City College of New York; Daniel Hartl, Washington University, St. Louis; Carl A. Huether, University of Cincinnati; Roger C. Johnson, Adelphi University; Roger Keller, University of Akron; Mary R. Murnik, Ferris State College; Leonard W. Storm, University of Nevada–Las Vegas; David Usher, University of Delaware; and William Wissinger, St. Bonaventure University. We also wish to acknowledge with thanks the help of our anthropologist friends, James Spuhler of the University of New Mexico, Albuquerque, and Claude Bramblett of the University of Texas at Austin, and our genetical-medical-legal friend, Margery W. Shaw of the University of Texas at Houston. They have given us confidence that our chapters on human evolution and variation and those on the social-medical-legal aspects of genetics are at least near the target.

No book happens without the very active involvement of editors. It has been a pleasure to work with Gregory W. Payne, senior editor at Macmillan, and Dora Rizzuto and Elizabeth Belfer, production supervisors.

Austin, TX   H. Eldon Sutton
Santa Fe, NM   Robert P. Wagner

1
The Paragon of Animals
2

2
The Genesis of Modern Genetics
8

Contents

3
The Chemical Origins and Basis of Life
24

4
The Structure and Function of Cells
54

5
Cell Cycles and the Continuity of Life
82

6
From Genotype to Phenotype
98

7
The Mechanics of Inheritance in Eukaryotes
120

8
Linkage and Recombination
144

12
Sex, Sex Determination, and Reproduction

214

13
Human Development and Gene Action

238

14
Bacteria, Viruses, Plasmids, and Genetic Engineering

254

15
Immunogenetics
270

16
Mutation and the Environment
290

17
Natural Selection
316

18
Behavioral Genetics
338

19
Human Evolution
356

20
Genetic Variations in Human Populations
394

21

The Breeding of Plants, Animals, and Microorganisms

412

22

Human Genetics Applied: Counseling, Screening, Engineering

442

23
Genetics, Eugenics, Public Policy, and Law

462

GENETICS

A Human Concern

This earth has several million species of plants and animals living on the land and in the sea. Among the million or so species of animals, one is constantly raising a ruckus, polluting the atmosphere and the oceans, destroying forests and grasslands, and generally creating a continuous ferment among the different members of its species. We need not go further in identifying that animal; you know who it is.

We humans are forever proclaiming the distinction between human and nonhuman life on this planet, and rightly so. We are unique. We are the only animal (of which we are aware) that considers itself detached from nature. What is more, we labor prodigiously to overcome and surpass nature. Since we consider ourselves apart from nature and have some control over its course, we feel great power and in thought and action proclaim ourselves paragons.

What a piece of work is a man! how noble in reason! how infinite in faculties! in form and moving how express and admirable! in action how like an angel! in apprehension how like a god! the beauty of the world! the paragon of animals!

Shakespeare (*Hamlet*, II, ii)

Our Biological Natures

But willy nilly we are animals, despite being paragons. We are subject to the same natural laws as all other animals and indeed all other forms of life. We have the same general constraints and limitations operating in and on our lives as on all other forms of life. What makes us unique is our capacity to cope with them. We can send people into space as far as the moon. No other form of life can do that without our help. Once beyond the dense layer of oxygen-containing gas that envelops the earth's surface, no form of life that we know could survive for long unaided, which means aided by us.

Our physical structure is only trivially different from many other animals. Basically we are constructed out of the same kinds of building blocks that are found in all other animals and higher plants. We belong to that group of organisms called *eukaryotes*. Eukaryotes are organisms made of one or more *nucleated cells*. For the present we can define a cell as a unit of life that can, by dividing, form more life. A eukaryotic cell contains a *nucleus* within which are *chromosomes* (Fig. 1.1). The chromosomes bear a specific kind of substance known as DNA (**d**eoxyribo**nu**-cleic **a**cid). The DNA contains the message (or the information) necessary to direct the activities of the cell or life itself and by its replication and distribution during cell division carry the information for the continuation of life into the next generation.

This flow of information via the chemical compound DNA from one cell generation to the next, and its expression in directing the activities of the cells in which it occurs, is the subject of the science of *genetics. The reason we humans are part of nature and not outside it is that our DNA places the same general constraints and limitations on us as on all other organisms.* The genetics of humans is essentially the same as the genetics of the other eukaryotes. It is a special subject,

1

The Paragon of Animals

FIGURE 1.1
A eukaryotic cell. All animals and plants consist of cells, or life units, that have nuclei containing chromosomes. The nucleus is separated from the remainder of the cell by a membrane. This cell, if it were a human egg, would have a diameter of 0.2 mm or about 0.0008 ins.

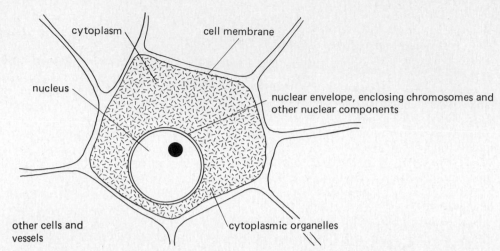

however, because it is a study of us, the paragons. Through it we learn a great deal about ourselves: how we function or why we sometimes dysfunction physiologically; why we sometimes behave as we do; what our limitations are individually and collectively; and most important what the imaginary limitations are that should be cast aside.

By means of an understanding of the fundamentals of genetics, we can view ourselves more objectively and rationally and avoid the emotional states that lead us to racism, sexism, hubris, and other obnoxious forms of "thinking" that seem to be special afflictions of paragons. The study of genetics is a humbling experience when undertaken by persons of reasonable intelligence.

Our Animal Forebears

Our relationship to two of the great apes, the gorilla and the chimpanzee, is so close as to make it certain that we are recently descended from the same ancestors (recent on the geological time scale). Ten to twenty million years ago an apelike creature existed whose descendants gave rise to the humanoid line and the contemporary ape lines. Some people find this difficult to accept. No good can come of ignoring the fact that we are somewhat hairless pseudo-apes that habitually walk on two legs instead of intermittently like the true apes.

So we are not only animals but we are reasonably certain about our position in the animal kingdom. We belong in the class Mammalia, order Primates, suborder Anthropoidea, superfamily Hominoidea, family Hominidae, according to our human view of how things should be classified. And the evidence to support the validity of this classification is considerable, as we shall make clear in succeeding chapters.

Our Uniqueness

Despite our similarity to other animals, especially the great apes, we are indeed very different in one respect. That is in the operation of our nervous system. Our brains, vocal cords, and hands enable us to communicate with each other by speech and writing and to build all kinds of contrivances ranging from buildings to computers and satellites. Our ability to communicate not only in the present but also with the past and the future has enabled us to build a culture that rests on the past while developing into the future. *We not only have a genetic inheritance through DNA like all other animals, but we also have a cultural inheritance.* In this lies our uniqueness. One may argue that chimpanzees can be taught to recognize words and signs and hence communicate with us and other chimps. Dogs and elephants can also be taught to walk or stand on their hind legs, but in neither case do they do it very well.

We are the only animals that have a highly organized system of communication between generations. This has enabled us to develop our culture(s) over a period of a million years or so. We are not born a *tabula rasa* (clean slate), but we are born with a slate in our brains that can be filled with information passed on from previous generations. Using this information, we can reason and create new information to be passed into the future. A sort of coevolutionary process is occurring in human society in which there is a continuous interaction between mind and culture with the result that culture is modified in succeeding generations. An obvious example of this is the rise of the industrial revolution in the eighteenth and nineteenth centuries. Now we are faced with as great changes in our society as a result of the application of silicon chips.

Some thinkers on this topic of mind and culture are suggesting that this interplay is, in fact, having a direct effect on the development of our minds themselves, hence the use of the term *coevolutionary*. Whether or not this is true is beside the main point, which is that we are unique among organisms because of the capacities of our brains to develop and sustain what we call culture.

Our Genetic Heterogeneity

The most important consequence of sexual reproduction is great genetic heterogeneity in all populations of plants and animals that practice it. And since most of them, like us, do, most are genetically heterogeneous. We are no exception. In fact, as we shall show in succeeding chapters, each of us is genetically unique unless we have an identical twin. The probability that anyone in times past or times future was or will be identical to us in his or her content of DNA is essentially zero.

This has important implications to every living human being. Because of our genetic heterogeneity reinforced in most instances by the environments (cultures) in which we develop, we have mental and physical differences among us varying from minor to major. Genetically we are *all* unequal, even though we may be (or should be) equal in the eyes of the law. We shall return to a fuller discussion of this most important point in several chapters.

Despite the fact that anthropologists, in recognition of the genetic heterogeneity of the human population, have divided us into races and sub-races, we are all members of one and the same species, *Homo sapiens* (wise man). This accolade that we have bestowed on ourselves may be an overstatement of our wisdom, but biologically it means that we are probably all descended from the same primitive, primordial nonhuman ancestor, and in the human population as a whole, no sexual isolation exists. What this latter means is that all matings between humans from whatever race or population are capable of producing offspring who are in turn fertile. This lack of sexual isolation is the crucial biological test for determining the existence of a single species. This means that such differences as skin color, body shape and size, and hair color are all rather insignificant biologically, even though we may place emphasis on them socially.

The Psychology of Genetics

As a final note to this introduction about us the paragons, let it be noted that the topic of human genetics is repugnant to many persons, even to those who consider themselves fairminded. The eminent British geneticist J. B. S. Haldane made this statement some years ago in a lecture given at the Eleventh International Congress of Genetics.

For a variety of reasons, men and women find it harder to accept the implications of genetics than those of many other branches of science. In the first place its findings may conflict with or reinforce two powerful emotions, the sexual and parental. Secondly they may conflict with or reinforce theories which support the class structure of human societies and the divisions between different societies, or alternatively with theories which regard class and national divisions as evil. . . . The findings of genetics may conflict with religious doctrines, but the conflict is probably less acute than those raised by the studies of astronomy and evolution. However, it is impossible to speak on the applications of human genetics without giving offense to someone.

The reasons for our attitudes toward genetics in general and human genetics in particular are many, as Haldane states. These attitudes are reinforced by the actions of irrational groups such as the Nazis who have created an illusion of a "master race," and by some eugenicists who dream of breeding a race of super humans. These deviations of the mind on the part of a minority have succeeded in convincing some members of our society that it is best not to mention that we are all different or unequal genetically. The fact is we are, and nothing is gained by denying it. And much is to be gained by recognizing it and using this knowledge about ourselves to forge a better society for all of us.

Reference and Further Reading

Haldane, J. B. S. 1965. The implications of genetics for human society. In *Genetics Today. Proceedings of the XI International Congress of Genetics*, vol. 2, xci. Pergamon Press, Oxford and London.

The history of the Western World since the sixteenth century has been punctuated by a series of revolutionary ideas that have influenced and in some places dominated thought. The most protrusive of these have been attempts to make logical descriptions and explanations of the physical world and the universe about us, such as the heliocentric theory of Copernicus (1473–1543), the foundations of experimental science by Galileo (1564–1644), the development of the mechanical explanation of the organization and operation of the universe by Newton (1643–1726), the synthesis of the theory of relativity by Einstein (1879–1955), and most recently the development of atomic theory by a number of physicists. The ideas of these persons have changed the human condition more in the last 400 years than in the previous 6000 or 7000 years that Western civilization has been thought to exist.

These advances in our interpretations of the physical world about us began several hundred years prior to equivalent developments interpreting the living, organic world. It was not until the nineteenth century when coherent theories and certain fundamental observations and experimental results began to be advanced that anything resembling modern biology started to develop. Charles Darwin (1809–1882) was among the greatest of the nineteenth century biologists (Fig. 2.1). His theory of evolution by means of natural selection was and still is the keystone of modern biology comparable in its influence to the work of Isaac Newton in physics. With him should stand Gregor Mendel (1822–1884), the founder of the particulate theory of inheritance, Rudolph Virchow (1821–1902), who showed us that cells are the ultimate units of life, and Louis Pasteur (1822–1895), who proved once and for all that life as we know it now cannot arise spontaneously and that much of our disease is caused by bacteria. All of these accomplishments were revolutionary insofar as they changed our outlook on the world in general and created the foundations of modern biology and scientific medicine.

Perhaps the most important idea philosophically was Darwin's thesis that humans have descended from lower forms of life, most recently from apelike creatures who were also the ancestors of the present-day great apes, such as the gorilla and the chimpanzee. This was a revolutionary idea because it brought humans down from a position *outside* nature to one *within* and *part* of nature. Our uniqueness lies not in our structure, strength, or ability to do the mile in less than four minutes, but in our brains. Otherwise, we are much like other animals, weaker than some and slower than many. This observation was and is still difficult for many to swallow, and some choose not to believe it. Or if they do believe it, they would rather it didn't become generally known.

The discoveries that were of more immediate significance to the development of human genetics, however, were those that had to do with cells and patterns of inheritance through cells. To learn about these we turn to the subject of cell biology. An understanding of cells is essential to any significant understanding of genetics.

2

The Genesis of
Modern Genetics

FIGURE 2.1
Four persons important in the founding of modern biology and genetics. (a) Charles Darwin (1809–1882), (b) Gregor Mendel (1822–1884), (c) Rudolph Virchow (1821–1902), (d) Louis Pasteur (1822–1895). (Photos a, b, and d: The Hunter Institute; photo c: Bettman Archive.)

(a)   (b)

(c)   (d)

The Beginnings of Cell Biology

One of the major scientific and philosophical questions that our ancient ancestors wrestled with was the origins of living things. In ancient Greece, where most of the thinking seems to have been done about this matter, there were two principal ideas current which are now most conveniently called (1) *spontaneous generation* and (2) *preformation.*

Spontaneous Generation

Spontaneous generation assumed that organisms arose from a formless ooze or mass of unorganized matter. One of the greatest thinkers and observers of the ancient world, Aristotle (384–322 B.C.E.), was a proponent, or perhaps better an adherent, of this explanation of origins, probably because he could think of nothing better. Recognizing that there had to be some link between generations, because after all dogs come from dogs and cats from cats, Aristotle proposed that the eggs from which animals arise receive an "essence" or "soul" giving them a self-realization or *entelechy* to develop into specific kinds of creatures. Hence, the link between generations was spiritual rather than physical.

This idea, despite its drawbacks, seems to have been the most popular one in the ancient world, continuing into the twentieth century, at least for the lower organisms or "vermin." It was finally dealt a death blow in this last century by Louis Pasteur and others who, working with bacteria and yeast, were able to show that when all the microorganisms in a culture are killed, a sterile medium is produced from which no living thing can arise unless it is again seeded with living organisms.

Preformation

If one rejected spontaneous generation, one had to assume a link between generations that caused the transfer of something that determined the characteristics of the species of that organism. What could it be? There was no knowledge on which to base an answer to that question prior to the nineteenth century, but valiant attempts to provide an explanation other than spontaneous generation were made and these centered around what we now call preformation. The adherents of this theory (or doctrine) believed that all organisms (except maybe the lower ones) arose from an egg within which was the potential to produce not only one generation but *all* future generations.

This theory deserves more than passing reference because it played an important role in thinking about the subject of descent all the way back in time to the ancient Greeks. The Greek philosopher Anaxagoras (ca. 500 B.C.E.) is generally credited with propounding the theory that was later elaborated upon by the theologians of the Middle Ages. The culmination of its development came with the encapsulation theory of the Swiss naturalist Charles Bonnet (1720–1793). According to Bonnet, the original individual of a species contained within it at the time of its special creation the fully *preformed* individuals of all following successive generations *one within the other.* Thus, there was no new creation; all formation of life forms was completed at the beginning of life on earth.

The microscopes were so bad in Bonnet's time that considerable imagination was needed to interpret what was seen in such things as sperm and eggs. As a result, some claimed to see a fully formed infant child in the human sperm head. It was called the *homunculus,* or fetus. Others saw the same thing in the egg. They were called the "ovists" in distinction to the former, who were called the "sperm-ists." These female and male chauvinistic attitudes did not last long, especially when one considered the obvious fact that children resemble both parents, not just one of them. The two schools argued for a period of years until the cell biologists of the last century finally put the matter to rest with better microscopes.

The Cell Theory

While spontaneous generation and preformation in the extreme form advocated by Bonnet were being abandoned, the *cell theory* began to fill the vacuum left by their demise. Early in the 1800s it was made clear that the structural and functional units of life were the cells first recognized by Robert Hooke (1635–1703), a seventeenth century contemporary and confrere of Isaac Newton (Fig. 2.2). Theodore Schwann (1810–1882) and M. J. Schleiden (1804–1881) are generally given most of the credit for bringing the cell into prominence as the life unit, but actually they were but two among many in the first 50 years of the nineteenth century who laid the foundations for this extremely important observation, which was to change the face of biology and medicine.

About the middle of the nineteenth century, a number of scientists began to appreciate that cells were more than structural units. This realization came about because of distaste for spontaneous generation. Schwann, in his attempts to explain the origins of new cells, proposed that they arise out of formless, unorganized masses in the tissues called blastemas. This was little more than spontaneous generation, and a quite unsatisfactory explanation. The correct answer was to come from one of the most eminent physicians of the nineteenth century, Rudolph Virchow, who, in his classic treatise *Cellular Pathology*, made the following statement (translated from the original German).

Where a cell arises there a cell must have previously existed (*omnis cellula e cellula*), just as an animal can spring only from an animal, a plant only from a plant. (Virchow, 1858)

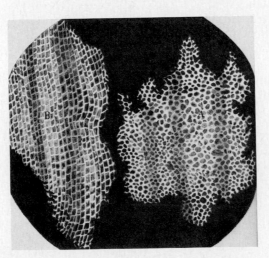

FIGURE 2.2
Cork cells as first observed by Robert Hooke (1635–1703). Hooke's observations were the first to show the cellular structure of living material. The illustration shows the cells as drawn by Hooke and published in 1665 in his book *Micrografia*. (Photograph from The Humanities Research Center, the University of Texas at Austin.)

Cells divide, and from one mother cell two daughter cells arise. The cell is more than a structural unit and a functional unit, it is also a *unit of reproduction*. By this we mean that it is the *smallest unit capable of reproducing more life*.

Soon after the appreciation of the significance of cells as life units by Virchow and his contemporaries, chromosomes were discovered, and the significance of the cell nucleus began to be recognized as the seat of heredity, or the bearer of some unknown substance (called the *idioplasm* by some) that is passed on to the offspring and determines what their characteristics are to be. It became obvious that the egg is a cell descended from cells. The early cell biologists of the latter part of the nineteenth century showed that what is in the sperm head and the egg are chromosomes with an associated material called cytoplasm. *We now know the egg and sperm contain preformed chromosomes that carry the instructions to make a child, not the child itself* (Fig. 2.3). In a sense the preformationists were important in the history of ideas about genetics, because they insisted that there had to be an explanation for the continuity of life. The question was settled by the cell biologists whose work led to the realization that this continuity is through the genetic

FIGURE 2.3
Sperm, egg, fertilized egg, and child. The sequence of events diagrammed here is general knowledge, even among children of our day, but it should be appreciated that our understanding of it is barely 100 years old. It was one of the most significant discoveries of the last century—the result of the observations of many cell biologists but principally those by O. Hertwig (1849–1922), E. Strasburger (1844–1912), and E. van Beneden (1846–1910). In the period 1880–1890, they announced that the cell nucleus carries the physical basis of heredity and the chromosomes of the offspring are derived in equal numbers from the nucleus of the sperm and the egg and hence equally from the two parents.

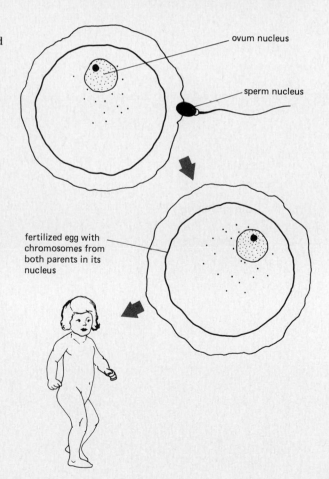

ovum nucleus

sperm nucleus

fertilized egg with chromosomes from both parents in its nucleus

material in the chromosomes. *The law of genetic continuity* for humans (and all other forms of life) states that what we inherit are our parents' chromosomes.

The cell is no longer a theory but a fact. The recognition of cells as the units of life is to be considered as revolutionary an advance in the history of biology as the recognition of the fact of evolution.

Mendelism

At the time that persons like Virchow were developing the modern cell theory, an obscure Roman Catholic monk by the name of Gregor Mendel (1822–1884) was growing pea plants in a monastery garden in the Czechoslovakian city of Brno (or Brünn) (Fig. 2.4), then part of the Austro-Hungarian Empire. Over a period of about 8 years, he made many crosses with his plants and came to certain conclusions that he presented at a scientific meeting in Brno in 1865 and published in 1866. His conclusions laid the foundations for the understanding of the *mechanics of inheritance.*

With his carefully selected pea plants, Mendel showed that by crossing plants with diverse opposite or alternative characteristics, such as pea or pod shape or color, one got results among the first, second, and succeeding generations of progeny that could best be explained by assuming the existence of *unit* or

FIGURE 2.4
The garden of the monastery in Brno, Czechoslovakia, in which Mendel carried out his experiments on hybridization of peas.

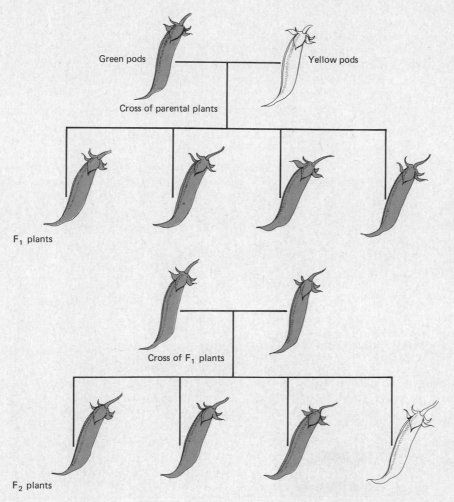

Green pods

Yellow pods

Cross of parental plants

F₁ plants

Cross of F₁ plants

F₂ plants

FIGURE 2.5
Mendel's technique of analysis. Two purebred strains of pea plants with different, alternative characters are crossed to get the hybrid F_1. In the example shown, the pods are either green or yellow in the parental plants. (The remaining parts of the plants are all green.) The F_1 hybrids all have green pods. Hybrids are crossed to one another to give the F_2 progeny. Among the F_2 are plants of *both* parental types, in distinction from the F_1 plants that are only of one type.

particulate characters in the pollen and egg cells of the plants. What these particulate or unit characters were he did not explain, for he did not know much about cells, let alone chromosomes. But by following their passage from one generation to the next among his progeny plants, he was able to draw several important conclusions that we now call *Mendel's Laws of Heredity*.

Mendel observed that when two *purebred* strains of an organism differing in contrasting, alternative unit characters are mated or "crossed" (Fig. 2.5), the hybrids (designated as the F_1 or first filial generation) usually express *one* of the two unit characters. When these F_1 hybrids are crossed (Fig. 2.5), *both* parental types are represented among the F_2 progeny, and they occur in a definite ratio. Mendel proposed as an explanation that the F_1 hybrids have within them the potential for both characteristics, even though only one is expressed (because it is *dominant*, as we explain later). He gave the letter A to one unit and a to the alternate unit. When the germ cells, that is, pollen and eggs, are formed, A and a separate or *segregate*, so that half are *A* and half *a*, as shown in Box 2.1. As the result of fertilization, one expects three different combinations: *AA*, *Aa*, and *aa*. *AA* and *aa* are the purebred parental types. By purebred, we mean that when crossed to one of the same kind, or selfed (*AA* × *AA* or *aa* × *aa*), they produce only progeny like themselves.

It was Mendel's genius that led him in the first place to prepare plants that were purebred, or homozygous as we would call them now, for specific characteristics, and then to cross these different purebreds. Even though the hybrids did not show the characteristics of both parents, individuals of subsequent generations did, as shown in Box 2.1. Hence, there was a *transmission of something* from generation to generation. That something we now call a *gene*.

Mendel's essential contribution is sometimes overlooked by emphasizing his technique of analysis of crossing inbred strains with alternative characteristics

── Box 2.1

Mendel's Explanation for the Results He Obtained

In his paper, "Experiments in Plant Hybridization," published in 1865, Mendel wrote these words (here translated from the German), which expressed the essence of his thinking.

Experimentally, therefore, the theory is confirmed that the pea hybrids form egg and pollen cells which, in their constitution, represent in equal numbers all constant forms which result from the combination of the characters united in fertilization.

He gave one characteristic the symbol *A* and its alternative *a*. Thus following his statement above, he designated the F₁ hybrids *Aa*. The purebred parents he designated simply *A* and *a*. The parents thus could only form either *A* or *a* pollen and eggs, but not both. When crossed,

only *Aa* hybrids could result. These, however, he pointed out can form two kinds of pollen and eggs in *equal* numbers, and when crossed, should produce three kinds of plants, *AA*, *Aa*, and *aa*, in a ratio of 1*AA* : 2*Aa* : 1*aa*.

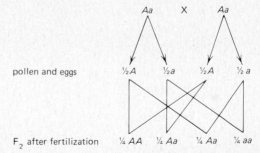

Here he was applying probability. He was probably the first person to apply probability in biology.

together. This was a coup to be sure, but the conclusion to be drawn from the result was that inheritance is the passage of something *particulate* from generation to generation. The particulate characteristic is *undiluted* by its passage. This may seem to be of no great significance to us now, but in the nineteenth century a *blending hypothesis of inheritance* occupied a prominent position in biology. It was even subscribed to by Charles Darwin, who apparently never became acquainted with Mendel's work. According to the blending hypothesis, when two individuals with contrasting characteristics are crossed, white hair versus black, for example, the offspring will be intermediate gray. If these are crossed to white-haired persons, the next generation will be still lighter, and so forth.

The Revolution of 1900–1903 and the Origins of Genetics

The fusion of cell biology and Mendelism resulted in a synthesis that transformed our understanding of life, human and otherwise, and led directly to the founding of modern genetics. In about 1900, the knowledge accumulated up to that time about cells and the newly rediscovered results and conclusions of Mendel were brought together. We say "rediscovered" because, although Mendel's work was in itself revolutionary, little attention was given it for 35 years, until three botanists, H. M. de Vries (1848–1935), K. E. Correns (1864–1933), and E. Tschermak (1871–1962), did experiments similar to Mendel's and then found that he had done the same years earlier and had come to the same conclusions. Soon after their results were published, cell biologists such as E. B. Wilson (1856–1939) and Th. Boveri (1862–1915) began to realize that what was known about the movement and distribution of chromosomes during the formation of eggs and sperm in animals and their subsequent union in fertilization might explain Mendel's and his rediscoverers' results with their plant crosses. This was proved to be the case by a student of Wilson, W. S. Sutton, in 1903.

Modern genetics began with these events. From this period on, activity in the laboratories, planting fields, and theoreticians' offices became furious. Box 2.2 lists the names of some of the principals in these early efforts that resulted in the genesis of the new science of genetics.

Human Genetics

The principles discovered by Mendel and elaborated upon by those persons listed in Box 2.2 apply to humans, other animals, and plants. One of the intriguing things about genetics is that its principles are essentially universal in their application, and much has been learned about human genetics by studying the genetics of the fruit fly *Drosophila* and the corn plant maize as well as mice, rats, and guinea pigs. It is not possible to carry out breeding experiments with humans as one can with these other organisms (and one hopes it never will be), but we can carry over with a high degree of confidence what we learn from these experimental plants and animals. Furthermore, when genealogical records are kept for human kinships, the inheritance of certain characteristics can be analyzed. This is especially true of the records kept for the lineages of European royalty, some of which go back 900 to 1000 years, or about fifty generations. Also, church records, especially those of the Mormon Church, can be of great value in tracing the inheritance of certain characteristics. A family tree starting with Queen Victoria of England (Fig. 2.6) illustrates how the inheritance of the blood disease hemophilia can be traced starting with Queen Victoria herself. Other families can be studied in a similar way of course, but few families have kept the detailed records available for royalty.

Genetics in Medicine

Most of what has been learned about the genetics of humans, however, has come from the observations of patients by physicians and subsequent collaboration with geneticists. (Many physicians now are also trained as geneticists, and more are selecting this combination.) A great deal can be learned about the inheritance of certain conditions by examining the families and other relatives of patients. Over the past 80 years, more than 3000 different clearly inherited conditions have been recognized. Most are classified as pathological, but many are not. In general, these conditions have what we call a simple pattern of inheritance, in which the passage of the trait can be easily followed through several generations. For example, such a simple pattern is found for hemophilia, as described in Figure 2.6. Not all traits show this simple pattern. For example, human diabetes has a very complex pattern of inheritance. As we say, "It tends to run in families," but it is impossible to predict accurately whether a newborn from a diabetic family will contract the disease later in life. Other even more complex traits, such as musical and artistic ability, mathematical ability, and so forth, have even more complex modes of inheritance. Reasons for these complexities will be discussed and many explained in the succeeding chapters of this book. For the present, it will be pointed out that an organism is the product of the activity of many genes, and the resultant adult individual has characteristics determined not only by the genes acting in concert but also by the environment within which the individual develops.

Despite these difficulties, genetic considerations have become an important part of modern medicine, and the physician who ignores them does so at his or her peril. We are all different genetically (although similar enough to be in the single species *Homo sapiens*), and as a consequence we respond differently to different kinds of drugs, anaesthetics, and other treatments. To take an obvious example, operating on a person with hemophilia requires precautions not necessary with a person who doesn't have hemophilia.

In addition, because a disease is inherited, its etiology or cause can frequently be investigated and understood more easily. Most diseases (in fact, perhaps all not caused by viruses and bacteria) are metabolic in origin. By comparing the metabolic pattern (the body chemistry) of the person who has a particular disease with that of a person who does not, one can frequently find a specific metabolic difference. In the case of inherited diseases, the differences generally can be traced to lack of a certain metabolic activity and hence to deficiency of a specific enzyme.

Box 2.2

Founding Fathers of Genetics After Mendel

1900   *H. de Vries*   Repeated in its essentials the work of Mendel
           C. Correns   and revived interest in his largely forgotten
           E. Tschermak   conclusions.

1901–1903   In this period the outlines of what we now call genetics were
                 sketched in by

           W. Bateson   Showed that Mendelian laws applied to animals
           L. Cuénot   as well as plants.

Bateson coined the word *genetics* and wrote one of the first popular genetics texts. Both he and Cuénot were the first proponents of the hypothesis that the hereditary determinants (genes) determined the presence or absence of enzymes.

           W. E. Castle   Showed Mendelian laws applied to humans.
           A. E. Garrod

Garrod pointed out the link between what is inherited and the ability to carry out certain reactions in the body.

           E. B. Wilson   Made the tie between cell biology and Mendel-
           C. E. McClung   ism, along with a number of other scientists of
           T. Boveri   this period.
           W. S. Sutton

1909   *W. Johannsen*   Coined the terms *gene, genotype, phenotype.*

1910–1916   *T. H. Morgan*   Working together in the same laboratory (the lat-
                 A. H. Sturtevant   ter three were students of Morgan), these scien-
                 H. J. Muller   tists put the genes on the chromosomes and
                 C. B. Bridges   showed they are arranged linearly.

                 R. A. Emerson   Started the science of modern corn genetics,
                 E. M. East   which was to influence all plant and animal
                                       genetics.

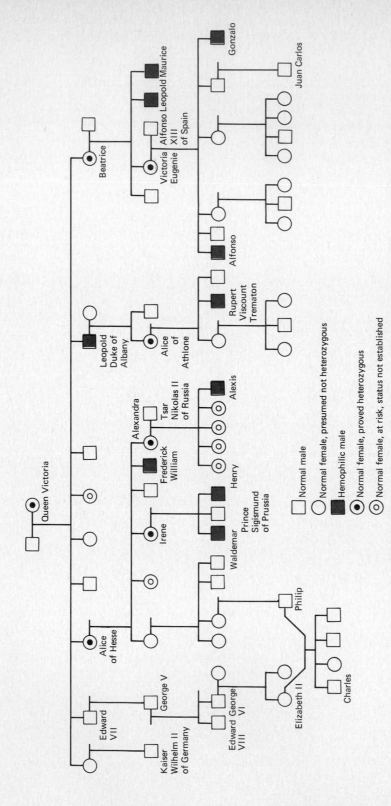

FIGURE 2.6 The inheritance of hemophilia among the descendants of Queen Victoria of Great Britain. Hemophilia is transmitted as a simple Mendelian trait on the X chromosome. Males, who have only one X chromosome, express the trait. The trait is recessive in females, who have two X chromosomes, so that heterozygous females do not express the trait, although they may transmit it.

20

Knowledge of the basic cause of the disease gives the physician a rational basis for treatment. Consider the disease galactosemia, which occurs in about 1 in 65,000 births. If untreated, the disease, which has a simple pattern of inheritance, results in severe mental retardation. Now we know that the cause is inability of the galactosemic person to utilize the sugar galactose, an important constituent of milk. The infant with this condition builds up high concentrations of galactose that affect the proper development of the brain. Treatment is obvious: take the galactose out of the infant's diet. This is easily done and the child develops normally. It is now routine in many hospitals to screen all infants for galactosemia and other similar metabolic diseases, an easy and inexpensive procedure. At the present writing about 32 different kinds of inherited metabolic disorders can be screened in infants. Once recognized, a disorder can often be treated before severe damage results to the afflicted person.

It wasn't too long ago, perhaps fifty years, that it was thought by most physicians as well as the general public that an inherited disease could not be treated. The attitude was fatalistic. By understanding the biochemical and metabolic basis of disease, the medical profession can now intelligently devise methods of treatment. We have come a long way from treatment by the exorcising of evil spirits and demons.

Genetics and Behavior

The extent to which the genes that we inherit from our parents affect our behavior is a highly controversial subject that we shall not delve into deeply. It is an extremely complex subject, difficult to analyze, and difficult to draw firm conclusions about. The fact is that we all do think of ourselves as paragons. In this lie our strengths and abilities to do remarkable things, like putting men on the moon and showing compassion for one another, but also in this lie our weaknesses, such as intolerance, cruelty, racism, and other modes of behavior that we generally tend to hide in polite society, except when we come out of our closets as members of extremist "true believing" groups. The fact is we all have our angelic and demonic sides.

What role do our genes play in these things? Are there genes for criminality, musical ability, mathematical ability, sensitivity toward others, religious feelings, and so on? Or does the way in which we are brought up determine these things? Or is it a combination of genes and environment, nature (genes) versus nurture (environment)? We shall grapple with these questions in later chapters, but we warn now that we have no clear answers on many issues of this type. We will, however, be able to make some straightforward statements about the inheritance of certain diseases and the actions of the genes that cause them. But before we do this we must develop a background about cells and their constituents, how they function, and what their structures are. This is basic to any understanding of modern genetics.

Review Questions

1. What were the essential differences between preformation and spontaneous generation as stated in the eighteenth century?
2. State the law of genetic continuity.
3. Does the law of genetic continuity contain elements of the concepts of spontaneous generation and preformation? If so, what?
4. The statement "omnis cellula e cellula," aside from its obvious meaning, conveys also the connotation that cells are the ultimate life units. Why?
5. What significant and original approach(es) did Mendel take in his experiments that enabled him to be successful in demonstrating his F_2 ratios?
6. Why was Mendel driven to the conclusion that inheritance is particulate?
7. What is the significant difference between blending inheritance and particulate inheritance?
8. What was the essential significance of the biological revolution of 1900–1903?
9. What group of geneticists proved that the Mendelian unit characters, called "genes" by Johannsen, actually were located on the chromosomes?

References and Further Reading

Eiseley, L. 1958. *Darwin's Century.* Doubleday, Garden City, N.Y. A very readable history of Darwin's period with emphasis on the impacts of his writings on society.

Irvine, W. 1955. *Apes, Angels and Victorians. The Story of Darwin, Huxley and Evolution.* McGraw-Hill, New York. A general treatment of Darwin and his period that emphasizes the personalities of the principal participants.

Mayr, Ernst. 1982. *The Growth of Biological Thought.* Belknap Press of Harvard University Press, Cambridge, Mass. A major contribution to the history of the development of our ideas about the living world. Highly recommended.

Olby, R. C. 1966. *Origins of Mendelism.* Constable, London. A discussion of the works of plant breeders immediately before Mendel as well as Mendel himself.

Stubbe, H. 1965. *History of Genetics* (trans. from 2nd German ed.). The MIT Press, Cambridge, Mass. Genetics beginning with the prehistory of human society. Valuable for pre-Mendelian thoughts on the subject.

Sturtevant, A. H. 1965. *A History of Genetics.* Harper & Row, New York. An entertaining and informative history of genetics by one of its founding fathers.

McKusick, V. A. 1983. *Mendelian Inheritance in Man*, 6th ed. The Johns Hopkins Press, Baltimore. A compendium of the known genetic variations in humans.

Wilson, E. B. 1925. *The Cell in Development and Heredity*, 3rd ed. Macmillan, New York. A classical treatise worth reading even today for its style of scientific writing, which has never been surpassed.

Virchow, R. 1863. *Cellular Pathology* (trans. from 2nd German ed.). Reprinted (1971) by Dover Publications, New York. The treatise in which Virchow proposes his radical ideas about the origin of cells.

Life began on this planet about 3 billion years ago. Whether it started here *de novo* or came originally from an extraterrestrial source cannot be answered with certainty. But even if the latter is true, as some propose, life had to start somewhere in the universe, and therefore it had to arise from the inanimate organic and inorganic substances present in one form or another all through the universe. These substances, which make up the matter of the universe, are all chemical elements or combinations of elements.

Before considering in more detail how life began and how cells function, it will be useful to review the physical and chemical nature of matter. Our genes and chromosomes, cells, and all that surrounds us are made of atoms and molecules. The behavior of genes in cell division and metabolism must ultimately be explained in terms of the molecules of which they are composed and on which they act.

The Atomic Basis of Matter

The Nature of Atoms

The ancient Greeks were the first to propose the existence of atoms. Democritus (ca. 420 B.C.E.) is usually identified as the originator of the "atomists" school of philosophy. According to Democritus and his followers, all matter ultimately consists of discrete particles, *atoms*, that are indivisible. The different qualities we perceive are due to different combinations of atoms. This very modern sounding idea was based purely on reason and not on experiment, and one should not attribute too much insight to the early atomists. Indeed, the great philosopher Aristotle (384–322 B.C.E.) did not accept this idea, with the result that the atomists were not a major influence. Their ideas persisted though and are included by the Roman Lucretius (ca. 95 B.C.E.) in his *De rerum natura*, a treatise on natural history.

The Atom and John Dalton.   The modern concept of atoms is due to the great English chemist John Dalton (1766–1844), who is also of interest to geneticists because he published a description of his own inherited colorblindness. During the seventeenth and eighteenth centuries, a number of *elements*, such as hydrogen and oxygen, had been discovered, and the amounts of these elements in various *compounds* had been measured. The theory that elements consisted of atoms that could not be further divided was widely accepted, but there was no general idea how atoms differ from each other and behave toward each other.

Dalton's theory, published in detail in 1808, proposed several properties of atoms that explained the existing experimental observations.

1. All matter consists of atoms.
2. Atoms are indivisible.
3. Different elements consist of different kinds of atoms that differ from each other in their masses.

3

The Chemical Origins and Basis of Life

4. In the course of a chemical reaction, atoms enter into different combinations but are never created nor destroyed.
5. Chemical compounds are simple combinations of small numbers of atoms, always with exactly the same number of each kind of atom.

These generalizations have withstood nearly two centuries of research without substantial change. To be sure, we now know that atoms can divide, some, such as radium, spontaneously and others with the help of cyclotrons. This exception will concern us only when we later address the question of the genetic effects of atomic radiation. We also know that compounds need not be simple. Some of great interest to us consist of millions of atoms. But even in these giant molecules, each atom bears a simple relationship to other atoms to which it is directly attached.

The Structure of Atoms.   All atoms consist of a dense *nucleus* that has a positive electrical charge. The nucleus is surrounded by shells of negatively charged *electrons*. The positive charge of the nucleus is due to *protons*, which have over 1800 times the mass of electrons. The number of electrons and protons in an atom ordinarily are equal when the atom is not chemically combined with some other atom. Also present in most nuclei are *neutrons*, particles with the same mass of protons but with no charge.

The simplest atom is hydrogen, which consists of a single proton in the nucleus and one electron in the surrounding shell (Fig. 3.1a). The mass of such a hydrogen atom is due essentially to the one proton of the nucleus. The next larger atom is helium with two protons in the nucleus and two electrons. In addition, helium has two neutrons, so its mass is four times that of hydrogen (Fig. 3.1b).

Helium illustrates another property of atoms. The electrons are distributed in

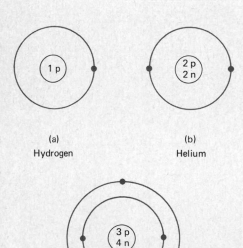

(a)
Hydrogen

(b)
Helium

(c)
Lithium

FIGURE 3.1
The structure of atoms. (a) The hydrogen atom. The nucleus consists of a single proton, giving this atom the lowest atomic weight. There is a single electron in the shell outside the nucleus. The negative charge of the electron balances the positive charge of the proton, giving a net charge of zero for the atom as a whole. (b) The helium atom. Since He has two protons and two neutrons in its nucleus, it has a mass approximately four times that of H. (c) The lithium atom. Li has three protons in its nucleus as well as four neutrons. The positive charge of the three protons is balanced by three electrons, two of which are in the inner K shell and one in the L shell.

concentric shells around the nucleus. Each shell has a maximum number of electrons that it will accommodate, and when that number is reached, the atom is extremely stable and will not react with other atoms. The inner electron shell holds only two electrons; hence, helium is very unreactive, a so-called *noble gas*. That is why it is preferred over hydrogen for lighter-than-air craft. Both are very light gases, but hydrogen burns explosively whereas helium will not burn at all.

Additional electrons must go into the second shell. Thus lithium, which has three protons in the nucleus, has two electrons in the first shell and one in the second (Fig. 3.1c). The second shell holds up to eight electrons. This would give a total of ten electrons, corresponding to ten protons in the nucleus. Such an atom would be neon, the second in the series of noble gases. The third shell also is stable with eight electrons, but in heavier elements, an additional 10 electrons can be added for a total of 18. Additional shells hold 18 or 32 electrons. The chemical properties of an atom are heavily dependent on the number of electrons in the outer shell, as described in a later section.

The Chemical Elements

Atoms differ from each other in the number of protons in the nucleus (which equals the number of electrons when the atom is neutral) and in the number of neutrons in the nucleus. The chemical behavior of an atom is determined ultimately only by the number of protons. The number of protons is called the *atomic number*, often symbolized by Z, and each atomic number corresponds to a different element. The term *element* is an old one that refers to the basic kinds of matter of which all else is composed. The four elements of classical civilization were earth, air, fire, and water. None of these is an element in modern chemistry. Instead, our modern elements are hydrogen, nitrogen, oxygen, uranium, and so forth.

Approximately 90 elements occur in nature. All others are too unstable to persist for the billions of years since the origin of the earth, if in fact they ever existed at all. Physicists have been able to create additional elements that are sufficiently stable for study, and some, such as plutonium, are stable enough to last for years. A list of all known elements is given in Table 3.1.

The Periodic Table. In 1869, the Russian chemist Dmitri Mendeleyev (1834–1907) discovered that if he arranged the elements by increasing atomic mass, certain periodicities in chemical properties occurred; that is, elements with similar properties were distributed at regular intervals in the array. With the discovery of new elements, additional support was provided for the periodic occurrence of similar elements. Finally, with the discovery of the structure of atoms in the twentieth century, the reasons for the periodicity became apparent.

A modern *periodic table of the elements* is shown in Figure 3.2. In group I are those elements with one electron in the outer shell; group II has two electrons, group III has three, and so forth. The transition elements are different. Typically they have two electrons in the outer shell. As the atomic number increases, the additional electrons of transition elements do not go into the outer shell. Rather, they go into inner shells. For example, the number of electrons in the shells of Ca ($Z = 20$) are 2K, 8L, 8M, 2N. For Sc ($Z = 21$), the electron formula is 2K, 8L, 9M, 2N.

Additional electrons go into the third (M) shell until its maximum of 18 is reached. Then filling of the fourth shell up to eight electrons resumes. Similarly, the so-called rare earths, 58–71 and 90–103, add their additional electrons to the second preceding shell rather than to the immediately preceding shell. Thus, the series beginning with cerium ($Z = 58$) adds electrons to the fourth (N) shell rather than to the outer sixth shell or the next inner fifth shell.

TABLE 3.1   The Chemical Elements

Atomic Number	Element	Symbol	Atomic Weight	Atomic Number	Element	Symbol	Atomic Weight
1	hydrogen	H	1.008	41	niobium	Nb	92.906
2	helium	He	4.003	42	molybdenum	Mo	95.94
3	lithium	Li	6.939	43	technicium	Tc	98.91
4	beryllium	Be	9.012	44	ruthenium	Ru	101.07
5	boron	B	10.811	45	rhodium	Rh	102.905
6	carbon	C	12.011	46	palladium	Pd	106.4
7	nitrogen	N	14.007	47	silver	Ag	107.868
8	oxygen	O	15.999	48	cadmium	Cd	112.40
9	fluorine	F	18.998	49	indium	In	114.82
10	neon	Ne	20.179	50	tin	Sn	118.69
11	sodium	Na	22.990	51	antimony	Sb	121.75
12	magnesium	Mg	24.305	52	tellurium	Te	127.60
13	aluminum	Al	26.982	53	iodine	I	126.904
14	silicon	Si	28.086	54	xenon	Xe	131.30
15	phosphorus	P	30.974	55	cesium	Cs	132.905
16	sulfur	S	32.064	56	barium	Ba	137.34
17	chlorine	Cl	35.453	57	lanthanum	La	138.91
18	argon	Ar	39.948	58	cerium	Ce	140.12
19	potassium	K	39.102	59	praseodymium	Pr	140.907
20	calcium	Ca	40.08	60	neodymium	Nd	144.24
21	scandium	Sc	44.956	61	promethium	Pm	(147)
22	titanium	Ti	47.90	62	samarium	Sm	150.35
23	vanadium	V	50.942	63	europium	Eu	151.96
24	chromium	Cr	51.996	64	gadolinium	Gd	157.25
25	manganese	Mn	54.938	65	terbium	Tb	158.924
26	iron	Fe	55.847	66	dysprosium	Dy	162.50
27	cobalt	Co	58.933	67	holmium	Ho	164.930
28	nickel	Ni	58.71	68	erbium	Er	167.26
29	copper	Cu	63.546	69	thulium	Tm	168.934
30	zinc	Zn	65.37	70	ytterbium	Yb	173.04
31	gallium	Ga	69.72	71	lutetium	Lu	174.97
32	germanium	Ge	72.59	72	hafnium	Hf	178.49
33	arsenic	As	74.922	73	tantalum	Ta	180.948
34	selenium	Se	78.96	74	tungsten	W	183.85
35	bromine	Br	79.904	75	rhenium	Re	186.2
36	krypton	Kr	83.80	76	osmium	Os	190.2
37	rubidium	Rb	85.47	77	iridium	Ir	192.22
38	strontium	Sr	87.62	78	platinum	Pt	195.09
39	yttrium	Y	88.905	79	gold	Au	196.967
40	zirconium	Zr	91.22	80	mercury	Hg	200.59

TABLE 3.1 (*continued*)

Atomic Number	Element	Symbol	Atomic Weight	Atomic Number	Element	Symbol	Atomic Weight
81	thallium	Tl	204.37	93	neptunium	Np	237.05
82	lead	Pb	207.19	94	plutonium	Pu	(240)
83	bismuth	Bi	208.980	95	americium	Am	(243)
84	polonium	Po	(210)	96	curium	Cm	(247)
85	astatine	At	(210)	97	berkelium	Bk	(247)
86	radon	Rn	(222)	98	californium	Cf	(251)
87	francium	Fr	(223)	99	einsteinium	Es	(254)
88	radium	Ra	226.03	100	fermium	Fm	(257)
89	actinium	Ac	(227)	101	mendelevium	Md	(258)
90	thorium	Th	232.038	102	nobelium	No	(255)
91	protactinium	Pa	231.04	103	lawrencium	Lr	(257)
92	uranium	U	238.03				

These various complications of the electronic structure of elements are of more interest to chemists than to biologists. The important points to remember are that the *outer* shell of an atom never has more than eight electrons, even though that same shell may accommodate more when it is an inner shell. Also, the configuration in which there are eight electrons in the outer shell (two in the case of the *K* shell) is extremely stable.

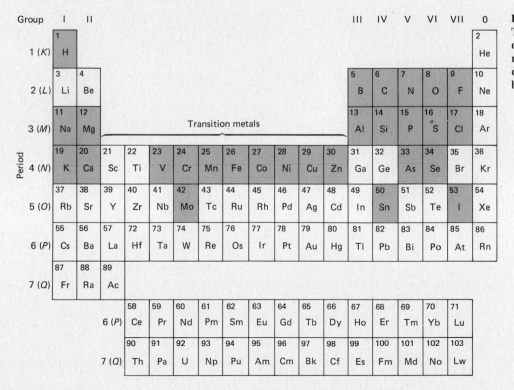

FIGURE 3.2
The periodic table of the chemical elements. Elements important to living cells are on a shaded background.

Atomic Weights and Isotopes. With the exception of the common form of hydrogen, all atomic nuclei have neutrons as well as protons. Ordinarily the number of neutrons is equal to or greater than the number of protons, and for elements of high atomic number, the ratio of neutrons to protons is relatively larger than for elements of low atomic number. For example, in the most common form of oxygen, there are eight protons and eight neutrons. Uranium has 92 protons and 146 neutrons.

The number of neutrons can vary within narrow limits, however. In addition to the common form with one proton and no neutrons, hydrogen also occurs in forms with one neutron and one proton and with two neutrons and one proton. Since the number of protons and, hence, the number of electrons do not change, the *chemical* characteristics of these three forms of hydrogen are identical. But the masses are different, being in a ratio approximately of $1:2:3$. These three forms of hydrogen are called *isotopes*. The isotopes of an element are the different forms that have varying numbers of neutrons. They are symbolized by placing the total number of neutrons plus protons to the upper left of the element symbol. Thus, the three isotopes of hydrogen are ^1H, ^2H, and ^3H. In the case of hydrogen, the two rare isotopes have special names: ^2H is known as deuterium and ^3H as tritium. All other isotopes are known only by their isotope number, for example ^{12}C is carbon-12 and ^{14}C is carbon-14.

The different combinations of neutrons and protons are not equally stable. Some, such as ^1H, ^{12}C, ^{14}N, and ^{16}O, are extremely stable. Others, particularly those with an excess of neutrons, are unstable. The nuclei have a tendency to fly apart, forming smaller nuclei and releasing energy and radiation. Such an isotope is said to be radioactive because of the radiation released. This happens continually in our bodies due to its content of ^{40}K, a relatively abundant form (0.01%) of radioactive potassium.

When atoms are bombarded with neutrons, the neutrons may be absorbed by the nuclei, become unstable, and undergo fission. The process of fission may release neutrons and other forms of subatomic particles and radiation. In such a case, one may have a "chain reaction." The first atomic bombs were based on the fission of ^{235}U, which is initiated by absorption of one neutron but releases several as products of the fission. If only a small amount of ^{235}U is present, most of the neutrons produced are absorbed by surrounding materials. However, if a sufficiently large amount is present (a "critical mass"), a chain reaction occurs. Given the right amounts of material and geometry, enormous numbers of atoms will split almost simultaneously, releasing the enormous energy of the atomic bomb instantaneously.

It should be obvious from the above that different atoms have different masses, which correspond approximately to the numbers of protons and neutrons. Although the reason for it was not known, the early chemists noted that the weights of the smaller elements entering into reactions with each other could usually be expressed by ratios of small whole numbers. Thus, hydrogen and oxygen in water are in a weight ratio of $1:8$. This led to the concept of atomic weights. Each atom has its characteristic mass. (*Mass* is the preferred term, since weight depends on gravity. A 75 kg man in an orbiting satellite still has his mass, even though he is weightless. The term *atomic weight* is established by tradition although not by logic.) The masses of the common elements are approximately whole-number multiples of hydrogen. This would be expected, since the mass is due to the number of protons and neutrons. However, the elements are mixtures of isotopes as they occur naturally. The atomic weight of an element is therefore an average value based on the naturally occurring mixture. The standard is the ^{12}C isotope of carbon, which is given a value of exactly 12. Table 3.1 gives the atomic weights of the naturally occurring isotope mixtures of the elements. One atomic weight unit is often called a *dalton* and is equal to 1.66×10^{-24}g. It is approximately equal to the weight of one hydrogen atom.

Chemical Elements in Biological Systems

In general, all cells have the same elemental chemical composition whether they be from bacteria or humans. Table 3.2 lists the chemical elements common to all kinds of cells. The main elements of organic matter are given in column A. The average amounts of these in the human body cells are given in percentages of atoms. Column B lists elements that are present in traces in cells, but not all are present in human cells. By trace is meant a few parts per million. These trace elements are essential to the organisms in which they occur, but only in traces. With a few exceptions, they can be extremely toxic if present in excess.

Hydrogen and oxygen are primary elements in the energy exchanges of living organisms. Calcium is present in abundance in humans and other vertebrates because of their bones. The trace elements are found in small quantities, but many are necessary to human life.

TABLE 3.2
The Elemental Composition of Cells

Primary Elements				Trace Elements	
	Element	Fraction in Human Body			
H	hydrogen	0.63	Mn	manganese	
O	oxygen	0.255	Fe	iron	
C	carbon	0.095	Co	cobalt	known to be present
N	nitrogen	0.014	Cu	copper	and necessary in
Ca	calcium	0.0031	Zn	zinc	human cells
P	phosphorus	0.0022	Mo	molybdenum	
Cl	chlorine	0.0008	I	iodine	
K	potassium	0.0006	Cr	chromium	
S	sulfur	0.0005	Se	selenium	
Na	sodium	0.0003			
Mg	magnesium	0.0001	Sn	tin	
		1.00	B	boron	
			Al	aluminum	present in cells of
			V	vanadium	other organisms
			Si	silicon	
			Ni	nickel	
			F	fluorine	

Structure and Function of Molecules

It is obvious from casual observation that we and our surroundings are not composed of pure elements. Rather, the atoms of various elements interact with each other to form *compounds* or *molecules* that consist of two or more atoms. Some biological molecules have hundreds of thousands of atoms. The term *molecule* may be defined as a stable grouping of one or more atoms held together by strong chemical bonds. A molecule of helium consists of a single atom, as do many metals in the pure elemental state. Single atoms of hydrogen or oxygen have a very transient existence, and the usual molecules of hydrogen and oxygen each consist of two atoms, symbolized H_2 and O_2.

Types of Chemical Bonds

Ionization and Ionic Bonds.   The stability of completed electron shells has already been noted. Atoms with outermost shells that are incomplete can acquire more stable complete shells in several ways. One way is to get rid of the electrons in the outer shell and rely on the next inner shell, which would be complete. Or electrons might be added to the outer shell to make it complete. In either case, the atom is no longer electrically neutral. It has formed an *ion* by the process of *ionization.* As an example, a hydrogen atom H consists of one proton and one electron. In theory, it could either gain one electron to fill its K shell or lose one electron. The latter is in fact what happens. This can be written as

$$H \rightarrow H^+ + e^-$$

The positively charged hydrogen ion is quite stable. The electron produced would need to go somewhere. One place would be to fill the outer shell of another element, such as chlorine.

$$Cl + e^- \rightarrow Cl^-$$

In this case, Cl has seven electrons in its outer shell and needs only one more to fill the shell. The chloride ion produced is negatively charged, since there are 17 protons in the nucleus and 18 electrons in the shells. Positively charged ions are often called *cations*, since they are attracted to a *cathode* (a negatively charged pole, as in a battery). Negatively charged ions are called *anions*, since they are attracted to the anode (Fig. 3.3).

One cannot have a flask with only hydrogen ions or chloride ions. Electrical neutrality must be preserved for a substance to be stable. We can achieve this by putting equal numbers of H^+ and Cl^- ions together, forming in this case hydrochloric acid, H^+Cl^-, sometimes also written just HCl, it being understood that both elements are ionized. The plus and minus charges attract each other to form a relatively stable arrangement called an *ionic bond*. It is characteristic of ionic bonds that the partners may exchange readily. Also, there is no specificity. Any pair of positive and negative ions are attracted to each other. The number of bonds that

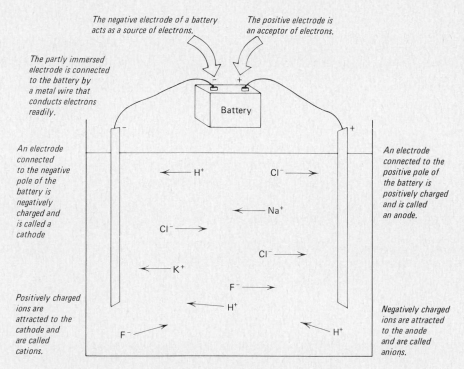

The negative electrode of a battery acts as a source of electrons.

The positive electrode is an acceptor of electrons.

The partly immersed electrode is connected to the battery by a metal wire that conducts electrons readily.

Battery

An electrode connected to the negative pole of the battery is negatively charged and is called a cathode

An electrode connected to the positive pole of the battery is positively charged and is called an anode.

H^+ Cl^-

Na^+

Cl^-

Cl^-

K^+

F^-

H^+

Positively charged ions are attracted to the cathode and are called cations.

F^-

H^+

Negatively charged ions are attracted to the anode and are called anions.

FIGURE 3.3
The formation and electrical behavior of cations and anions. When an atom gains or loses one or more electrons, it is no longer electrically neutral. If it is in an electric field, it will be attracted to the electrode of opposite charge. Thus H, Na, and K become H^+, Na^+, and K^+ and are attracted to an electrode that has a negative charge. A negative electrode is called a cathode, and ions attracted to it are called *cations*. Cl and F readily gain an electron to become Cl^- and F^-. They are attracted to a positive electrode (an anode) and are therefore *anions*.

an atom can form is its *valence.* An ion such as Fe^{3+} would have a valence of three.

Hydrochloric acid is a very strong acid. Sodium chloride is neither acidic nor alkaline but is a neutral salt. The difference is due to the presence of hydrogen ions in H^+Cl^-. The acidity of a substance depends entirely on the concentration of hydrogen ions. If the concentration is greater than in water, the substance is *acid.* If less than in water, it is *basic* or *alkaline.* The acidity of a substance is often expressed as its pH. Very pure water is pH 7.0. Values below 7.0 are acidic and above 7.0 are alkaline. Blood is usually about pH 7.35, and substantial deviations from this value can create grave problems. The usual range of pH values is 1–14, although acids and bases outside this range are routinely used in laboratories.

The group I *alkali metals* (lithium, sodium, potassium, rubidium, cesium, and francium) are very prone to give up their single outer electrons, forming the ions Li^+, Na^+, K^+, and so forth. Of these, Na^+ and K^+ are especially important in biological systems. The compound sodium chloride, Na^+Cl^-, is common table salt, and the substantial amounts of sodium and chlorine listed in Table 3.2 are primarily in this form. In general, the elements of groups I and II, the heavier elements of groups III–VI, the transition metals, and the rare earths form positive ions by giving up one or more electrons, more often two or three.

The group VII *halogens* (fluorine, chlorine, bromine, iodine, and astatine) have a very strong tendency to pick up an electron to form negative ions. This accounts for the great reactivity of these elements. Only chloride and iodide ions are of biological significance, although fluoride affects tooth hardness. The iodide is required in trace amounts for thyroid function. Other of the biologically important major elements (carbon, nitrogen, oxygen, phosphorus, sulfur) do not ionize except in special situations that will not be of concern to us.

Covalent Bonds.   Another way for atoms to acquire completed outer shells of electrons is to share electrons with each other. If two atoms of hydrogen pool their electrons, then there are enough to fill a K shell. If the electrons are symbolized by dots, the reaction may be written

$$H \cdot + H \cdot \rightarrow H : H$$

or
$$2 \ H \rightarrow H_2$$

Such sharing provides a very stable *covalent bond* between the atoms. Common hydrogen gas consists of molecules, each with two atoms of hydrogen bound covalently, as does oxygen, O_2. However, oxygen is short two electrons in its outer shell and must therefore share four.

$$2 \ : \ddot{O} : \rightarrow \ : \ddot{O} : : \ddot{O} :$$

In this case, a *double bond* binds the oxygen atoms together in O_2. Two atoms of nitrogen must share six electrons, forming a *triple bond*, in order to form N_2.

Each pair of shared electrons is a single covalent bond, and the number of such bonds that an atom can form, its valence, is quite constant. The valence of hydrogen is one, of oxygen and sulfur is two, of nitrogen is three, and of carbon is four. Keeping the valences in mind, one can assemble these atoms much like building blocks to form an unlimited number of molecules that are stable.

Carbon forms covalent bonds very readily, with itself and with other elements. Molecules that include carbon are called *organic* molecules, because it was formerly believed that they could only arise from living organisms. This is not true, but most of the carbon compounds do come from biological sources.

The simplest carbon compound is the common CO_2, in which one carbon atom shares its electrons with two oxygen atoms: $O\!=\!C\!=\!O$. Since hydrogen can form only one bond, four are required to *saturate* a single carbon atom.

$$
\begin{array}{c}
H \\
| \\
H\!-\!C\!-\!H \\
| \\
H
\end{array}
$$

methane

Methane, the gas commonly used in heating and cooking, is the simplest of the *hydrocarbons*, or compounds of hydrogen and carbon. Others are

$$
\begin{array}{c}
H \quad H \\
| \quad\; | \\
H\!-\!C\!-\!C\!-\!H \\
| \quad\; | \\
H \quad H
\end{array}
\qquad
\begin{array}{c}
H \quad H \quad H \\
| \quad\; | \quad\; | \\
H\!-\!C\!-\!C\!-\!C\!-\!H \\
| \quad\; | \quad\; | \\
H \quad H \quad H
\end{array}
\qquad
\begin{array}{c}
H \quad H \quad H \quad H \\
| \quad\; | \quad\; | \quad\; | \\
H\!-\!C\!-\!C\!-\!C\!-\!C\!-\!H \\
| \quad\; | \quad\; | \quad\; | \\
H \quad H \quad H \quad H
\end{array}
$$

ethane   *propane*   *butane*

Oxygen, with a valence of two, can be added to form a variety of familiar compounds.

<pre>
    H   H   H   O   H   H   H   H
    | | | || | | | |
H—C—C—O—H   H—C—C—O—H   H—C—C—O—C—C—H
    | | | | | | |
    H   H   H   H   H   H   H

      ethanol   acetic acid   diethyl ether
   (ethyl alcohol)   (the active agent   (an anaesthetic)
                               in vinegar)
</pre>

Virtually any combinations are possible, so long as hydrogen forms only one bond, oxygen two, and carbon four. Nitrogen can form three covalent bonds, as it does in ammonia, NH_3. One molecule of ammonia can also react with a proton (H^+) to form an ammonium ion, $NH_4{}^+$.

$$
\begin{array}{ccc}
\text{H} & & \text{H}_+ \\
\text{H:N:} + \text{H}^+ & \rightarrow & \text{H:N:H} \\
\text{H} & & \text{H}
\end{array}
$$

A common nitrogen-containing molecular grouping in biological molecules is the amino group, —NH_2.

Hydrogen bonds.   The third type of bond we must consider is the *hydrogen bond*. This is a very weak bond but one that is quite important in biology. It is formed when hydrogen that is covalently bound to oxygen or nitrogen is attracted by unshared electrons on other oxygen or nitrogen atoms. The bond can form only if the atoms are lined up properly. The electrons of the atoms can be represented as follows.

$$
\text{H:O:} + \text{H:O:} \rightarrow \text{H:O:H:O:} \\
\;\;\text{H} \qquad \text{H} \qquad\;\; \text{H} \;\;\; \text{H}
$$

Hydrogen bonds are only about $\frac{1}{10}$ to $\frac{1}{20}$ as strong as covalent bonds and are easily disrupted by heat. In the example above, two molecules of water are joined by a hydrogen bond. Many additional molecules could be added to these two, and in liquid water most of the molecules are involved in such weak aggregates. Were this not so, water would be a gas at ordinary temperatures, and there would have been no primordial oceans.

Other Molecular Interactions.   One additional interaction between molecules is important in biological systems, especially in membranes. Many organic compounds, including diethyl ether, hydrocarbons, fats, and oils, will not dissolve in water or will do so only to a very limited extent. If mixed vigorously together, as in a salad dressing, they will separate, the oil layer on top because it is less dense and the water layer on bottom. Inorganic salts, such as sodium chloride, will always be dissolved in the water layer. Some organic compounds are primarily water soluble (*hydrophilic*) and some are fat soluble (*hydrophobic*). (The term "fat" is used in a general sense for all oils and similar organic solvents, such as ether and benzene.)

Compounds with a high oxygen and nitrogen content tend to be hydrophilic. Those that are mostly hydrocarbon are hydrophobic. Many molecules are both; that is, one portion of the molecule is hydrophobic and the other portion is hydrophilic. Such a molecule will tend to line up in the water–oil interface, with one foot in each camp. Soaps and detergents operate on this principle. Cells are mostly water, although many molecules in cells have strongly hydrophobic portions. Such hydrophobic areas tend to stick together in what is sometimes referred to as a *hydrophobic bond*. Such a bond is not a strong specific relationship between atoms, but it is important in the stability of many large molecules, such as proteins, and in the formation of membranes.

All atoms have some attraction for each other if they approach very closely. This attraction is called *van der Waals* force. While it is undoubtedly important in many biological systems, the force is weak and nonspecific. The effect of van der Waals forces is small compared to other atomic and molecular interactions described.

Chemical Reactions

With the information given above, one may write out many chemical reactions that ought to occur. Whether they will or not depends on several factors. Foremost, perhaps, is the relative levels of energy associated with each state. Just as water has a strong tendency to flow downhill, those molecules at lower energy levels are more favored than those at higher levels. A second factor is the energy barriers between different molecular states; that is, is it possible to get from here to there?

Chemical Reactivity and Energy. We are all familiar with the reaction of methane with oxygen.

$$CH_4 + 2\ O_2 \rightarrow CO_2 + 2\ H_2O + \text{heat}$$

Once started, the reaction goes spontaneously and produces heat to warm our houses and cook our food. Energy, in the form of heat, is one of the products of the reaction. This is typically the case when a hydrocarbon or any other substance is *oxidized*. The energy is stored in the form of C—H bonds and is released by combination with oxygen. When methane is burned, the heat released is sufficient to energize unreacted methane so that more is oxidized in a chain reaction.

Many chemical reactions have energy as one of the products. When energy is supplied from external sources, many of these same reactions will reverse and form products that have stored energy. Many reactions in our cells ought to go opposite from the way that they do, based purely on chemical considerations. This happens because energy is supplied from other reactions to force the reaction to go the wrong way. Water doesn't flow uphill by itself, but it can be pumped uphill if there is energy to run the pump. Many of the chemical reactions in cells require energy obtained through the metabolic oxidation of food.

Catalysis. When water flows downhill, it does so because there is a stream bed. If a barrier is encountered, the flow stops until sufficient water has backed up to raise the level to the top of the barrier. If a barrier that intervenes between two water levels is higher than the upper level, no flow can occur, although it would do so were there a break in the barrier.

Every chemical reaction has an energy barrier, even if the reaction produces more energy than it consumes. One can mix hydrogen and oxygen together quite safely, provided no sparks occur in the mixture. A spark would put in sufficient energy to "excite" some of the molecules and cause them to react and produce more energy in a chain reaction. True spontaneous combustion is a rare event indeed. It depends on the input of energy by some mechanism.

Another way to cause reactions to occur is to reduce the energy barrier; that is, reduce the amount of external energy required to initiate and maintain a reaction. Substances that can do this are called *catalysts. Most gene products are catalysts.* A substance such as glucose can undergo hundreds of reactions in a biological environment. But they would occur very slowly—taking hundreds of years in some instances. Yet glucose reacts very rapidly in a biological system, because the presence of catalysts (enzymes) causes the energy barriers of certain reactions to be lowered greatly. What we inherit is, in a sense, an array of catalysts that makes each of us unique.

Some Molecules of Biological Importance

Inorganic Molecules

There are tens of thousands of different molecules in any living organism. Many fall into one of several major classes that are important to heredity and to cellular function. Foremost perhaps are the inorganic molecules, of which sodium chloride, NaCl, has already been mentioned. Potassium is also very important and occurs in a completely ionized form K^+, as does sodium, Na^+. When CO_2 dissolves in water, carbonic acid is formed.

$$H_2O + CO_2 \rightarrow \begin{matrix} HO \\ \diagdown \\ \diagup \\ HO \end{matrix} C{=}O \rightarrow 2\,H^+ + \begin{matrix} {}^-O \\ \diagdown \\ \diagup \\ {}^-O \end{matrix} C{=}O$$

Under physiological conditions, one of the hydrogen ions is ordinarily replaced with a Na^+ to give sodium bicarbonate, $NaHCO_3$ (sold in the grocery stores as baking soda). Although the equation is written to show two ionizations, one of the hydroxyl (—OH) groups is often in the covalent rather than the ionized form. Phosphate is another inorganic compound of great importance. The phosphate ion is formed by combination of one phosphorus atom with four oxygen atoms.

$$\begin{matrix} & O & \\ & \| & \\ {}^-O- & \!\!P\!\! & -O^- \\ & | & \\ & O_- & \end{matrix}$$

As in the case of bicarbonate, one or more of the oxygens would combine with a hydrogen ion to form an unionized hydroxyl group. The extent to which this occurs depends on the pH and hence on the concentration of H^+. Phosphoric acid, H_3PO_4, has a strong tendency to ionize, producing hydrogen ions; hence, it is a strong acid.

Proteins

Proteins are actually *macromolecules*, meaning they consist of very large numbers of atoms. They also are *polymers*, meaning that they are built up by hooking together smaller molecules. The smaller molecules in proteins are *amino acids*, twenty of which are found regularly in the proteins of the cells of all organisms. Figure 3.4 lists these and gives their structures.

Certain features of amino acids are important to an understanding of protein structure and function and should be noted.

1. All amino acids except proline have the structure $H_2N—\overset{\overset{\text{H}}{|}}{\underset{|}{C}}—COOH$. The H_2N— group is the *amino* group, and the —COOH (an abbreviated form of $—\overset{\overset{\text{O}}{\|}}{C}—OH$) is the *carboxyl* or acid group, hence the name amino acid.
2. The rest of the amino acid molecules (to the right of the vertical line in Fig. 3.4) constitute their *side chains*. These are all different and in fact differentiate one amino acid from the other.
3. With the exception of methionine and cysteine, amino acids contain only C, H, N, and O. These four elements constitute the greater part of the mass of all organisms. The two exceptional amino acids also contain S (sulfur).

A protein is a chain or polymer of amino acids called a polypeptide. Figure 3.5 illustrates a polypeptide chain with five amino acids. Note that the amino acids are tied together through their carboxyl carbons and amino nitrogens by *peptide bonds*. (A peptide bond is an ordinary covalent bond given a special name because of the chemical groups that it joins.) Associated with the formation of a peptide bond is the loss of one H from the amino group and one OH from the carboxyl group.

$$H_2N—\overset{\overset{\text{R}}{|}}{\underset{\underset{\text{H}}{|}}{C}}—\overset{\overset{\text{O}}{\|}}{C}\text{—OH} \quad H—\overset{\overset{\text{H}}{|}}{N}—\overset{\overset{\text{R}}{|}}{\underset{\underset{\text{H}}{|}}{C}}—COOH \rightarrow H_2N—\overset{\overset{\text{R}}{|}}{\underset{\underset{\text{H}}{|}}{C}}—\overset{\overset{\text{O}}{\|}}{C}\blacksquare\overset{\overset{\text{H}}{|}}{N}—\overset{\overset{\text{R}}{|}}{\underset{\underset{\text{H}}{|}}{C}}—COOH + H_2O$$

(R = side chain; ■ = peptide bond)

The result of this is that the amino acid side chains stick out from the polypeptide chain. Also note that a polypeptide has an *amino end* and a *carboxyl end*. This is true of all polypeptides without exception. The peptide bond is a strong covalent bond not easily broken. To reduce a polypeptide to its constituent free amino acids, it is necessary to break the peptide bonds by boiling in strong acid. We also

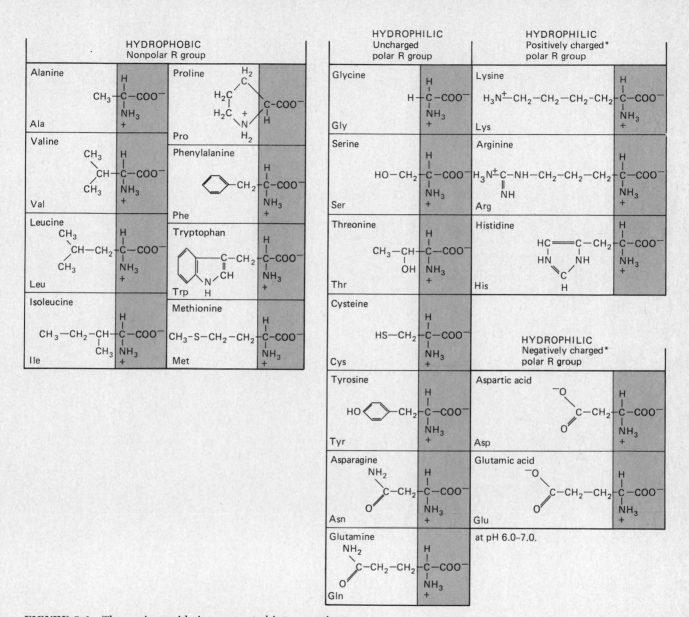

FIGURE 3.4 The amino acids incorporated into proteins.

do the job ourselves at body temperature with digestive enzymes (catalysts) in our stomachs and small intestines.

Every organism has many thousands of different kinds of proteins in its cells, all of which are constructed from the twenty amino acids listed in Figure 3.4. These

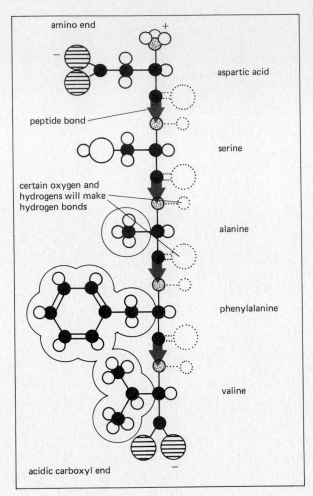

amino end

aspartic acid

peptide bond

serine

certain oxygen and
hydrogens will make
hydrogen bonds

alanine

phenylalanine

valine

acidic carboxyl end

FIGURE 3.5
A polypeptide consisting of five
amino acids joined by peptide
bonds.

proteins differ in the character and arrangement of the possible twenty amino acids in their polypeptide chains. Figure 3.6 illustrates this. The amino acids are represented by their abbreviations as given in Figure 3.4. They are numbered from left to right by convention starting with the amino end. Polypeptides A and B have the same amino acids present *in the same proportions but in a different linear sequence.* Hence, A and B are two different polypeptides and will have different functional properties in a cell. Polypeptides B and C obviously are related in structure because they have much the same sequence of amino acids, but they do differ in that the Glu in position 4 in polypeptide B is replaced by a Phe in polypeptide C. This again may be sufficient to give B different properties from C as we shall discuss in future chapters. *The important thing to remember at this point is that the specificity (character) of a protein is conferred by the particular sequence of the amino acids in its chain.*

FIGURE 3.6
A comparison of three different polypeptides that have the same number of amino acid residues but that would differ in function in a cell. A and B have the same amino acid compositions, but the linear orders differ. B and C have similar but not identical compositions. Their sequences of amino acids are identical except for the different amino acids in position 4.

Amino Acid Sequence Numbers			
1	2	3	4

Polypeptide A: Glu — Ala — Ser — Gly — · · ·

Polypeptide B: Ser — Ala — Gly — Glu — · · ·

Polypeptide C: Ser — Ala — Gly — Phe — · · ·

Proteins rarely if ever have as few amino acids as we have used for illustration. Generally their chains have fifty or more amino acids in sequence. Some have as many as 600 or more amino acids in a single chain, but the average probably lies between 100 and 200.

Twenty amino acids can form many different polypeptides of 100 amino acids. All the millions of different proteins that exist in the organic world now, and that have existed in the past, can be accounted for by assuming the participation of only twenty amino acids in their structure. Theoretically, twenty amino acids can form 10^{130} different polypeptides of 100 amino acids each!

The specific sequence of amino acids in a polypeptide is its *primary structure* and, as we have tried to emphasize, is of primary importance in determining the nature of the protein. However, few if any proteins actually function in the form of simple chains. The chains of virtually all proteins are coiled into an α-helix to form a configuration of a type shown in Figure 3.7 for the protein ribonuclease. The ribonuclease polypeptide consists of 124 amino acids that are first formed into an α-helix, which, in turn, is folded as shown in Figure 3.8. Cysteines in the chain may

FIGURE 3.7
The structure of ribonuclease. The polypeptide chain coils into an α-helix with consecutive turns of the helix held together by hydrogen bonds.

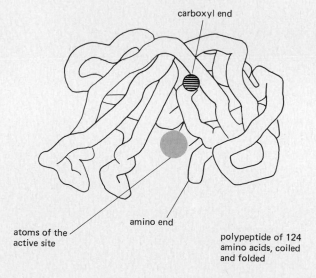

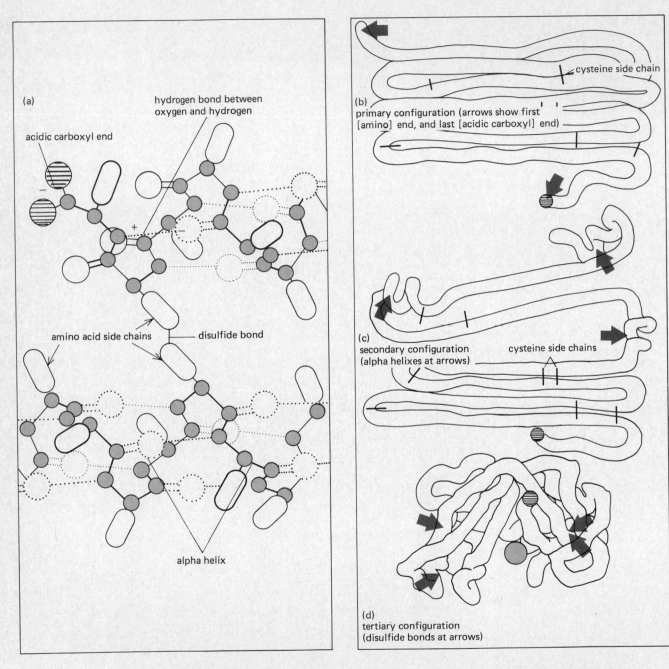

(a)

hydrogen bond between
oxygen and hydrogen

acidic carboxyl end

amino acid side chains — disulfide bond

alpha helix

(b)
primary configuration (arrows show first
[amino] end, and last [acidic carboxyl] end)

cysteine side chain

(c)
secondary configuration
(alpha helixes at arrows)   cysteine side chains

(d)
tertiary configuration
(disulfide bonds at arrows)

FIGURE 3.8   The secondary and tertiary structure of a protein. Only part of the protein is shown. The α-helix constitutes the *secondary* structure. It, in turn, is folded to form the *tertiary* structure.

form *disulfide bonds,* which give the molecule more stability than it might otherwise have.

The α-helix form of a polypeptide chain is called its *secondary structure,* and the folded three-dimensional structure formed from this is called its *tertiary structure.* As a result of the folding, an *active site* is formed in the tertiary structure. This is the site at which the protein molecule does its things as is explained in later chapters. At this point, just remember that each protein in the body has a specific function to perform, and this function is made possible by the tertiary structure, which, in turn, is determined by the primary structure—the nature and sequence of the amino acids in the chain.

Carbohydrates

Carbohydrates include simple sugars, generally consisting of a chain of five or six carbon atoms with approximately one H and one OH group per carbon. They also include polymers of these simple sugars, that is, sugars hooked together with covalent bonds into very large molecules, such as glycogen, starch, and cellulose. Examples are given in Figure 3.9. Carbohydrates are used as energy storage depots intermediate between the immediate energy source ATP (adenosine triphosphate) and the long-term energy storage in the form of fat. Many small sugar polymers are attached to the surfaces of cells and give them some of their characteristics.

Fats

Fats, also called *lipids,* generally consist of a molecule of glycerin to which are attached three fatty acids through covalent ester bonds (Fig. 3.10). In some fats, the glycerin is replaced by more complex derivatives that can bind fatty acids in much the same way. In some, only two fatty acids are bound; the third position is occupied by a phosphate group.

One of the characteristics of fats is their high energy content. Most of the fatty acid portion is pure hydrocarbon in nature, comparable in energy content to petroleum products such as gasoline. The excess calories that we consume can be efficiently stored as fat. A second property is the very hydrophobic nature of the fatty acid chain and the hydrophilic nature of the glycerin or similar structure. This means that in an aqueous medium the hydrocarbon "tails" will tend to stick together and create a small hydrophobic pocket. Biological membranes consist of a double layer of fat molecules aligned so that the hydrophobic tails are on the interior and the hydrophilic parts face the aqueous regions of the cell (Fig. 3.11).

Nucleic Acids

Like protein molecules, nucleic acids are macromolecules, but their building blocks are not amino acids. Rather, they are the three kinds of molecules shown in

44

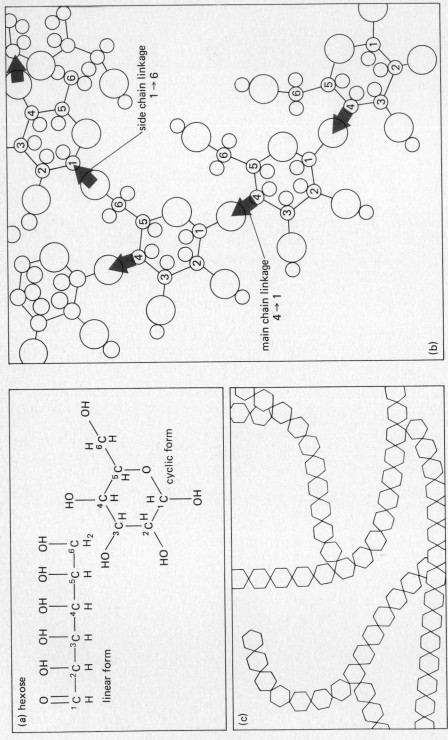

(a) hexose

linear form

cyclic form

(b)

side chain linkage
$1 \rightarrow 6$

main chain linkage
$4 \rightarrow 1$

(c)

FIGURE 3.9 Some examples of carbohydrates found in cells. (a) A representative hexose or six-carbon sugar. Each of the carbon atoms is part of a carbon chain. One of the carbons is attached by a double bond to an oxygen atom, forming a carbonyl group. Hydroxyl groups and hydrogen atoms are attached to each of the remaining carbons. The particular hexose shown has all the hydroxyls on the same side of the carbon chain. Exchanging the positions of the hydroxyl groups and hydrogen atoms on a carbon results in a different sugar. Thus there are a number of different hexoses, all with the same number of C, H, and O atoms but with different arrangements of hydroxyl groups and hydrogen atoms. Such a simple sugar is called a **monosaccharide** and can exist either in a linear form or in a cyclic form. Cyclic forms are able to join by covalent bonds to form **disaccharides** (two monosaccharide units), **trisaccharides**, or **polysaccharides**. (b) Diagram of part of a polysaccharide, showing the covalent attachment of monosaccharide units. (c) Glycogen, the principal form of energy storage in liver, consists of as many as 25,000 glucose units joined into a branched chain.

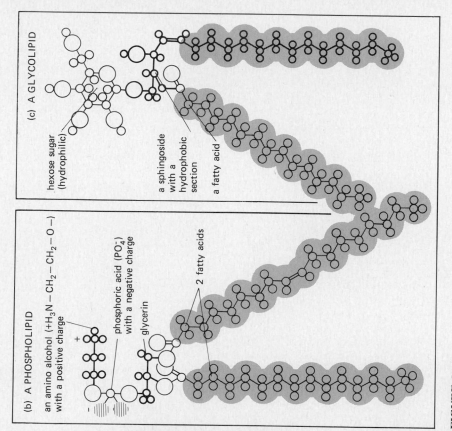

(a) LIPID STRUCTURE

glycerin

acid group

a fatty acid

a lipid molecule

(b) A PHOSPHOLIPID

an amino alcohol ($^+H_3N-CH_2-CH_2-O-$) with a positive charge

phosphoric acid (PO_4^-) with a negative charge

glycerin

2 fatty acids

(c) A GLYCOLIPID

hexose sugar (hydrophilic)

a sphingoside with a hydrophobic section

a fatty acid

FIGURE 3.10

(a) A typical lipid molecule. This lipid is made of a glycerin molecule to which three fatty acids are attached by covalent bonds. Hence, it is called a *triglyceride*. The part of the molecule where the glycerin occurs is somewhat hydrophilic because of the cluster of oxygen atoms, but the fatty acid "tails" are very hydrophobic because the structures are essentially hydrocarbons. (b) A phospholipid. One of the fatty acids is replaced by a highly polar hydrophilic phosphate group to which is also attached a hydrophilic organic structure, in this case ethanolamine ($-O-CH_2-CH_2-NH_3^+$). (c) A glycolipid. *Glycolipids* are so named because they contain sugar, usually glucose. In this case, glycerin is replaced by a related compound, sphingosine, to which the hydrophilic hexose is attached.

45

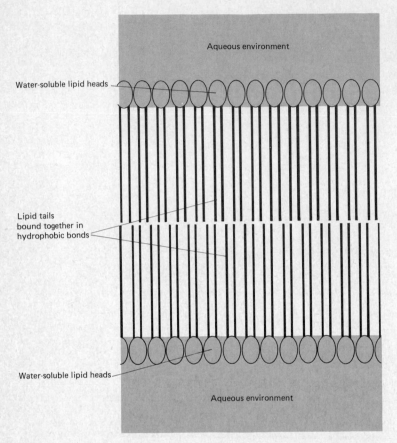

Water-soluble lipid heads

Lipid tails bound together in hydrophobic bonds

Water-soluble lipid heads

Aqueous environment

Aqueous environment

FIGURE 3.11
The bilayer structure of a biological membrane. The lipid molecules form a sandwich, the oily hydrophobic "tails" facing inward and the hydrophilic heads facing out to the aqueous medium of the cell interior or exterior.

Figure 3.12: *phosphate* or phosphoric acid, a *sugar*, and an organic *base.* Two kinds of five-carbon sugars participate: ribose and deoxyribose. These sugars differ in the number of hydroxyl groups they contain. Five kinds of bases are found represented by three kinds of pyrimidines—cytosine (C), uracil (U), and thymine (T)—and two kinds of purines—adenine (A) and guanine (G). Henceforth, we shall refer to these bases as C, U, T, A, and G.

The phosphate, sugars, and bases are joined by covalent bonds, as indicated in Figure 3.12, to form *nucleotides.* The four DNA nucleotides are shown in Figure 3.13. Certain pairs of nucleotides can associate with each other by formation of hydrogen bonds between the bases. The pairs guanylic acid (G)/cytidylic acid (C) and thymidylic acid (T)/adenylic acid (A) are the only pairs in DNA. When joined by hydrogen bonds, the nucleotide pairs have a flat hydrophobic interior due to the bases and a hydrophilic deoxyribose-phosphate on each end.

The nucleotides are the building blocks of the nucleic acids comparable to the amino acids in the structure of proteins. Figure 3.14 shows how the nucleotides are tied together to form *nucleic acids*, which are nothing more than *polymers* of nucleotides. RNA is generally composed of single chains, but DNA nearly always

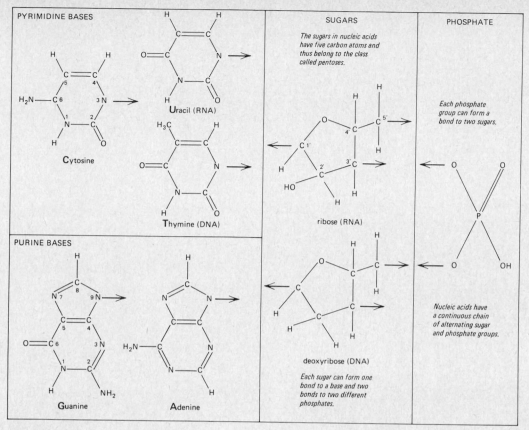

FIGURE 3.12
The building blocks of nucleic acids.

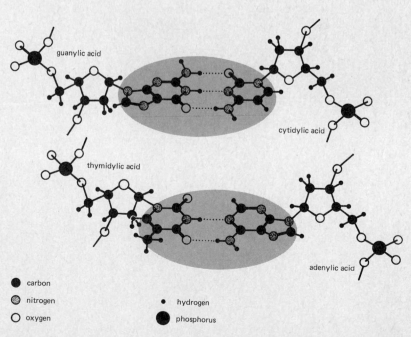

FIGURE 3.13
Nucleotides commonly found in DNA. The basic structure of all nucleotides is the same: base–sugar–phosphate. In this drawing, the nucleotides are shown paired by hydrogen bonds as they would exist in DNA. The bases (identified by the tinted background) form a rigid planar structure. The remaining portions of the nucleotide molecules are more flexible and are drawn opened out to reveal the molecular structure better.

FIGURE 3.14
A portion of a DNA molecule (portrayed opened out rather than in the usual helical conformation). DNA is double stranded except in a few viruses. The phosphate of one nucleotide forms a covalent bond to the deoxyribose of an adjacent nucleotide, whose phosphate forms a bond to the next nucleotide, and so forth, forming a continuous covalent structure that can be millions of nucleotides long. The two strands are said to be complementary because the nucleotide sequence of one strand determines what sequence can be on the other. The two strands are held together by hydrogen bonds, indicated by the row of dots in the figure.

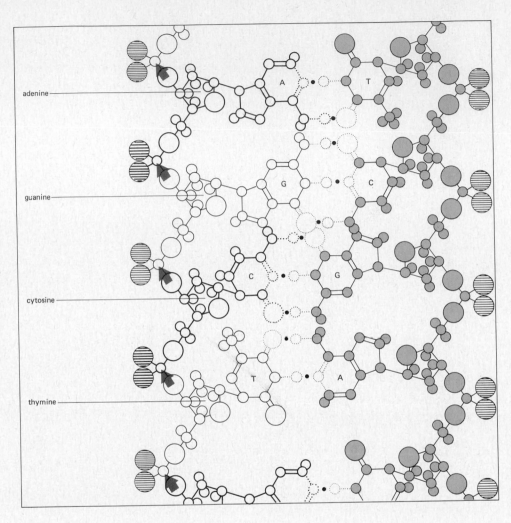

consists of two chains held together by hydrogen bonds between the purines and pyrimidines, as shown in the figure.

The double polynucleotide structure shown in Figure 3.14 is actually formed into a helical structure called the double helix because two chains are involved, as depicted in Figure 3.15. The two chains are *complementary* and of *opposite polarity*

FIGURE 3.15   (*opposite*) The two principal nucleic acids. (a) A portion of an RNA chain. (b) The DNA double helix, as described by James D. Watson and Francis Crick in 1953. DNA ordinarily has this structure in what is called its "native state" in chromatin. The Watson–Crick double helix has a right-handed twist. Another form of double helical DNA with a left-handed twist, the Z form, has been found in some regions of chromatin. In order for DNA to replicate or to act in the synthesis of RNA, it presumably must unwind partially, as described in Chapters 5 and 6.

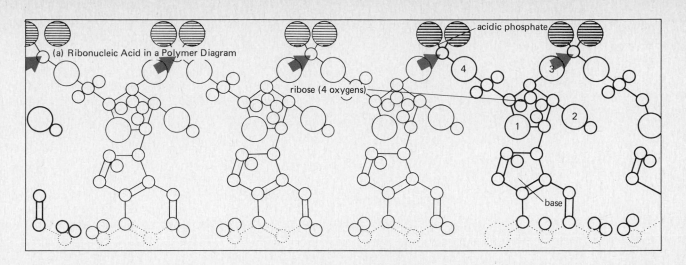

(a) Ribonucleic Acid in a Polymer Diagram

acidic phosphate

ribose (4 oxygens)

base

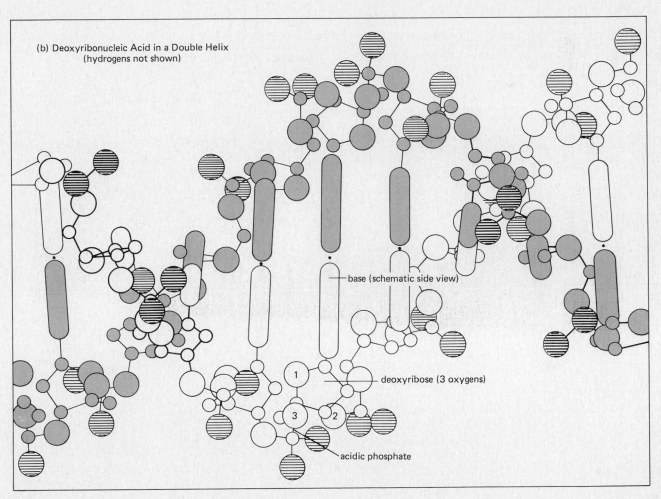

(b) Deoxyribonucleic Acid in a Double Helix
(hydrogens not shown)

base (schematic side view)

deoxyribose (3 oxygens)

acidic phosphate

49

in any given DNA. By complementary we mean that the hydrogen bonding between the purine and pyrimidines is limited to A bonding only with T, and C bonding only with G. The result is that the sequence of bases in one chain is determined entirely by the sequence in the other.

$$5'—G—T—C—A→3'$$
$$3'←C—A—G—T—5'$$

The two complementary chains will always be of opposite polarity as shown in Figure 3.13. By this we mean that in one chain the phosphate bonds to the sugars are $3'←5'$, while in the other chain they are $5'→3'$. This has important consequences, as we shall see later.

The chromosomes of the eukaryotic organisms contain large amounts of the double-stranded DNA illustrated in Figures 3.14 and 3.15. In these chromosomes, the DNA is generally coiled, folded, and bound to certain proteins. The total conglomerate is called *chromatin*. The DNA of this chromatin constitutes the genetic material.

The Beginnings of Life

No one knows exactly how life began on this planet. When the earth was first formed, it may have been a hot mass, but even this is not certain. This probably was about 4.5 billion years ago. By about 3.5 billion years ago, life may have begun; by 3 billion years ago, life had reached a stage of development adequate to leave a fossil record for us to see in the present rocks.

The Atmosphere

When life first began, the atmosphere must have been much different from what it is now. We can only make educated guesses as to what it was actually like. Table 3.3 gives the probable composition of the earth's atmosphere 4 BYA (billion years ago), 3 BYA, and 2 BYA. Obviously these estimates are crude, as to both time and composition. The types of compounds for each stage are listed in the order of abundance. Thus, methane gas (CH_4) was most abundant 4 BYA, but nitrogen (N_2) was a minor component. By 2 BYA, N_2 had replaced methane as the most abundant gas. Today methane is present as a trace amount in our atmosphere, although large amounts are still trapped in pockets below ground. The natural gas (mostly pure methane) below ground at present mostly arose from decomposition of organic materials and is found in association with larger molecular weight hydrocarbons (oil), also products of decomposition of plants and animals.

Another significant change that has apparently occurred over the past 3 to 4 billion years is the appearance of free oxygen (O_2). At present, it makes up 21% of our atmosphere, but 3 to 4 BYA it was virtually nonexistent. CO_2, also relatively abundant now at about 0.03%, may not have been present 4 BYA. Why these

TABLE 3.3
Probable Composition of the
Earth's Atmosphere in the Past as
Compared to Present

Only the major components are listed in the order of their abundance.

4.0 BYA[a]	3.0 BYA	2.0 BYA	Present[b]	
methane	N_2	N_2	N_2	78.1%
H_2	H_2O	O_2	O_2	21.0
N_2	CO_2	H_2O	Ar	0.9
H_2O		CO_2	CO_2	0.03
		Ar		

[a]Billions of years ago.
[b]Values are given in percent by volume.

changes? As we shall see, trying to answer this question gives us some important clues about what actually occurred at least during the first 2 billion years of the evolution of life from some very primitive bacterial form to the present highly developed plants and animals.

The Oceans

The consensus among cosmological geologists is that water was present and was in the liquid state very early in the earth's existence. The oceans are extremely old in any case, and from the beginning they contained substances such as the gases listed in the first column of Table 3.3 and many different kinds of inorganic salts made of the elements listed in Table 3.2. Life almost certainly started in the ocean.

It has been demonstrated in many laboratories that by using high-voltage electrical discharges or intense ultraviolet light, organic molecules such as amino acids can be made in the absence of life from mixtures of methane (CH_4), ammonia (NH_3), H_2, and H_2O. Heating mixtures such as these to temperatures of 950° C will also cause amino acid synthesis, as will high-intensity X rays. The essential constituents of DNA have also been synthesized de novo from simple mixtures, as have other essential compounds, such as fatty acids. These sorts of experimental results show that if life did start on earth, it could have done so in the ocean after a long period of chemical synthesis of its essential constituents from the very simple compounds present in solution.

The First Cells

However life did start, cells of a very simply type, which we now call prokaryotic cells, were the first to appear, probably about 3–4 BYA. These were called prokaryotic because they contained no nuclei. Their present-day descendants are the bacteria and blue-green algae. They obviously did not need or use free oxygen at first, because there was none. They were *anaerobic* organisms in contrast to *aerobic* or oxygen-requiring organisms such as we are.

Over a period of time, the anaerobic organisms or bacteria caused the accumulation of carbon dioxide (CO_2), and an important, or more accurately, a revolutionary event occurred. Some of these organisms that had been using the dissolved chemical matter in solution in the ocean for food and energy developed the ability to use the sun's light energy to react carbon dioxide and water together to form free oxygen and a form of sugar. This can be written in chemical shorthand as

$$CO_2 + H_2O \rightarrow C_xH_yO_z + O_2$$

This process is *photosynthesis*. It makes food in the form of sugar compounds designated by the general formula $C_xH_yO_z$. Over millions of years, these photosynthesizing organisms released more and more O_2 into the atmosphere and set the stage for the next giant step in evolution—the eukaryotic organisms that can function only in the presence of oxygen.

Review Questions

1. In what ways are atoms and cells similar? In what ways are they different?
2. Does a molecule that can reproduce itself exist?
3. How many different kinds of elements are known to exist?
4. How many different kinds of elements are found in living cells?
5. What is an element?
6. What is a molecule?
7. What happens to an atom that loses an electron?
8. Distinguish between protons, electrons, and neutrons.
9. Define atomic number.
10. What is a dalton?
11. Distinguish between a stable and an unstable isotope.
12. What is atomic fission?
13. What causes an element to be radioactive?
14. What radioactive isotope of a common element found in our bodies is present as a relatively high percentage of its stable isotopic form?
15. K^+ is an ion. What does this mean, and how is it formed from the elemental K?
16. What does pH measure?
17. What distinguishes an organic molecule from an inorganic molecule?
18. Compare covalent bonds and hydrogen bonds with respect to their characteristics important to life.
19. What is the difference between hydrophilic and hydrophobic chemical groups?
20. What are the three characteristic parts of all amino acids?
21. What two parts of amino acid molecules enter into the formation of covalent bonds to form peptides?
22. What basic feature determines the specificity of a protein?
23. A polypeptide chain becomes a functional protein upon doing what?
24. What are the essential constituents of a biological membrane?
25. Amino acids are to proteins as _____ are to nucleic acids.
26. What are the five purine and pyrimidine bases found in nucleic acids?
27. Which pyrimidines are found in only one kind of the two nucleic acid types, RNA and DNA?
28. What is meant by the complementarity of the two chains of a DNA molecule?
29. The opposite polarity of the two chains of a DNA molecule is the result of what?
30. What important gas necessary for the life of all eukaryotes and most prokaryotes was

probably missing from the earth's atmosphere of 3 to 4 BYA? What kind of organisms produce this gas, replacing that used up by so many organisms?

References and Further Reading

Bloomfield, M. 1980. *Chemistry and the Living Organisms*, 2nd ed. Wiley, New York. An elementary introduction to the structure of molecules and their function in biological systems.

Dickerson, R. E. 1978. Chemical evolution and the origin of life. *Scientific American* 239(3):70–86. An authoritative article written in easy to understand language.

Kenyon, D. H., and G. Steinman. 1969. *Biochemical Predestination*. McGraw-Hill, New York. A rather advanced treatise on the subject of the origins of life on earth. But the first few chapters are worth trying even for those with a meager chemistry background.

Miller, S. L., and L. E. Orgel. 1974. *The Origins of Life on the Earth*. Prentice-Hall, Englewood Cliffs, N.J. 229 pp. Two of the foremost scientists working on the origin of life summarize current ideas.

Oparin, A. I. 1957. *The Origin of Life on the Earth*. Academic Press, New York. The classic in the field. Start with this if you are interested in the subject.

Sackheim, G. I., and D. D. Lehman. 1981. *Chemistry for the Health Sciences*, 4th ed. Macmillan, New York. A basic textbook that covers the elements of chemistry and biochemistry.

Cells are made of molecules organized in various ways to form specific kinds of structures with specific functions. Two fundamentally different kinds of cells exist, and the distinction between them is the basis for dividing the living world into two distinctly different categories: those organisms called *prokaryotes* and those called *eukaryotes*.

Bacteria and blue-green algae are prokaryotes, and all other organisms are eukaryotes. The root *karyo* refers to *nucleus*. Thus prokaryotes are before (*pro* = before) nuclei and have none, whereas eukaryotes have "true" nuclei (*eu* = good), which consist of chromosomes suspended in a liquid surrounded by a membrane. The rest of the eukaryotic cell outside the nucleus is called the *cytoplasm*.

Human beings, all other animals, and all plants are eukaryotes because of their cell structure. Hence we shall be considering *eukaryotic genetics* principally in this book. However, we shall also have something to say about *prokaryotic genetics* because of the important role that bacteria play in human affairs.

The Physical Structure of Cells

Bacterial Cells

Figure 4.1 depicts a bacterial cell. It may be considered the simplest kind of cell, consisting as it does of a *cell membrane* and cell wall surrounding an inner mass containing different kinds of proteins and nucleic acids mostly suspended in a dilute salt solution. Also present, of course, are many other kinds of molecules. Nucleic acids are represented by at least three different kinds of RNA, that will be discussed in more detail in Chapter 6, and a circle of double-stranded DNA that is generally called a *chromosome*. This chromosome is simpler in structure than a eukaryotic chromosome and is not enclosed in a membrane. Hence, there is no nucleus as distinct from cytoplasm.

In addition to the chromosomal DNA, that is about 1.1 mm in length in the common bacterium *Escherichia coli*, bacteria may also have smaller rings of DNA floating about called *plasmids* (Fig. 4.1). Plasmids ordinarily have 1/1000 to 1/100,000 of the DNA found in the *E. coli* chromosome. These plasmids are not generally necessary for the bacterium's existence, but they can confer properties on the bacterium that it would otherwise not have. Plasmids replicate themselves just as does the host bacterial cell's chromosome. Some plasmids stay independent of the host cell chromosome. Others can integrate themselves into the host chromosome, in which case they are called *episomes*.

Bacteria may also play host to *bacteriophage*, often just called *phage*. These are simply *viruses* that infect bacteria. They resemble plasmids in having a relatively small circle of DNA, but they differ significantly in also possessing a coat of protein surrounding the DNA, as shown in Figure 4.2.

Bacteria, plasmids, and phage are considered in more detail in Chapter 14.

Eukaryotic Cells

A diagram of an animal cell is given in Figure 4.3. In addition to being several orders of magnitude larger than the bacteria, it is much more complex internally. Its DNA

4

The Structure and Function of Cells

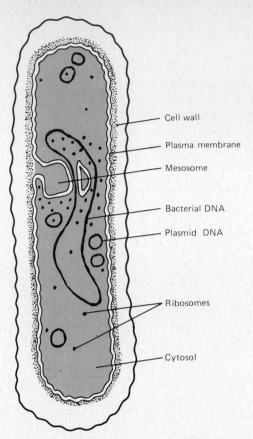

FIGURE 4.1
A bacterial cell.

Cell wall

Plasma membrane

Mesosome

Bacterial DNA

Plasmid DNA

Ribosomes

Cytosol

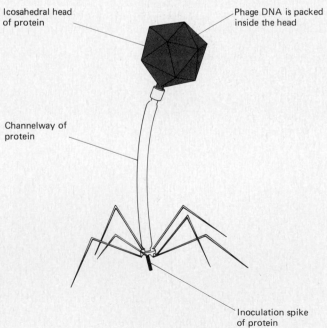

Icosahedral head
of protein

Phage DNA is packed
inside the head

Channelway of
protein

Inoculation spike
of protein

FIGURE 4.2
A bacteriophage particle.

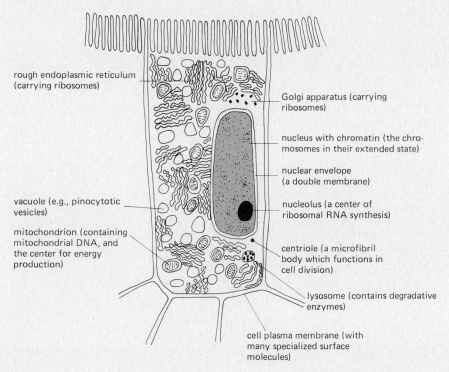

FIGURE 4.3
An animal cell.

rough endoplasmic reticulum
(carrying ribosomes)

Golgi apparatus (carrying
ribosomes)

nucleus with chromatin (the chromosomes in their extended state)

nuclear envelope
(a double membrane)

vacuole (e.g., pinocytotic
vesicles)

nucleolus (a center of
ribosomal RNA synthesis)

mitochondrion (containing
mitochondrial DNA, and
the center for energy
production)

centriole (a microfibril
body which functions in
cell division)

lysosome (contains degradative
enzymes)

cell plasma membrane (with
many specialized surface
molecules)

is organized into chromosomes enclosed inside a *nuclear membrane* to form a *nucleus*. The rest of the cell, enclosed within a *plasma membrane*, forms the *cytoplasm*. The cytoplasm is a complex of different kinds of structures such as *mitochondria, endoplasmic reticulum, centrioles*, and the *Golgi apparatus*. These *organelles* have specific roles to play in the functioning of animal and plant cells and are not found in bacterial cells.

Plant cells are similar to animal cells in the complexity of their structure and have the same organelles as animal cells plus one more type, the chloroplasts (Fig. 4.4). In addition, plant cells have a rather rigid *cell wall* composed of cellulose outside the plasma membrane.

Both plant and animal cells harbor viruses and plasmids in cytoplasm and nucleus. These entities and their effects on eukaryotes are discussed further in Chapter 14.

Eukaryotic cells vary enormously in size but they are uniformly larger than bacteria and viruses. Box 4.1 gives some idea of the ranges in dimensions of various kinds of cells and cellular components.

The Nucleus and Its Chromosomes.   The nuclei of eukaryotic cells have chromosomes as their major constituents. The chromosomes bear the hereditary determinants that are passed from generation to generation through sperm and eggs. These determinants are called *genes*. The kinds of genes an individual carries on his or her chromosomes constitute the *genotype* of that individual. The genotype determines the course of development from egg to adult under the

—Box **4.1**

Cellular Distances

The complete range of eukaryotic cell sizes is enormous. Animal eggs may range up to 2–3 cm or more, and marine plant cells may exceed 5 cm. The range in size (or length) from smallest to largest is over 10 million fold.

Average Size	Entity
2nm	Width of DNA molecule.
5,000 nm	Length of human mitochondrial DNA.
16,000,000–82,000,000 nm	Average length of human chromosomal DNA.
1,980,000,000 nm	Total length of human diploid genome DNA.
10–50 nm	Diameter of plasmid.
50–100 nm	Diameter of virus.
1.5 μm	Length of bacterium.
1–2 μm	Length of mitochondrion.
1–2 μm	Length of chloroplast.
0.01–1 mm	Diameter of eukaryotic cells (but nerve cells may have fibers that extend up to several meters).

influence of the various environmental factors to which the organism is subjected. The result of this interaction between genotype and environment is the *phenotype*—the appearance, functioning, and behavior of the organism.

The nucleus has been described as the seat of heredity because it contains the chromosomes. We now know that mitochondria, chloroplasts, and plasmids that may be present possess a few genes of their own. The number of these is small compared to the tens of thousands of genes in eukaryote nuclei.

Chromosome Structure. The chromosomes of all animals and plants have the same basic structure: a double helical strand of DNA extending their length and which is intimately associated with a number of different kinds of protein. The DNA, plus a small amount of RNA, and a large amount of protein are the principal components of chromosomes. The combination is referred to as *chromatin* because of its high affinity for colored stains that make the chromosomes visible in the microscope.

Chromatin is not a haphazard aggregation of protein molecules and DNA, but rather is a highly organized conglomerate containing two principal kinds of protein—the *histones* and the *nonhistone proteins*. The histones are the major protein component in the chromosomes (Table 4.1). There are five kinds: H1, H2A, H2B, H3, and H4. Some of the characteristics of these are listed in Table 4.2. Note that all are basic proteins, because they are rich in the basic amino acids arginine and lysine.

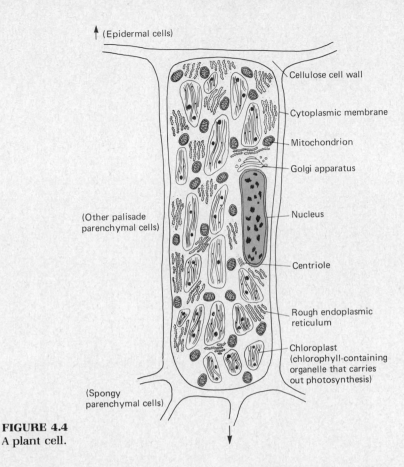

(Epidermal cells)

Cellulose cell wall

Cytoplasmic membrane

Mitochondrion

Golgi apparatus

Nucleus

(Other palisade parenchymal cells)

Centriole

Rough endoplasmic reticulum

Chloroplast (chlorophyll-containing organelle that carries out photosynthesis)

(Spongy parenchymal cells)

FIGURE 4.4
A plant cell.

Four of these histones, H2A, H2B, H3, and H4, are organized in chromosomes as *nucleosomes*, which occur at intervals along the DNA strand as shown in Figure 4.5. Each of the four kinds of histones is represented in the nucleosome by two molecules, forming a histone octomer that is intimately associated with the DNA strand wound about it (Fig. 4.5). The H1 histone is shown in Figure 4.5 and is present in the intact chromosome, where it presumably binds nucleosome to

TABLE 4.1
The Chemical Composition of Chromatin

The values given are for calf thymus chromosomes. Values may vary somewhat in other organisms.

Component	Percent in Chromatin
DNA	39
Histone	45
Nonhistone protein	13
RNA	3

TABLE 4.2
Types of Histone Found in Chromatin

Histone Type	Approximate Molecular Weight	Characteristics of Basic Amino Acids
H1	21,000	lysine rich
H2A	14,000	lysine rich
H2B	14,000	lysine rich
H3	15,000	arginine and lysine rich
H4	11,000	arginine and lysine (as well as the neutral glycine) are well represented

nucleosome. Other related kinds of histones exist in embryonic animals, but the histones described in Table 4.2 are the ones we have by the time we are born.

The functional chromosomes also include an undetermined number of other kinds of proteins, the nonhistone proteins, many of which are highly acidic in contrast to the basic histones. The arrangement of these is not understood, but

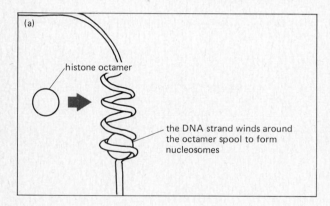

(a) histone octamer

the DNA strand winds around the octamer spool to form nucleosomes

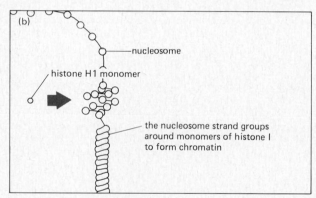

(b) nucleosome

histone H1 monomer

the nucleosome strand groups around monomers of histone I to form chromatin

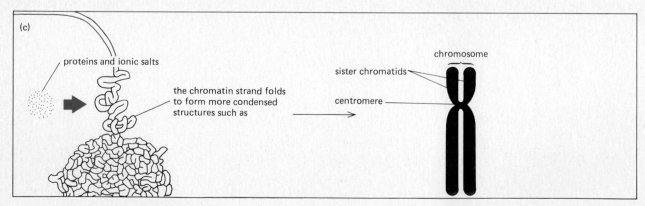

(c) proteins and ionic salts

the chromatin strand folds to form more condensed structures such as

chromosome

sister chromatids

centromere

FIGURE 4.5
The interaction of histones and DNA to form nucleosomes.

some of them undoubtedly function in the processes of DNA replication and in the formation of RNA.

Chromosomes can change their appearance considerably with the phases of the cell cycle. We deal with cell cycles in the next chapter, but here we need to point out that at a stage of the cell's existence called *interphase*, its chromosomes are quite invisible with the light microscope, and even with the electron microscope, they are difficult to observe clearly. It is only when a cell is about to divide that its chromosomes become visible. Then they may appear as shown in Figure 4.6, when they are said to be in the *condensed* state.

During interphase, the chromosomes are stretched out and greatly extended. As they approach the period of cell division, they begin to coil more and more and then fold as shown in Figure 4.6 until they appear as shown in Figure 4.6C. As they coil and fold, they become shorter and thicker and are visible under the light microscope. When stained, they appear as shown in Figure 4.6D. Here each appears as a double figure because the arms have replicated but are still attached at one point, the *centromere*. This area plays a very important role in the distribution of chromosomes at the time of cell division. The centromere may be located at any point along a chromosome, but it is always at the same position for a particular chromosome.

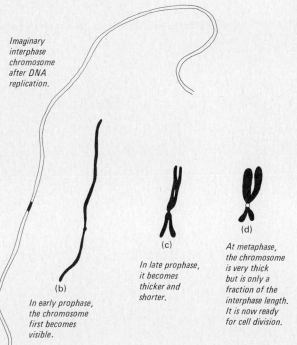

FIGURE 4.6
The appearance of chromosomes at different stages of the cell cycle.

Imaginary interphase chromosome after DNA replication.

(a)

(b)
In early prophase, the chromosome first becomes visible.

(c)
In late prophase, it becomes thicker and shorter.

(d)
At metaphase, the chromosome is very thick but is only a fraction of the interphase length. It is now ready for cell division.

The Human Chromosome Complement.   Under ordinary circumstances, all the cells in our bodies have nuclei containing 46 chromosomes. If examined at the time of maximum condensation, called *metaphase*, they have the general appearance of those shown in Figure 4.7. The display in Figure 4.7a can be organized better by cutting out the chromosomes in the photograph with scissors. They can then be arranged in a standard sequence with those that are alike in appearance next to each other (Fig. 4.7b). The procedure for analyzing human chromosomes is described in Box 4.2. A standard array of chromosomes is called a *karyotype*. The karyotype is therefore the chromosome complement of an individual, including

FIGURE 4.7
A set of human chromosomes at metaphase. (a) A set from a single nucleus as it appears on a microscope slide after preparation and staining. (b) A set after the chromosomes are photographed and cut out and the similar chromosomes (homologs) are arranged in pairs to give the 22 pairs of *autosomes* (numbered 1 to 22) and the pair of sex chromosomes (labeled X and Y). [(a) Provided by Dr. Patricia N. Howard-Peebles, The University of Texas Health Science Center at Dallas; (b) provided by Dr. Frances Arrighi, The University of Texas System Cancer Center, Houston.]

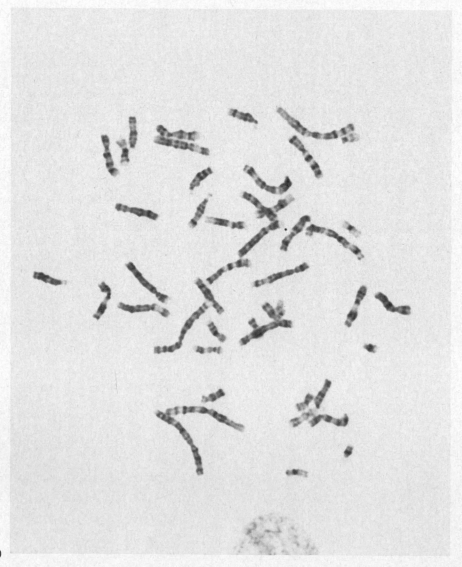

(a)

FIGURE 4.7
(continued)

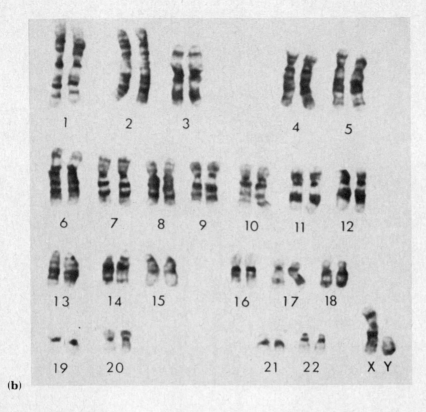

(b)

the number of chromosomes and the morphology or appearance of the chromosomes. For example, note that there are 23 pairs of chromosomes. In most of them the centromeres are in the middle or near the middle so that the two *arms* are quite easy to see. These are described as *metacentric* or *submetacentric* chromosomes. If the centromere is near one end, as it is for chromosomes 13, 14, 15, 21, 22, and the Y, the chromosome is said to be *acrocentric*.

By means of various staining techniques, *bands* can be developed in the chromosomes. The location and size of these bands are characteristic of the particular chromosome in virtually all members of the species and indeed of homologous chromosomes in closely related species. The chromosomes in Figure 4.7a are stained to show G-bands (named after the Giemsa stain). Other staining techniques show the same bands, but the different bands have different relative intensities. Careful analysis of the bands permits the construction of an *idiogram*, a diagram showing the chromosome bands. Figure 4.8 shows an idiogram of the human chromosome complement. The bands do not represent genes but are stretches of chromatin. Certain chromatin regions have a greater affinity for dyes for reasons that are only partially understood.

As noted above, the banding pattern of a particular chromosome is the same in nearly all members of the species. Unusual patterns in a chromosome may occur in some individuals, but these are usually associated with abnormalities, either in

The Preparation of Human Chromosomes for Analysis

Chromosomes can be observed in any cell that is dividing. In most human tissues, very few cells are dividing at any one time, and a search would yield few cells in metaphase, when the chromosomes are highly condensed and most readily visible. Earlier studies were limited to biopsies of the testes, since the germ cells of males do divide continuously. Such cells would be undergoing meiosis. Testicular biopsies obviously are not suitable for routine chromosome analysis.

A major advance in chromosome analysis was provided by the development of methods for culturing human cells in the laboratory. A variety of cell types can be cultured: bits of skin, hair follicles, white blood cells, and cells from amniotic fluid. All have their special uses. In the most common procedure, a few milliliters of blood are withdrawn from an arm vein and the white cells are separated from the red, although the separation step may be omitted. The white cells are placed in a culture medium, along with a *mitogen* (a substance that induces mitosis). Several plant extracts are excellent mitogens. The lymphocytes respond to the mitogen by growing and dividing. After 2–3 days, the chemical colchicine or its derivative colcemid is added. This arrests cell division in metaphase. After 2h, the cells are harvested. This consists in treating them with a dilute salt solution, transferring them to microscope slides, and staining them. Those in

metaphase will have their chromosomes spread as in Figure 4.7. The chromosomes can then be counted and arranged into karyotypes.

The analysis of chromosomes in cells from skin or other tissues is carried out in much the same way except that no mitogen is required and the cells may be cultured for several weeks if very few cells are available initially.

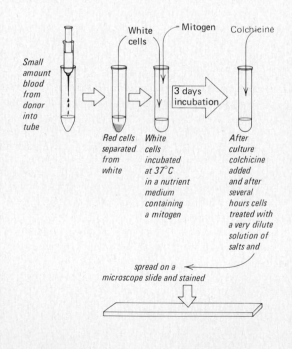

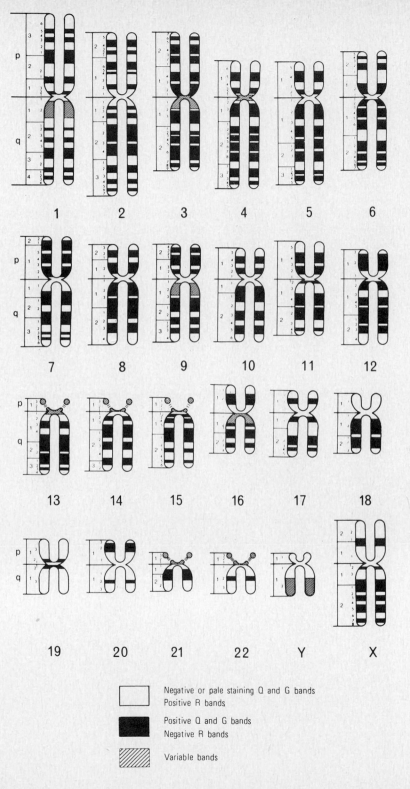

FIGURE 4.8
An idiogram of the human chromosome complement. This shows a male karyotype with both X and Y chromosomes. (From Paris Conference, 1971; reproduced by permission of the March of Dimes Birth Defects Foundation.)

Negative or pale staining Q and G bands
Positive R bands

Positive Q and G bands
Negative R bands

Variable bands

the individual or in offspring of that person. Our close evolutionary relatives, the great apes, have karyotypes that are almost identical to ours (Chapter 19). Although the apes have 48 chromosomes (24 pairs) rather than the 46 (23 pairs) that we humans have, nearly all the chromosome segments can be matched with the banding patterns. Human chromosome 2 corresponds to two of the ape chromosomes (see Fig. 19.8).

Of the 23 pairs of human chromosomes, 22 are numbered. The remaining pair are *sex chromosomes.* The two kinds of sex chromosomes are called X and Y. Females have two X chromosomes, and males have one X and one Y. The karyotype of Figure 4.7 is from a male and shows two different sex chromosomes. The chromosomes that occur in identical pairs in males and females and are numbered 1 through 22 are called *autosomes.* The *karyotype formula* of a person is a brief symbolic representation of the karyotype that indicates the total number of chromosomes, the sex chromosome complement, and any unusual chromosomes that may be present. A normal human female is 46,XX and normal male is 46,XY.

The Structure of Cytoplasm. The cytoplasm is an active part of the cell in which occur most of the cell's chemical processes, collectively called metabolism. Scattered throughout the cytoplasm is what appears under the electron microscope as a network of tubules, vesicles, and fibrils. This, the *endoplasmic reticulum,* is the site where 99% of the cell's proteins are synthesized.

Also prominent in the cytoplasm when viewed under the high magnification of the electron microscope are the *mitochondria* scattered about between the strands of the *endoplasmic reticulum.* These organelles (Figs. 4.9 and 4.10) have a

FIGURE 4.9
An electron micrograph of an ultrathin section of a human cell showing mitochrondria (Mt) and a portion of the nucleus. (Furnished by Dr. Donald R. Robberson, The University of Texas System Cancer Center, Houston.)

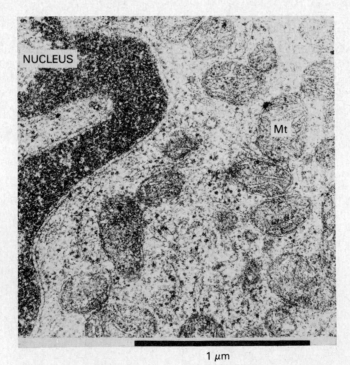

NUCLEUS

Mt

1 μm

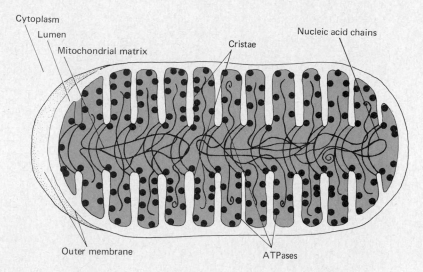

Cytoplasm
Lumen
Mitochondrial matrix
Cristae
Nucleic acid chains
Outer membrane
ATPases

FIGURE 4.10
A mitochondrion in section, showing some of its internal structure.

complex structure consisting of membranes, proteins associated with these membranes, and nucleic acids of both types, DNA and RNA.

Other organelles in the cytoplasm that bear mentioning are the *lysosomes* and the *Golgi apparatus* (Fig. 4.3). Lysosomes produce (or secrete) enzymes involved in the breakdown (lysis) of proteins, fats, and many other substances, while the Golgi body or apparatus is instrumental in the secretion of a number of proteins such as milk proteins and collagen, as well as certain proteins in membranes.

Cell Membranes.   Membranes constitute an integral part of cells and are found around and throughout them. The *plasma* or *cell membrane* encloses the cell; the endoplasmic reticulum is a membrane within the cytoplasm; the mitochondria consist principally of membranes, as do lysosomes and Golgi bodies, and the nuclei have a double membranous nuclear envelope separating the nuclear contents from the cytoplasm.

The basic structure of the cell membranes is shown diagrammatically in Figure 4.11. The fatty or lipid part of the membrane is made up of an assortment of lipid molecules, organized into a bilayer as described in Figure 3.11. A membrane is a very fluid structure, with the components constantly moving about. The proteins of membranes "float" in the lipid part. The kinds of proteins present vary with the kind of membrane. Generally the cell-membrane proteins have carbohydrate chains attached to the end that protrudes to the outside. These play an important role in the reactions of the cell with molecules external to the cell. For example, the blood groups A, B, and O are determined by the kinds of sugars associated with the membranes of the red blood cells. These sugar proteins, or *glycoproteins* as they are called, function as sites of recognition for other molecules delivering messages, such as hormones. They are therefore also called *receptors*. Different kinds of cells will have different kinds of receptors in their membranes and hence respond to different kinds of signals from molecules that can bind to the receptors. Many membrane proteins function in the transport of specific metabolites from the exterior to the interior of cells, forming a highly selective gateway for substances needed in cellular metabolism and function.

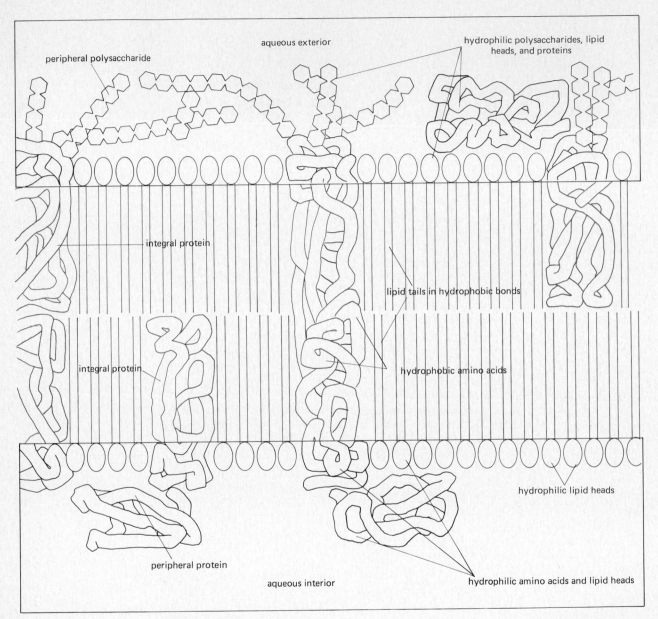

FIGURE 4.11
The basic structure of a cell membrane.

Cells and Metabolism

Cells are the functional units of the body. They carry out the chemical reactions that constitute metabolism, and they reproduce. The term *metabolism* is applied to the totality of the chemical reactions that occur in cells. Literally thousands of these reactions take place, and we can only consider them here in the most general way.

The biochemist who studies the various metabolic reactions frequently classifies them into two categories: (1) the *catabolic* and (2) the *anabolic*. Figure 4.12 outlines the three principal stages of each and shows how they are related. The main difference between the two categories is that the catabolic reactions are *degradative*, while the anabolic are *synthetic*. It is important to arrive at some understanding of these principal pathways because so much of what we will have to say about genetics in succeeding chapters is intimately related to metabolic processes.

Catabolism and the Transformation of Energy

The catabolic pathways are indicated by the black arrows in Figure 4.12. They consist in the breakdown of the complex macromolecules such as proteins, simple carbohydrates and polysaccharides, and lipids to their constituent building blocks (stage 1) and the further reduction (stages 2 and 3) of these to the simple endproducts NH_3 (ammonia), H_2O, and CO_2, which are excreted through the kidneys, lungs, and skin. (In humans, NH_3 is principally excreted as urea through the kidneys.) An important part of the catabolic process involves the production of energy in usable form for the anabolic processes indicated by the gray arrows.

Most of the energy generated in cells is from the catabolism of sugars, lipids, and amino acids in stage 3 of Figure 4.12. The breakdown of these is principally in the mitochondria of the cytoplasm via what is called the *tricarboxylic acid cycle* (or Krebs cycle) and the *electron transport system*. The net result of these metabolic activities is the production of H_2O, CO_2, and ATP and the utilization of O_2. These metabolic reactions constitute what is familiarly called *respiration* and may be written in the overall reaction as

$$P_i + O_2 + ADP + C_xH_yO_z \rightarrow CO_2 + H_2O + ATP$$

The ADP in this reaction is a nucleotide with the structure given in Figure 4.13. It is given the acronym ADP derived from its chemical name **a**denosine **dip**hosphate and differs from the adenine nucleotide in RNA only by having an additional phosphate. ADP occupies a central role in the metabolism of cells because it combines with another phosphate (P_i) to form ATP or **a**denosine **trip**hosphate in the presence of available chemical energy. ATP is *energy rich* and gives up its energy to drive anabolic reactions as shown in Figures 4.12 and 4.13.

The $C_xH_yO_z$ in the above chemical equation represents either sugar or fat. What the equation says is that the complete catabolism of these compounds to CO_2 and H_2O (Fig. 4.12) results in the transfer of energy from the chemical bonds in the

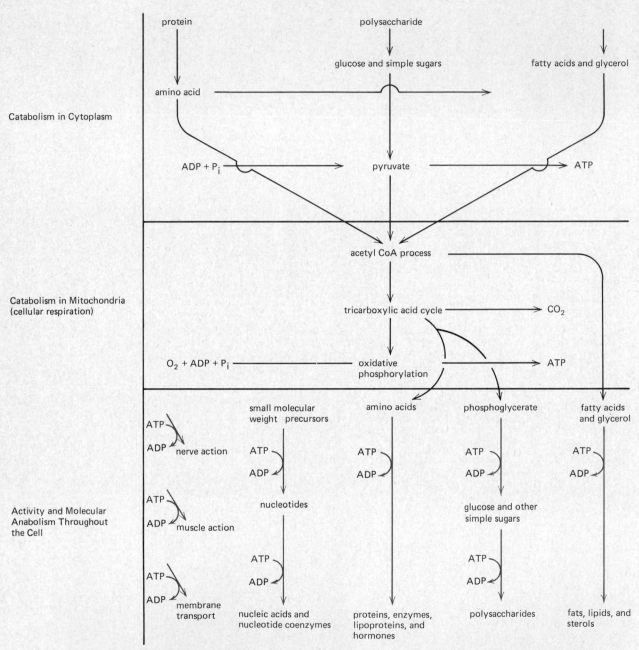

FIGURE 4.12
Catabolic and anabolic metabolism.

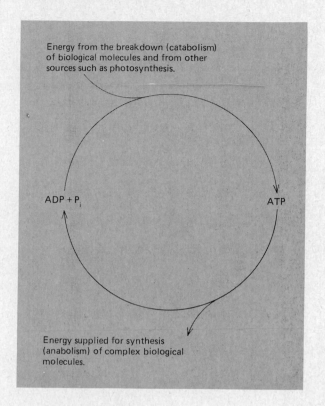

This drawing of ATP shows three phosphate groups; ADP is shorter by one phosphate.

FIGURE 4.13
ADP, ATP, and their role in metabolism.

$C_xH_yO_z$ molecules to ADP with the production of ATP. Energy is neither created nor destroyed but transformed from $C_xH_yO_z$ to ATP. In ATP it is readily available to be used in the anabolic and other life processes, such as muscle contraction. As ATP energy is used, ADP is formed, and the process must be repeated to regenerate the ATP from the energy derived from the $C_xH_yO_z$ compounds coming in as food.

Amino acids are also catabolized in the body to produce urea, $H_2N-\overset{\overset{\displaystyle O}{\|}}{C}-NH_2$, and in the process ATP is generated from ADP in the presence of inorganic phosphate (P_i).

Anabolism and Photosynthesis

The anabolic reactions of cells are the synthetic processes. They use the energy provided by the consumption or catabolism of $C_xH_yO_z$ to synthesize proteins, nucleic acids, lipids, and the numerous other molecules that form the structural and functional elements of cells. These anabolic activities are outlined briefly in Figure 4.12. Obviously in order for anabolism to continue so the cells may grow and

FIGURE 4.14
Photosynthesis. The sun's energy, in the form of light, is converted into chemical energy. Photons of light are absorbed by chlorophyll and related pigments. The absorbed energy is converted into chemical energy by a series of complex reactions. Without this key process, life could not be sustained over a long period.

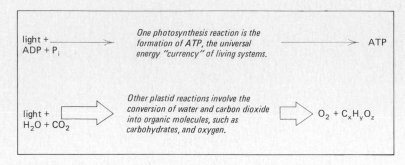

function, there must be a continual supply of ATP and hence $C_xH_yO_z$ in the form of food. There must also be oxygen, O_2. The O_2 and $C_xH_yO_z$ come from the photosynthetic activities of green plants (Fig. 4.14). As we have already mentioned on page 52, early in the history of life, a type of prokaryote arose that could use sunlight, CO_2, and H_2O to make $C_xH_yO_z$ plus O_2. We are thus totally dependent ultimately on the green plants to carry out these reactions.

Proteins and Metabolism

The diagram in Figure 4.12 is highly simplified. It must be appreciated that catabolism and anabolism involve many thousands of chemical reactions. Furthermore, with few exceptions, none of these reactions would proceed sufficiently rapidly to support life, were it not for the presence of *enzymes*, which are proteins that function as catalysts.

As noted in the previous chapter, a catalyst is a substance that increases the rate of a chemical reaction. Most proteins function as catalysts. They typically are highly specific; that is, they will catalyze only one or a very few reactions. Each kind of chemical reaction in cells has an enzyme specific for that reaction. Hence, since there are thousands of different kinds of reactions, there must be at least thousands of different kinds of enzymes (proteins). These proteins differ from one another in their amino acid sequences, as we have described in the previous chapter.

The great catalytic power of enzymes can best be appreciated by realizing that they can increase the rate of chemical reactions above the rate in the absence of enzymes by 10,000 to 100,000,000,000,000 times! They are essential to life because cells and hence organisms can only function if their chemical, metabolic reactions are carried out rapidly.

Not all proteins are enzymes in the strict sense. Some carry out transport roles. For example, the hemoglobin in our blood carries oxygen to our cells for use in respiration. However, it can be considered a catalyst of this transport, for without it oxygen would diffuse so slowly that our cells would die from inadequate amounts of this essential substance that functions as an acceptor for hydrogens produced in catabolism, as shown in Figure 4.12. Without oxygen, the hydrogens resulting from the breakdown of carbohydrates would have no place to go.

Some proteins are basically structural. One example is *collagen* found in the connective tissues of animals. It is perhaps the most abundant protein in our bodies, being about ⅓ of our total protein. It plays the important role of holding us together as tendons and by carrying out protective roles as the cornea of the eye, which is almost pure collagen.

It is important to note that in conformance with the high degree of organization of the eukaryotic cell, the distribution of the cell's enzymes is also highly organized. Indeed, most enzymes are *compartmentalized* within specific organelles or attached to specific membranes. The mitochondria provide the best example of this. The outer membrane of mitochondria is very similar to the other membranes of the nonmitochondrial cytoplasm, and such enzymes as are associated with it are generally found in other parts of the cytoplasm. The inner mitochondrial membrane is a highly organized entity, as shown in Figure 4.10. It contains many different kinds of enzymes arranged in a precise order to carry out the important function of converting the carbon fragments produced by *glycolysis* in the cytoplasm to CO_2 and H_2O with the production of ATP. Both the electron transport system and the tricarboxylic acid cycle are arranged inside the mitochondria enclosed by or attached to the inner membrane. The amount of ATP made outside the mitochondria is miniscule compared to that produced within them. As a result, poisoning of the mitochondrial functions, for example with cyanide, results in almost instantaneous death.

To a considerable extent we are what our proteins are. By virtue of their control of the rates of the chemical reactions constituting our metabolism and by their functioning in our structure, they determine not only that we are humans but that each of us is a special kind of human different from other humans.

Complex Organisms and the Division of Labor Among Cells

Organisms can be single cells or multicellular, made of a few or very many cells. When they are multicellular, as of course we are, they nearly always have different kinds of cells. The different kinds of cells are organized into tissues and the tissues into organs. The organism is the conglomerate of organs, tissues, and cells.

All cells in the body are not the same in structure or function. During the process of development from the fertilized egg (or *zygote*), the differentiation of cells occurs. The result is the production of cells with specialized functions, such as nerve cells, muscle cells, skin cells, and liver cells (Fig. 4.15). These different kinds of cells differ primarily in the kinds of proteins (enzymes) they contain. Proteins make the cells different and give them different structures and functions.

Although differentiated cells have different enzymatic capacities for carrying out their specific functions (liver cells can make glycogen from glucose, for example, whereas kidney cells cannot), they maintain those organelles necessary for staying

FIGURE 4.15
Some of the different kinds of cells found in the human body. All cells are drawn to the same scale. These different cell types illustrate the variety of cell shapes and sizes found in descendants of the fertilized egg. The functions of the cells are equally diverse.

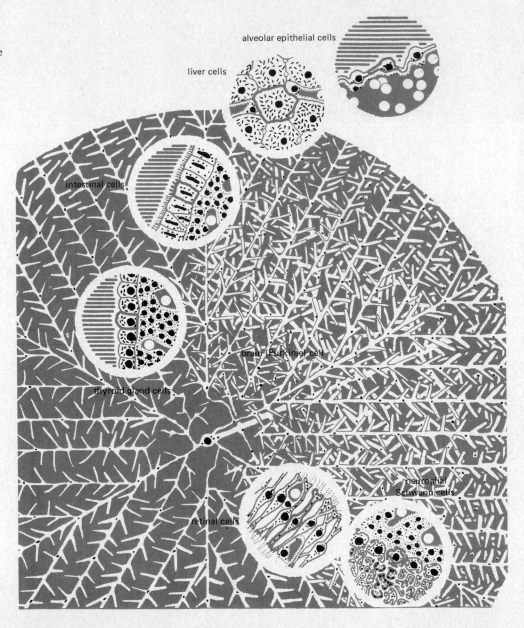

alive. Exceptions to this rule are our red blood corpuscles (erythrocytes), which lose their nuclei during their development in our bone marrow. (But red cells of birds and many other species retain their nuclei when they mature.) As a consequence, they also lose their capacity to reproduce. But in return, they have much more room for hemoglobin, which comprises about 95% of their total protein. Red cells must continually be replaced by new cells produced in the marrow of our long bones, sternum, and skull.

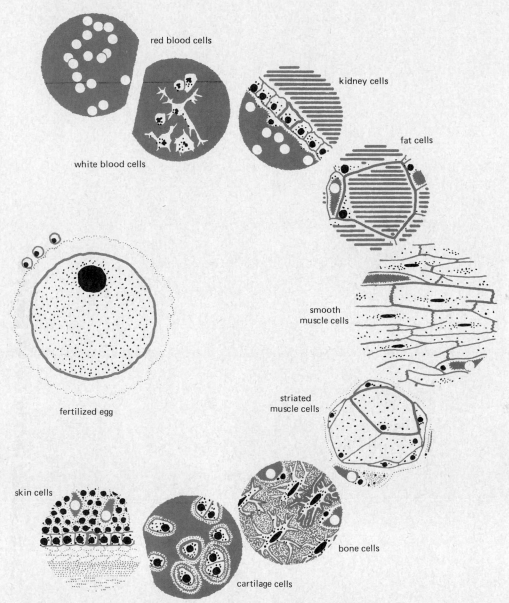

FIGURE 4.15   (continued)

During the process of differentiation in development, the cells, although changing in structure and function, retain the genetic identity of the fertilized egg. By this we mean that they generally contain the same complement of chromosomes as the fertilized egg. We will elaborate on this in the following chapter.

The organization of cells into tissues and the tissues into organs is illustrated in Figure 4.16, which shows how the various kinds of cells in tissues are organized in one particular organ, our small intestine.

FIGURE 4.16
A cross section of the small intestine, showing the organization of different types of cells into tissues.

(a) Cell Functions

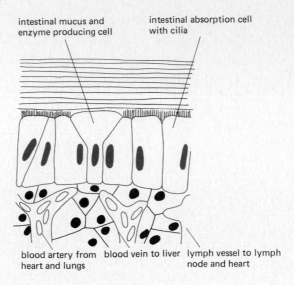

(b) The Digestive Tube

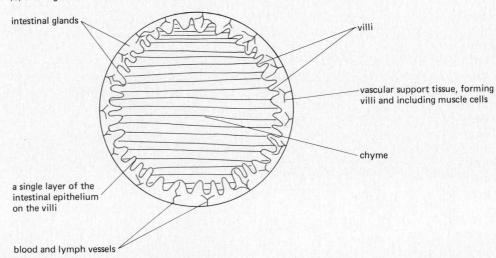

Reproduction of Cells

The other important function of cells aside from metabolism is reproduction. Cells are not only units of metabolism but also units of reproduction. This reproductive function is accomplished by cell division during the cell cycle. Whereas proteins are the important elements in metabolism, the nucleic acids are the important elements in reproduction, as we describe in the following chapter. Protein molecules cannot reproduce themselves, but nucleic acid molecules can. Since they in turn control the synthesis and structure of proteins, it is quite easy to see that the nucleic acids are the "king pins" of the cell.

The Origin and Evolution of Cells

From the foregoing, it should be apparent that cells are very complex in structure and function. Even the simplest bacterial cell has thousands of genes, and plant and animal cells have a variety of cytoplasmic organelles in addition. One can imagine a primordial soup teeming with nucleotides, amino acids, simple sugars, and innumerable other chemicals formed during hundreds of millions of years of sterile existence of the earth. But one cannot imagine that all of a sudden the right molecules happened to come together to form anything like a modern cell capable of metabolism and reproduction. The earliest precursors of life must have been molecules that were self-replicating, that is, that could select building blocks from the primordial soup and catalyze their assembly into additional copies of themselves. One may speculate that it was a small polynucleotide, but the arguments and evidence are beyond the scope of this text.

However it happened, a sort of primitive, prebiotic metabolism must have existed based on the energy recovered from the accumulated compounds. Through enlargement and mutation, the self-replicating units became more sophisticated, being favored by natural selection over their less well adapted relatives. At some point, perhaps as the soup became thinner, one of the replicating units acquired a membrane, giving it an enormous advantage by keeping together the components of its increasingly complex metabolism. Further evolution of this primitive cell gave rise to a more efficient prokaryote.

At this stage, the energy was still entirely derived from breakdown of high energy chemical substances in the environment. The metabolism was anaerobic. It was similar in principle to the glycolysis that occurs today in most organisms, including our own cells. At some stage, one of these early cells developed the ability to use sunlight as a source of energy, which enabled it and its progeny to carry out the reaction

$$H_2O + CO_2 + \text{light energy} \rightarrow C_xH_yO_z + O_2$$

This ability to transform light energy into chemical energy by photosynthesis was another giant step forward in the evolution of life. It permitted the direct use of part of the sun's energy, and it produced O_2 as a product. The presence of O_2 allowed the aerobic organisms to arise and become the dominant forms of life. At first these aerobes were protokaryotes that used the O_2 to carry out respiration to form ATP and CO_2. They, in turn, provided the CO_2 needed by the photosynthetic prokaryotes. They therefore lived together in a kind of mutual aid system or *symbiosis*.

It has been suggested that at the time this symbiotic relationship was being established, perhaps 1.5–2 BYA, a third kind of cell was in existence that was still anaerobic, living off the dissolved foods in the ocean. This kind of cell had to be quite sluggish in its activities because it could not photosynthesize, and more importantly it could only get part of the energy out of the $C_xH_yO_z$ compounds about it. Thus, it could not carry out respiration,

$$P_i + ADP + O_2 + C_xH_yO_z \rightarrow ATP + CO_2 + H_2O,$$

which catabolizes $C_xH_yO_z$ completely to CO_2 and H_2O, transforming the maximum amount of energy to ATP. Instead, it could only carry out the anaerobic process of glycolysis in which $C_xH_yO_z$ is only partially catabolized in the absence of O_2 and with only a limited synthesis of ATP. The significance quantitative advantage attained by *aerobiosis* over *anaerobiosis* can be appreciated best by noting that for *each* molecule of glucose catabolized via glycolysis only *two* molecules of ATP are made. What remains from the glucose in the form of lactic acid still contains 93% of the available energy of the original glucose (Fig. 4.17). If, however, the lactic acid is further catabolized all the way to CO_2 and H_2O via the tricarboxylic acid cycle and the electron transport system with the utilization of O_2, the remaining 93% of the energy in glucose is converted to ATP (Fig. 4.17). Obviously the aerobic life style is much to be desired from the standpoint of energy efficiency. If we have a cell that can do both glycolysis anaerobically as well as respiration, it can get a total of 32 molecules of ATP from one molecule of glucose.

It is believed that such kinds of cells arose 1–2 BYA. One kind was a partnership between a prokaryotic cell like a present-day bacterium, which was able to carry out the aerobic phase of catabolism (respiration), and an anaerobic cell that could carry out the glycolytic phase. The prokaryote lived *within* the anaerobes as shown

FIGURE 4.17
Glycolysis and the formation of ATP, followed by oxidative phosphorylation and the formation of more ATP.

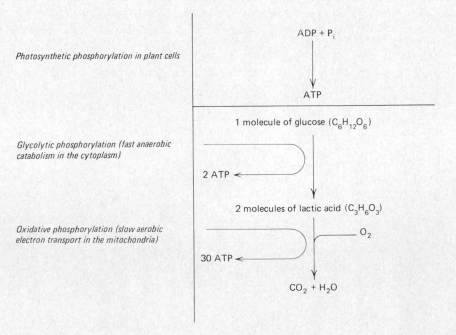

Photosynthetic phosphorylation in plant cells

ADP + P_i

ATP

1 molecule of glucose ($C_6H_{12}O_6$)

Glycolytic phosphorylation (fast anaerobic catabolism in the cytoplasm)

2 ATP

2 molecules of lactic acid ($C_3H_6O_3$)

Oxidative phosphorylation (slow aerobic electron transport in the mitochondria)

O_2

30 ATP

$CO_2 + H_2O$

FIGURE 4.18
The primordial eukary-
otic cell. This represents
what is currently the
most logical explanation
for the origin of the
eukaryotes about one to
two billion years ago.

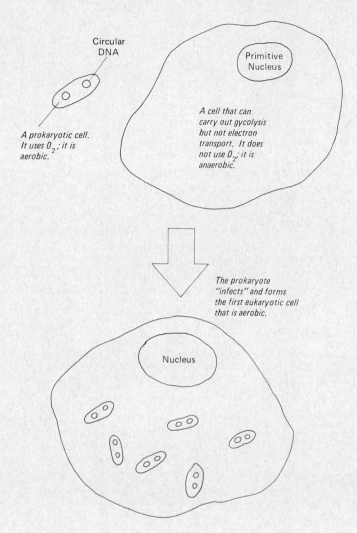

Circular
DNA

Primitive
Nucleus

*A cell that can
carry out gycolysis
but not electron
transport. It does
not use O_2; it is
anaerobic.*

*A prokaryotic cell.
It uses O_2; it is
aerobic.*

*The prokaryote
"infects" and forms
the first eukaryotic cell
that is aerobic.*

Nucleus

in Figure 4.18 and became the first eukaryotic cells. We eukaryotes, according to
this theory, are therefore composite organisms! The invading prokaryotes became
our *mitochondria*, which to this day provide the major portion of the energy in the
form of ATP required by our cells.

Review Questions

1. List and briefly describe the significant differences between the prokaryotes and
   eukaryotes.
2. What do the terms genotype and phenotype refer to?
3. What are the main constituents of chromatin?
4. How are these constituents organized?

5. List the significant parts of an animal cell, and briefly describe their functions.
6. Relate the anabolic, photosynthetic activities of green plants to the catabolic respiratory activities of animals to show how one depends on the other.
7. What are the primary sources of ATP in human cells?
8. What is meant by the term differentiation as applied to cells?
9. What are the three chief functions of the DNA in cells?
10. Contrast the aerobic condition and the anaerobic condition of life. Which probably arose first in the evolution of life?

References and Further Reading

Avers, C. J. 1981. *Cell Biology*, 2nd ed. Van Nostrand, New York. An elementary treatise that will answer many questions about cell structure.

Dyer, A. F. 1979. *Investigating Chromosomes*. John Wiley, New York. A short dissertation on the preparation of chromosomes for microscopic examination. Mostly about plants, but some applies to animals.

Gall, J. G., K. R. Porter, and R. Siekevitz. 1981. Discovery in Cell Biology. *J. Cell Biol.* 91 (2). An attempt to bring together facts bearing on some of the most important discoveries in cell biology and their consequences. Rather advanced, but readable.

Hsu, T. C. 1979. *Human and Mammalian Cytogenetics. An Historical Perspective.* Springer-Verlag, New York/Heidelberg/Berlin. An authoritative and amusing description of the discovery of the techniques used to make human and other mammalian chromosomes fit for analysis of their characteristics.

Kornberg, R. D., and A. Klug. 1981. The nucleosome. *Scientific American* 244:52–64. A well-illustrated article by two pioneers in the study of nucleosomes.

Paris Conference (1971): Standardization in Human Cytogenetics. 1972. *Birth Defects: Original Article Series* 8 (No. 7). The National Foundation–March of Dimes, New York. 46 pp. The official guide to human chromosome nomenclature.

Therman, E. 1980. *Human Chromosomes: Structure, Behavior, Effects.* Springer-Verlag, New York. 235 pp. A concise textbook of human cytogenetics.

Watson, J. D. 1968. *The Double Helix.* Atheneum, New York. A personal account of the exciting events leading to the description of the Watson–Crick model of DNA and the activities of the players in the drama surrounding the events.

White, M. J. D. 1973. *The Chromosomes*, 6th ed. Chapman and Hall, London. 214 pp. A simple introduction to chromosomes by one of the world's foremost cytogeneticists.

Living cells never rest: They metabolize and they reproduce. When they rest, they are dead. With some exceptions, the cells in our body undergo a cycle consisting of a period of growth followed by cell division to form two daughter cells. Some of our cells, such as nerve cells, never divide after we are born. Others, such as those in our skin, digestive tract lining, liver, and kidney, divide continually and reproduce at varying rates. Cells in the red marrow of our bones divide constantly to provide red and white blood corpuscles.

The period of most active cell division in our lives occurs after the sperm meets the egg and forms the fertilized egg or *zygote* that becomes us. Cell division after this conception is very rapid and occurs in all parts of the embryo and fetus. The general process slows down after we are born and continues to slow down until our early twenties, when it comes to a halt except in areas such as those mentioned.

Simple cell division such as occurs in most of our body cells is called *mitosis*. As a result of mitosis, a mother cell will produce two daughter cells, each with the same number of chromosomes, 46, and with the same kind of chromosomes as the mother cell (Fig. 5.1). In other words, the human karyotype shown in Figure 4.7 would have been replicated exactly had that cell continued to divide.

Another kind of cell division occurs in the gonadal tissues of our bodies: in our ovaries if we are female and in our testes if male. These organs contain our *germ cells*, which upon undergoing *meiosis* become eggs ready to be fertilized or sperm ready to fertilize. Meiosis is a specialized type of cell division that in fact consists of two divisions and results in the reduction of the chromosome number by ½. Thus, a cell with 46 chromosomes will produce eggs or sperm with 23 chromosomes. The reduction in number is not random, however, as we shall show.

Cell division ensures not only an increase in numbers of cells but also an increase in numbers of individual organisms. It is the basis for the continuity of life. To understand genetics, we must begin with some understanding of both mitosis and meiosis. We start with the mitotic cycle, which incorporates mitosis as the stage of active division.

The Mitotic Cycle and DNA Replication

The mitotic cycle is diagrammed in Figure 5.2. It is divided into four distinct phases or stages: G_1, S, G_2, and M. Collectively the G_1 + S + G_2 phases are called the interphase, and M stands for mitosis. As we have noted previously, the chromosomes are essentially invisible during the interphase (page 61). In G_1, the extended chromosomes of the cell have DNA present in them as a single molecule of double-stranded DNA extending their length. In the S (synthesis) phase, the DNA of the chromosomes begins to replicate chemically; by the end of this phase, chemical replication of the chromosomes is complete. However, the replicas remain together attached by their centromeres. At this stage, they are called *sister chromatids*. The sister chromatids remain together through the G_2 phase. In G_2, the amount of DNA per nucleus is double that in the G_1 phase because of its replication in the S phase, but the number of chromosomes has not changed. Figure 5.3 shows how DNA is thought to be synthesized by replication in the presence of the enzyme *DNA polymerase*. Because the two strands of DNA are complementary, the net

5

Cell Cycles and the
Continuity of Life

FIGURE 5.1
The consequences of simple cell division by mitosis. Each daughter cell receives an equal share of genetic material.

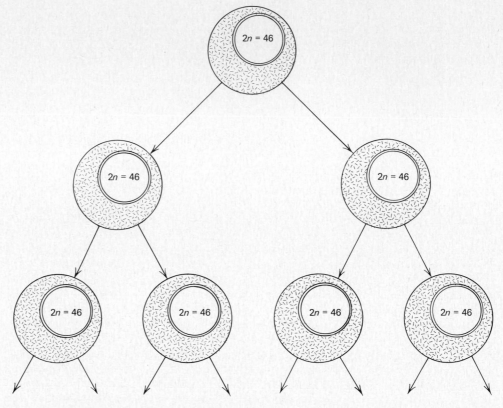

result of the replication is two identical DNA molecules. Since the genetic message is contained in the sequence of bases in the DNA strands, the resultant two molecules must therefore have the same messages.

DNA replication can be considered the most fundamental of biological processes. It not only ensures that we develop from an egg, but also that we can have offspring related to us. The continuity of life is through DNA. Although superficially quite simple, the process is, in fact, quite complex. Many enzymes are involved, and by no means are all of the details of their activities known. The molecule of DNA is an average human chromosome is about 3.8 cm in length. If each chromosome had only one point at which its DNA were being synthesized, it would take at least 12 h for an average chromosome to be replicated. Human cells, especially in early embryos, divide more frequently than that. Therefore, their chromosomes must replicate more rapidly. This rapid synthesis of long strands is accomplished by many replicative points starting along the length of the DNA molecule (Fig. 5.4). Each of these (we shall call them *bubbles*) has two replicative forks. Note that the DNA helix must unwind in order for the replicative process to occur. The helix then re-forms as shown in Figures 5.3 and 5.4

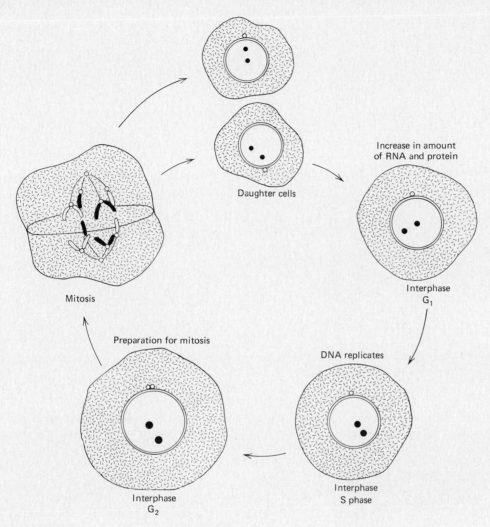

FIGURE 5.2
The mitotic cell cycle.

Daughter cells

Increase in amount
of RNA and protein

Mitosis

Interphase
G₁

Preparation for mitosis

DNA replicates

Interphase
G₂

Interphase
S phase

Chromosome Replication

While the DNA is being replicated, more histones and other chromatin proteins become associated with the new DNA molecules. As a result, *at the end of the S phase, each original chromosome has become two identical sister chromatids and the amount of DNA in the nucleus has doubled.* Also the amount of histone doubles so that each chromatid is essentially a new chromosome at this point, although it is not called that since sister chromatids are still held together by an undivided centromere or two attached centromeres. Also, the sister chromatids are still in close physical association throughout G_2. The G_2 phase following the replicative S phase is one in which nothing of consequence seems to be happening. But this is probably illusory. Its end is signaled by the onset of mitosis.

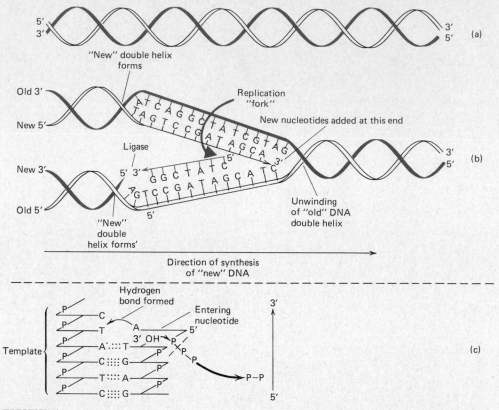

FIGURE 5.3

The replication of DNA. Outwardly, the replication is quite simple. The two polynucleotide chains of the double helix separate, and new chains form that are complementary to the old chains. Actually, this seemingly simple process is quite complex, and many details are not well understood. This figure attempts to show in a highly diagrammatic way some of the more important elements of the process. A DNA molecule (a) begins replication by unwinding as shown in (b) and the separated strands shown in the figure as straight serve as templates. The enzyme(s) that add the nucleotides to form the new chains at the "replication fork" are called DNA polymerase. They add 5'-deoxytriphosphonucleosides (dATP, dCTP, dGTP, and dTTP) to a growing chain with the elimination of two phosphates as shown in (c). The polymerases add only at the 3' end. Hence, the new chains grow only in the 5' → 3' direction. Ligase enzymes connect the segments of the new DNA chains by forming phosphodiester bonds. This is necessary in the case of the chain forming away from the replication fork, as shown in the bottom part of (b). It may or may not happen in the other chain. Double helices reform after synthesis to give two new DNA daughter molecules, each with one old and one new chain.

FIGURE 5.4
The replication of a chromosome.

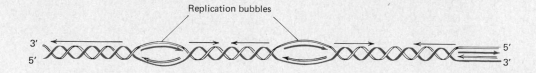

Mitosis

In the M (mitosis) phase, the chromosomes in the nuclei become fully visible and the sister chromatids formed in the S phase separate and go to the opposite poles of the cell. Mitosis is arbitrarily divided into four major stages by cell biologists: *prophase, metaphase, anaphase,* and *telophase.* The G_2 phase chromosomes enter prophase in preparation for mitosis by beginning to coil and fold. As they do this, they become visible under the light microscope because they are shorter and thicker. The sister chromatids derived from the original mother chromosomes by replication remain attached at their centromeric regions, as shown in Figure 5.5,

FIGURE 5.5
The distribution of chromosomes in mitosis.

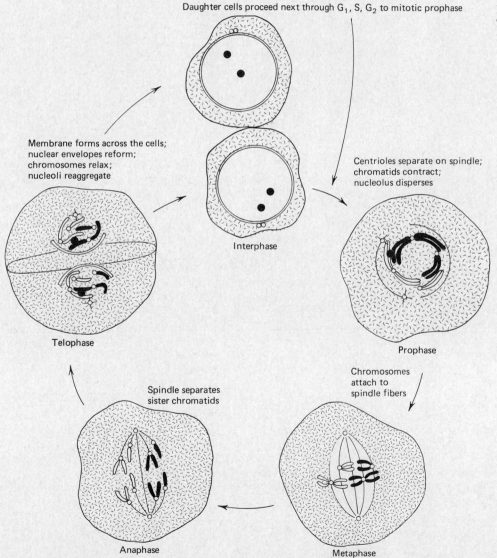

Daughter cells proceed next through G_1, S, G_2 to mitotic prophase

Membrane forms across the cells; nuclear envelopes reform; chromosomes relax; nucleoli reaggregate

Centrioles separate on spindle; chromatids contract; nucleolus disperses

Interphase

Telophase

Prophase

Chromosomes attach to spindle fibers

Spindle separates sister chromatids

Anaphase

Metaphase

well into the next stage, metaphase. During metaphase, the chromosomes arrange themselves into a single plane, and the nuclear membrane breaks down and is no longer visible. A large number of fibrils called *spindle fibers* now become visible. These converge at two points at opposite poles of the cell, as shown in Figure 5.5. Many of them attach to the chromosomes at the centromeric region, and the next stage, anaphase, begins. In anaphase, *the sister chromatids separate* at their centromeric regions and move to opposite poles, following the spindle fibers to their points of confluence. With this division of the centromeres, the chromatids now become chromosomes by definition.

Telophase begins with the two identical sets of chromosomes now completely separated and the cytoplasm of the mother cell beginning to divide. New nuclear membranes are formed around each of the two new chromosomes. When cytoplasmic division is completed, mitosis is completed. Figure 5.5 is a diagram of this process and shows that the two daughter cells produced are identical with respect to chromosomal content, though not necessarily with respect to cytoplasm. At the end of mitosis, the daughter cells enter the G_1 phase again, and the cycle is repeated.

The Meaning and Consequences of Mitosis

When a fertilized egg begins its journey toward the production of a new individual human or other kind of animal or plant, its successive cell divisions or mitoses result in a mass of cells, all of which have the same number and kind of chromosomes. After this initial period, called *cleavage* by embryologists, the cells begin to *differentiate*. This means that the early embryonic cells, which all look the same, change in form and function and start to become the muscle cells, nerve cells, liver cells, kidney cells, and so forth, of the fully developed infant (Fig. 4.15). This process of differentiation is accompanied by mitosis. The result is that cells are formed that all have the same chromosome constitution although they look and function differently. By same constitution we mean that the daughter cells have the same karyotypes as the mother cell. Refer to Figure 4.7 again and note that the human set of 46 chromosomes can be arranged into 23 sets of two. The chromosomes in each pair are identical in appearance. As we shall see, they are in fact closely related genetically because they bear the same kinds of genes. They are called *homologues* or *homologous* chromosomes. Thus, the human chromosome complement consists of 23 pairs of homologues, and mitosis occurs in such a way as to preserve this situation. Not just any 46 chromosomes are passed on to the daughter cells.

One of the consequences of differentiation of the embryo is that, certain cells arise that become localized in the gonads. These become the germ cells that, in addition to the ordinary cell cycle with mitosis, undergo a process called *meiosis*. Meiosis results in the formation of the male and female *gametes*, the sperm and the eggs, respectively. A clear understanding of the meiotic process is absolutely essential to an understanding of the genetics of the eukaryotes, including humans.

Meiosis and the Continuity of Life Through Sperm and Eggs

Whereas genetic continuity of our body cells is maintained by mitosis, continuity from our parents and to our offspring is preserved by the process of meiosis. The male gonads or testes produce sperm, and the female gonads or ovaries eggs. Both processes involve meiosis, and in both cases the meiotic mechanism is basically the same. Therefore, we shall describe meiosis first without reference to either sperm or eggs.

The diagram in Figure 5.6 presents the elements of what happens in meiosis. Note that it consists of two successive cell divisions. Germ cells about to enter meiosis go through the S and G_2 phases as if they were about to undergo mitosis, but then things change. The onset of meiosis is significantly different from the initial stage or prophase of mitosis. To understand this difference we must first refer back to what is meant by *homologous chromosomes*.

At the the time of mitosis, each of the 46 chromosomes in our nuclei can be seen to consist of two chromatids. Hence, the cells contain $2 \times 46 = 92$ chromatids at the beginning of mitosis because of the doubling that occurred in the preceding S phase. When mitosis is completed, each daughter cell now has 46 chromosomes (formerly chromatids), since the members of each pair of sister chromatids separate and become incorporated into separate daughter nuclei. The chromosome number thus remains unchanged from one cell generation to the next.

But in the case of meiosis, pairs of sister chromatids, in turn, pair or *synapse* to form *tetrads* consisting of four chromatids (Fig. 5.6). Those chromosomes (pairs of sister chromatids) that pair are said to be *homologous*. Homologous chromosomes are related chromosomes, since they carry essentially the same kind of genetic material. One set is paternal and the other maternal in origin. For this reason, the human cell with 46 chromosomes can be said to contain two sets of 23 chromosomes. The pairing phenomenon is therefore not random; *pairing occurs only between genetically related paternal and maternal (homologous) chromosomes*.

In the tetrads, the sister chromatids remain together attached by a common centromere. While in synapsis, they may, and generally do, exchange segments equilaterally and reciprocally with the *nonsister chromatids* of their homologues. This exchange is called *crossing over*. We will consider it in more detail in Chapters 7 & 8. For the present, it is more important to recognize that after the tetrads form, the pairs of sister chromatids of the tetrads separate and go to opposite poles, as shown in Figure 5.6. This step results in two daughter cells that are different from the daughter cells formed by mitosis, since *they contain only one member of each homologous pair*.

The first meiotic division is followed by a second division that resembles mitosis, since the "sister chromatids" separate and go to opposite poles. This second division is not exactly the same as a mitotic division, however, because if crossing over does occur in the prophase of the first division between nonsister chromatids, the resulting crossover chromatids are no longer identical to their "sisters" to which they are attached by their centromeres.

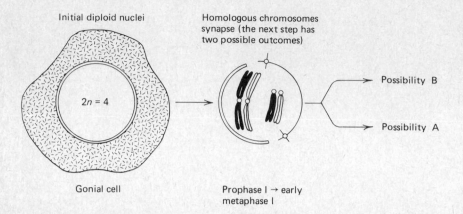

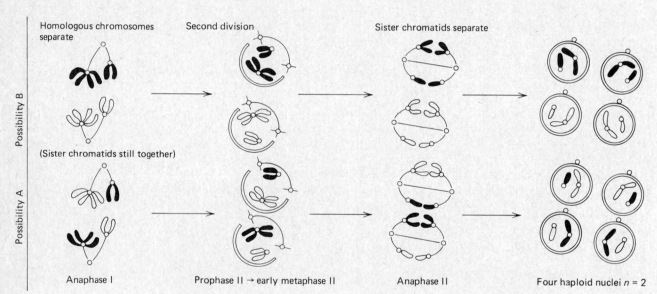

FIGURE 5.6
Meiosis. The initial germ cell has two pairs of chromosomes ($2n = 4$) and is diploid.
The products of meiosis, the gametes, are haploid with only one copy of each
chromosome.

Each of the cells resulting from meiosis will contain 23 chromosomes, consisting
of one representative of each homologous pair present in the original germ cell. If
this is not the case, serious consequences may result. One way of looking at
meiosis is that it is a process in which *two* cell divisions occur but only *one*
chromosome replication occurs.

The human chromosome number of 46 is generally referred to by geneticists as the *diploid number* and symbolized by 2*n*. After meiosis, the gametes have the *haploid number* of 23 designated *n*. Animals, including humans, are generally diploid and produce haploid gametes. Exceptions to this are found in a few animals and in many plants, as we shall describe in Chapter 9. For the present, consider the haploid number as the basic chromosome number. It is a complete single set of chromosomes called the *genome*. All members of the set are *non-homologous* with respect to each other.

Gametogenesis

Human sperm and eggs arise from diploid germ cells present in the testis and ovary. These cells are referred to as the *spermatogonia* and *oogonia*, respectively. They become functional haploid sperm and eggs by the meiotic divisions we have described plus other changes having to do primarily with the cytoplasm. The conversion of spermatogonia and oogonia to sperm and eggs is given the general name of *gametogenesis* and the specific male and female processes are referred to as *spermatogenesis* and *oogenesis* (Fig. 5.7).

The sperm are formed in the testes in the tubules. By the meiotic process, *spermatids* are formed as shown in Figure 5.7. These haploid cells then become transformed by *spermiogenesis* into functional sperm with tails to move them.

The diploid *oocytes* in the ovaries all go through the first meiotic prophase during embryogenesis and then stop further meiotic activity. When an egg leaves the ovary, in what is called *ovulation*, it enters the Fallopian tube. If it is fertilized by a sperm, it completes meiosis, and the fusion of the haploid egg nucleus and sperm nucleus occurs. Only one functional haploid nucleus is formed as the result of meiosis of a diploid oogonium. The other three haploid products are cast out as polar bodies with essentially no cytoplasm, as shown in Figure 5.7.

The sperm contributes a nucleus to the zygote and probably nothing more. Thus the fertilized egg starts its development with maternal cytoplasm only. This has interesting consequences, which we discuss in Chapter 7.

The Human Organism and the Reproductive Cycle

After the formation of sperm cells and egg cells by meiosis, *fertilization* takes place, as shown in Figure 5.8. A sperm penetrates the egg, and the sperm nucleus fuses with the egg nucleus. Since both sperm and unfertilized egg contain a haploid set of 23 chromosomes in humans, the fertilized egg regains the diploid number of 46 after fusion of the haploid nuclei.

FIGURE 5.7
Spermatogenesis and
oogenesis: meiosis and
germ cells.

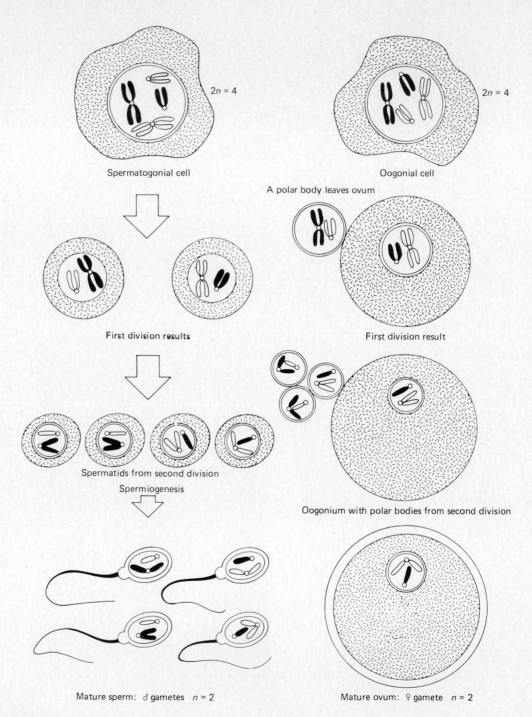

Spermatogonial cell

Oogonial cell

A polar body leaves ovum

First division results

First division result

Spermatids from second division

Spermiogenesis

Oogonium with polar bodies from second division

Mature sperm: ♂ gametes $n = 2$

Mature ovum: ♀ gamete $n = 2$

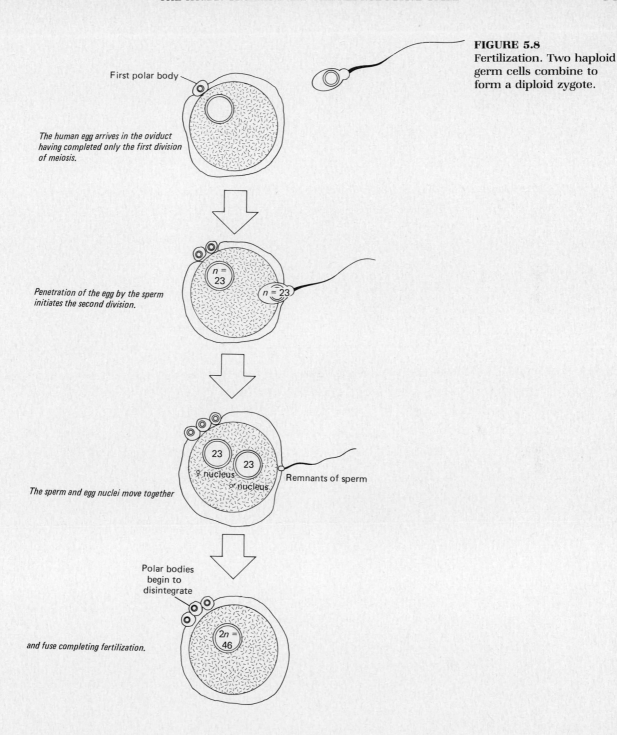

FIGURE 5.8
Fertilization. Two haploid germ cells combine to form a diploid zygote.

First polar body

The human egg arrives in the oviduct having completed only the first division of meiosis.

Penetration of the egg by the sperm initiates the second division.

$n = 23$

$n = 23$

The sperm and egg nuclei move together

23

♀ nucleus

23

♂ nucleus

Remnants of sperm

Polar bodies begin to disintegrate

and fuse completing fertilization.

$2n = 46$

FIGURE 5.9
Early divisions of a fertilized egg.

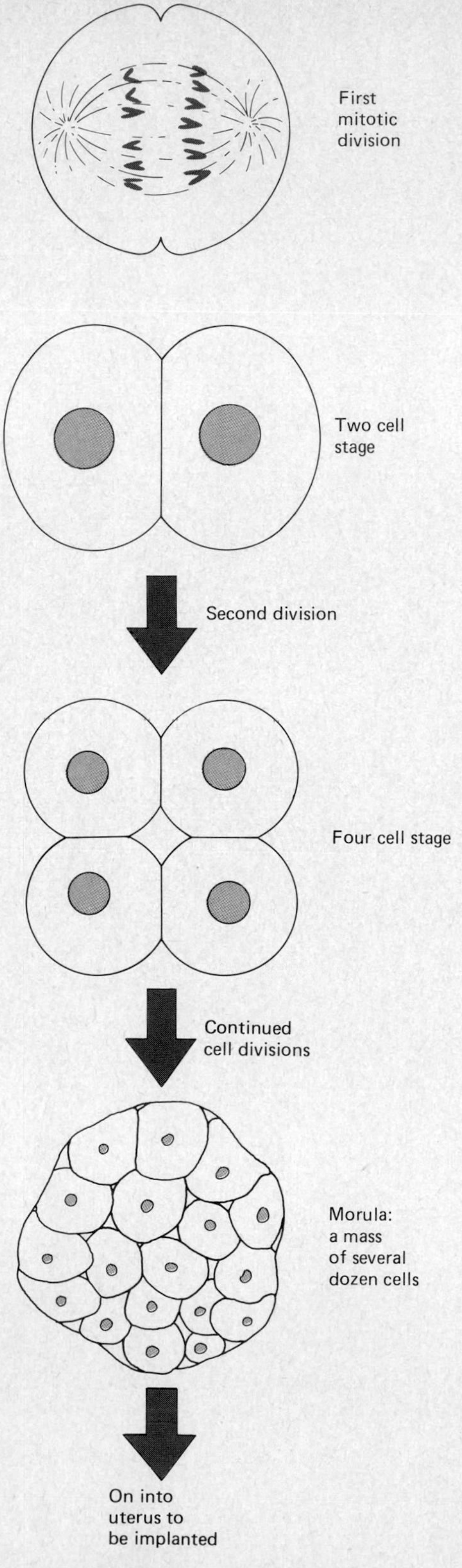

First
mitotic
division

Two cell
stage

Second division

Four cell stage

Continued
cell divisions

Morula:
a mass
of several
dozen cells

On into
uterus to
be implanted

94

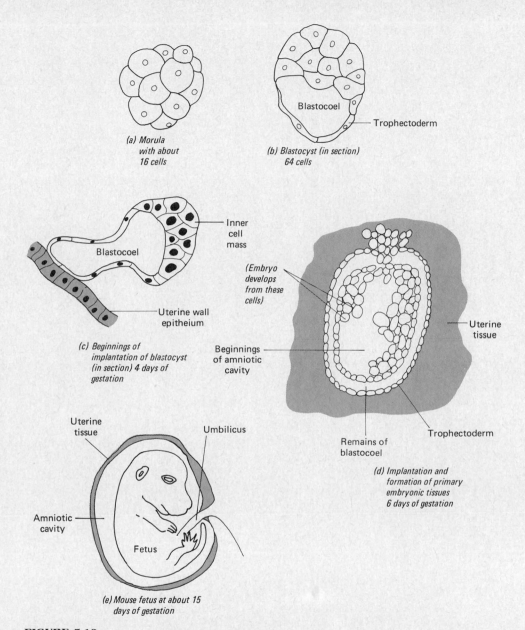

(a) Morula
with about
16 cells

(b) Blastocyst (in section)
64 cells

Blastocoel

Trophectoderm

Inner
cell
mass

Blastocoel

Uterine wall
epitheium

(c) Beginnings of
implantation of blastocyst
(in section) 4 days of
gestation

(Embryo
develops
from these
cells)

Beginnings
of amniotic
cavity

Uterine
tissue

Remains of
blastocoel

Trophectoderm

(d) Implantation and
formation of primary
embryonic tissues
6 days of gestation

Uterine
tissue

Umbilicus

Amniotic
cavity

Fetus

(e) Mouse fetus at about 15
days of gestation

FIGURE 5.10

The early stages of embryonic development in a mouse. The fertilized egg of the mouse develops very rapidly. By the time it is ready to implant in the uterus, the embryo consists of hundreds of cells. At ten days after fertilization, the embryo begins to look like a mammal, and by fifteen days it begins to look mouse-like. After 19–20 days of gestation, a mouse is born.

Ordinarily the human female has only one egg available at a time for fertilization. This becomes surrounded by many millions of sperm from the male, but generally only one sperm enters the egg. An automatic mechanism shuts off further penetration by additional sperm once the first one enters. (Occasionally two sperm get in and combine their chromosomes with those of the egg. The zygotes so formed do not produce fetuses that can come to term. They have three sets of chromosomes and hence are designated *triploid* or *3n*.)

With the onset of cleavage, a *morula* or clump of cells is first formed as shown in Figure 5.9. Then begins the process of differentiation that we have discussed previously. The very early development stages of a human are difficult to study for obvious reasons, but they are probably quite similar to what occurs in the mouse. Early stages of mouse embryogenesis are given in Figure 5.10.

Review Questions

1. List the four phases of the mitotic cell cycle and briefly describe the events occurring in each of them.
2. Briefly describe the main events in the replication of DNA.
3. Outline the sequence of events in mitosis and meiosis, drawing diagrams if necessary.
4. Contrast the mechanics of mitosis and meiosis and the endproducts.
5. Criticize this statement: "The only essential result of meiosis is the formation of gametes with half the number of chromosomes found in the mother cell."
6. Describe and define homologous chromosomes.
7. How many chromatids does a human cell contain in the prophase of mitosis? In the prophase I of meiosis?
8. How many tetrads are formed in human spermatogonia preparing to undergo spermatogenesis?
9. An animal with a diploid chromosome number of 48 has what haploid number?
10. How many chromatids are there in a cell in G_1 phase that has 23 chromosomes? In the G_2 phase?
11. What are the significant differences between oogenesis and spermatogenesis in humans?

References and Further Reading

Karp, G. 1979. *Cell Biology.* McGraw-Hill, New York. Good for further information about mitosis and meiosis.

Hamerton, J. L. 1971. *Human Cytogenetics.* Vol. 1, *General Cytogenetics.* Academic Press, New York. 412 pp. A general guide to chromosome behavior, written for the person with considerable background.

Hultén, M., and J. Lindsten. 1973. Cytogenetic aspects of human male meiosis. *Adv. Human Genet.* 4:327–387. A summary of investigations carried out in humans.

We all start from an egg containing 23 chromosomes from our fathers and 23 from our mothers. These 46 chromosomes contain about 1.8 meters of DNA. In addition, the egg to become us contains several thousand mitochondria, each with one or more circles of DNA about 5 μm in circumference. This DNA *in toto* was and is our *genotype.* It contained the directions for our development from the egg. Together with the environment in which we develop, it determined and is still determining our *phenotype,* which is us, our structure and our function, physiologically and behaviorally.

The DNA in our egg was replicated and by successive mitoses formed the thousands of billions of cells that constitute our bodies. Theoretically each cell contains the same quota of DNA as the original fertilized egg cell. This is a reasonable generalization. But like most generalizations, it is not entirely true because there are certain important exceptions we will describe in later chapters. For the present, we can state categorically that the DNA within us, which we have inherited from our parents, determines not only that we are to be human, but for the most part what kind of human. This process, the determination of the phenotype by the genotype, is an extremely complex one that we are far from understanding in its entirety. In this chapter we begin to discuss some of the basic factors that operate in the process of becoming human from a single cell.

The Organization of the Genotype

Our genotype is organized into chromosomes or, to put it more precisely, into two *genomes* of 23 pairs of homologous chromosomes. A genome is the genetic constitution of a haploid set of chromosomes. Each chromosome in turn is organized into units of function, our genes.

Genes Defined

Genes are segments of DNA containing particular sequences of bases of nucleotides that act as functional units in controlling the operations of the cell. They are replicated precisely during the replication of the chromosome of which they are a part. Thus, they are not only units of function, but also units of heredity, since by their replication the information they contain, used to accomplish their specific functions, is transmitted by cell divisions from one generation to the next.

Genes Classified

Genes are classified into a number of different categories according to their functions. These functions, so far as we understand them, center around a process called *transcription*. Transcription is a fundamental biological process by which the information contained in a base sequence of a segment of a single chain of DNA is copied or *transcribed*. The copy or transcript is a complementary chain of RNA. A diagram of the essentials of the process is given in Figure 6.1.

Several different kinds of RNA are transcribed from DNA. The genes we know something about are classified according to the type of RNA they transcribe. *Structural* genes transcribe an RNA called *messenger RNA*, usually written as

6

From Genotype to Phenotype

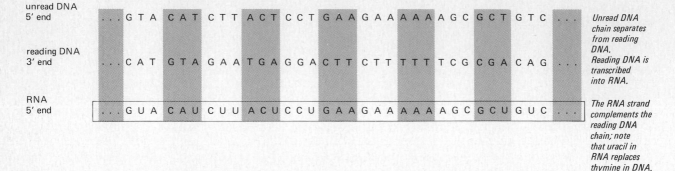

FIGURE 6.1
The essentials of transcription from DNA to RNA.

mRNA. The sequence of nucleotides in mRNA is *translated* into a sequence of amino acids in a polypeptide chain (Fig. 6.2). Thus, structural genes are directly identified with the instructions for sequences of amino acids. The proteins formed then function as enzymes and structural elements. For each kind of protein made in the cell, there are one or more genes transcribing mRNA specific for that protein.

Other genes transcribe RNA molecules that play essential roles in the process of translation, but they do not specify the sequence of amino acids in the polypeptide synthesized. The two kinds of RNA are designated *transfer RNA* or *tRNA* and *ribosomal RNA* or *rRNA*. Hence, tRNA genes and rRNA genes are distinguished from the structural genes we have already described. The basic functions of these three types of genes are summarized in Figure 6.2.

FIGURE 6.2
The essentials of translation from RNA to protein.

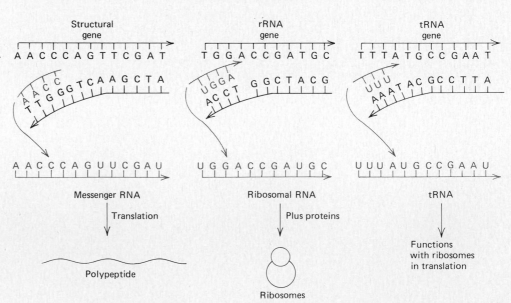

FIGURE 6.3
Examples of repetitive DNA.

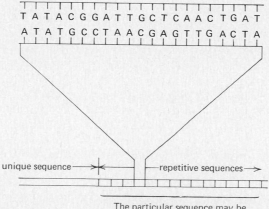

T A T A C G G A T T G C T C A A C T G A T
A T A T G C C T A A C G A G T T G A C T A

unique sequence ⟶ ← repetitive sequences ⟶

The particular sequence may be
repeated in tandem many times.

A fourth group of genes, the *regulatory* genes, can be described only in general terms, as very little is known about their function, especially in eukaryotes. The transcription of structural genes is regulated by DNA sequences outside the translated regions of structural genes. We are only now beginning to learn something about how this occurs.

A large part of the DNA in our chromosomes is not involved in the transcription of identifiable mRNA, rRNA, or tRNA, all of which have functions that are rather well understood. This DNA may comprise up to 90% of the total DNA in some animals. Some of it is highly repetitive; that is, it consists of sequences that are repeated over and over again as shown in Figure 6.3. Some sequences may be repeated thousands of times. In contrast, the genes coding for rRNA and tRNA may be present in hundreds of copies. Structural genes usually are present as only one or occasionally a very few copies, but they may have repeating sequences associated with them.

The main portion of our DNA has no identifiable function. What does all this DNA do? Some geneticists think of it as "junk" that just goes along for the ride and has no function. It is in a sense parasitic and has even been described as "selfish." But this point of view is an exaggeration with no support from experiments. Most if not all of this DNA probably does function in regulation, or in other ways we presently do not comprehend.

Transcription

DNA in its native state consists of two complementary polynucleotide chains held together by hydrogen bonds between the purine and pyrimidine bases, as described in Chapter 3 (Fig. 3.14). The pairing between these bases is not random: adenine bonds to thymine and guanine to cytosine. Thus, the sequence in one chain determines the sequence in the other. The result is that during the replication of DNA, as described in Chapter 5, two identical DNA molecules are formed in the presence of the DNA polymerase and associated enzymes (Fig. 5.3).

The transcription process bears a close similarity to the replication process. Polymerase enzymes are also involved, but they are called *DNA-dependent RNA polymerases*, or simple *RNA polymerases*, since they produce RNA rather than DNA. These polymerases, of which there are three kinds, act in the presence of single-stranded DNA and triphospho*ribo*nucleotides rather than triphospho*deoxyribo*nucleotides. RNA has a somewhat different base composition from DNA as well as the difference in sugar (Fig. 3.12). RNA contains no thymine, but in its place contains the pyrimidine uracil. When RNA is transcribed from DNA, uracil pairs with adenine and adenine with thymine (Fig. 6.1). Thus, an RNA transcript of the DNA sequence ATTCGAC will be UAAGCUG.

Transcription in human cells occurs throughout the interphase of the cell cycle. Nearly all of it ceases or is much reduced during mitosis. This is not unexpected because during this period of the cycle the chromosomes are coiled, folded, and condensed to such an extent that the RNA polymerases probably cannot act on such single-stranded DNA as may be formed.

Only a small part of the DNA of the chromosomes is transcribed at any given time in the interphase—perhaps only 2% or less. Furthermore, the pattern of the DNA transcribed in one cell type may be quite different from that transcribed in another cell type. Both rRNA and tRNA are probably transcribed continuously in all living cells, but the kinds of structural genes transcribed will differ from one cell type to another.

Several different kinds of rRNA are transcribed from specialized parts of the chromosomal DNA. The longer chains of rRNA (1000–5000 bases long) are transcribed in structures called *nucleoli* (Fig. 6.4) associated with certain of our chromosomes. The shorter chains of rRNA (100–200 bases long) are transcribed in regions on other chromosomes. All of the regions transcribing rRNA are repeated many times.

Protein synthesis requires the presence of all these different kinds of rRNA, and it is necessary to produce large amounts of the several kinds to keep up with the demands of protein synthesis, especially when the cells are multiplying and growing rapidly. Thus, the repetition of the rRNA genes provides multiple transcribing units within the nucleoli and elsewhere in the chromosomes.

Of the three known RNA polymerases, RNA polymerase I acts in the transcription of the major rRNAs in the nucleoli, while RNA polymerase III makes the minor rRNA and tRNA transcripts. The third polymerase, RNA polymerase II, is responsible for the synthesis of *heterogeneous nuclear RNA*, generally designated *hnRNA*. This RNA is the precursor of messenger RNA (mRNA). HnRNA is transcribed from the structural genes.

The RNA Genetic Code

Protein synthesis requires the reading of the genetic code in the structural genes represented in the mRNA. The code in the mRNA is called the *RNA genetic code*, and it is complementary to the code in the chain of DNA in the structural gene from which it is transcribed.

FIGURE 6.4
A Chinese hamster cell showing nucleoli in an early prophase nucleus. (a) The nucleoli indicated by the arrows are easily visible after staining, but in (b) they are more visible in the same cell shown in (a). This cell was allowed to incorporate radioactive ³H-uridine before it was prepared for observation. Then a thin photographic emulsion was placed over it, and the emulsion was exposed to the radioactivity for several weeks before being developed. The black spots show the location of incorporated ³H-uridine. The nucleoli have incorporated large amounts as compared to the rest of the nucleus, thus demonstrating that nucleoli are centers of RNA synthesis. (Provided by Dr. Frances E. Arrighi, The University of Texas System Cancer Center, Houston.)

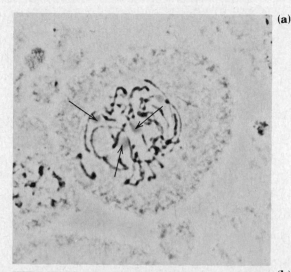

(a)

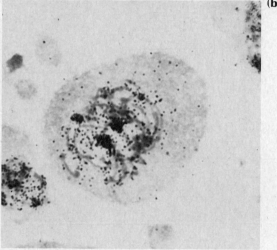

(b)

The RNA code is a triplet code. By this is meant that a sequence of three adjacent nucleotides provides the bit of information necessary to deliver a message. The message can be either to direct a tRNA to deliver a particular amino acid during translation of the mRNA or to start or stop translation. Each triplet of bases in the RNA code, such as UCG, AAG, and AUG, given in Table 6.1, is a message called a *codon*. Messenger RNA thus consists of a sequence of codons.

The following example will give some idea how the message is translated. The hypothetical nucleotide sequence for a short strand of DNA from a human chromosome is shown in Figure 6.5. This represents part of the structural gene that codes for glucagon, a small protein with only twenty amino acids and a molecular weight of 3483. Glucagon is a hormone synthesized in the pancreas. It causes the liver to release glucose into the blood when the blood glucose falls below a certain level. It acts in the opposite manner to insulin, which causes the liver to store glucose in the form of glycogen.

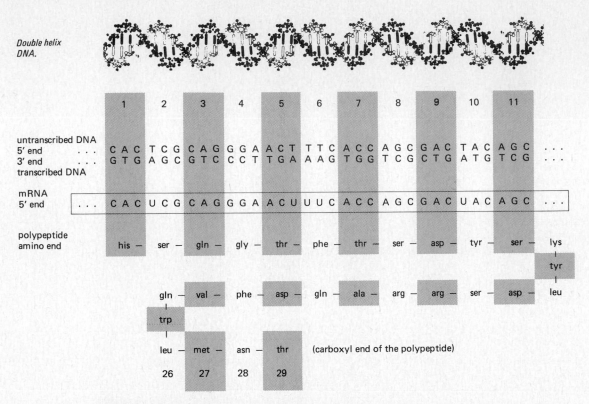

FIGURE 6.5
The structural gene for glucagon is transcribed into messenger RNA and then translated into a polypeptide that becomes the protein glucagon. Human glucagon (illustrated) has 29 amino acids, beginning with histidine at the amino terminal end. The nucleotide sequences shown for DNA and messenger RNA are the most probable sequences based on studies in other species, but the structure of the human glucagon gene has not yet been determined directly.

Figure 6.5 gives the nucleotide sequence of the glucagon "gene" for 33 consecutive base pairs. Below the double helix, the two strands are shown spread out so that they can be read. RNA polymerase proceeds along the transcribed DNA strand, faithfully producing an RNA copy, which becomes the messenger RNA (mRNA). The mRNA is thus complementary to the transcribed DNA strand and identical to the untranscribed DNA strand, except that it is composed of ribonucleotides and uridine is present in mRNA in place of thymidine in DNA.

The code in the mRNA is *translated* into the sequence of amino acids in the glucagon polypeptide shown at the far right. The rest of this polypeptide is also shown, but the messenger RNA and the DNA structural gene coding for it is not shown for reasons of space. The important thing to note is that each amino acid is determined by its codon, and *the sequence of codons determines the sequence of amino acids.* The DNA code determines the RNA code, which is read as triplets. Ordinarily the RNA code is the one shown because it directly determines which amino acids will be built into the polypeptide chain.

The three letters represent a sequence of three RNA nucleotides. The DNA codons would be complementary to the RNA codons. The corresponding amino acids are indicated by three-letter abbreviations.

TABLE 6.1
The Genetic Code

Phe	UUU	Ser	UCU	Tyr	UAU	Cys	UGU
	UUC		UCC		UAC		UGC
			UCA				
Leu	UUA		UCG	Ter[a]	UAA	Ter[a]	UGA
	UUG				UAG		
	CUU	Pro	CCU			Trp	UGG
	CUC		CCC	His	CAU		
	CUA		CCA		CAC	Arg	CGU
	CUG		CCG				CGC
				Gln	CAA		CGA
Ile	AUU	Thr	ACU		CAG		CGG
	AUC		ACC				
	AUA		ACA	Asn	AAU	Ser	AGU
			ACG		AAC		AGC
Met	AUG						
		Ala	GCU	Lys	AAA	Arg	AGA
Val	GUU		GCC		AAG		AGG
	GUC		GCA				
	GUA		GCG	Asp	GAU	Gly	GGU
	GUG				GAC		GGC
							GGA
				Glu	GAA		GGG
					GAG		

[a]Ter = termination codon.

Table 6.1 gives the RNA triplet codons for the twenty amino acids that occur in proteins. The same code is used by all organisms; viruses, bacteria, plants, and animals. The only exceptions are some found in mitochondrial DNA.

Several significant properties of the code should be noted. The first is that most amino acids have more than one codon. For example, arginine (Arg), leucine (Leu), and serine (Ser) each have six different codons. Several amino acids (alanine, glycine, proline, valine, and threonine) have four codons each, and isoleucine has three codons. The remaining amino acids have two codons each with the exception of methionine and tryptophan, which each have only one. This synonymic property of the code for most amino acids is called *degeneracy* or, if you wish to use a more elegant word, *redundancy.* The full significance of this redundancy is not understood. Since there are 64 different codons and only 20 amino acids, why are not the codons distributed more evenly so that each amino acid has approximately an equal number? We don't know, but it is probably significant that some have as many as six and some only one.

A second major feature of the code is that three of the codons, UAA, UAG, and UGA, do not code for amino acids. They perform the important function of signaling the termination of translation of codon sequences.

A close study of the RNA code reveals a third interesting feature: The third base in the codons seems quite unimportant in many cases. For example, CCX codes for proline, no matter what the third base, X, may be. The redundancy in the code is not random. For the most part, the redundancy involves only the third base.

To conclude this brief discussion of the genetic code, it should be emphasized that what is generally called the code is an RNA code, but that this code must necessarily always be complementary to the code in the DNA of structural genes.

Genes, RNAs, and Translation

The process of translation is in fact an extremely complex one. But the essentials are quite easy to understand, provided one has somewhat more understanding of the three forms of RNA that we have discussed.

rRNA and Ribosomes

The various kinds of rRNA do not act alone in translation. They are incorporated into bodies called *ribosomes*, along with fifty to sixty kinds of proteins. These *ribonucleoprotein* organelles consist of two elements, as shown in Figure 6.6, which come together and attach to the messenger RNA to start the process of translation. In cells in which protein synthesis is proceeding at a high rate, the ribosomes may contain 80–90% of the total cellular RNA.

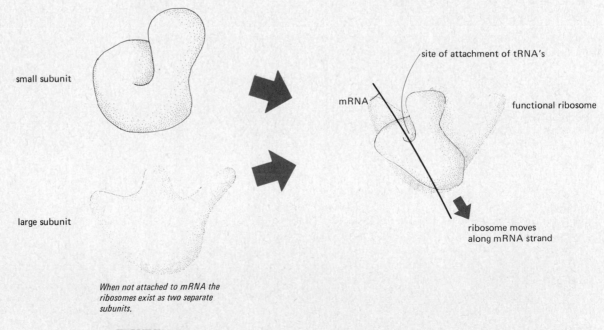

small subunit

large subunit

site of attachment of tRNA's

mRNA

functional ribosome

ribosome moves
along mRNA strand

*When not attached to mRNA the
ribosomes exist as two separate
subunits.*

FIGURE 6.6
A ribosome consists of RNA and protein. These different kinds of molecules form the complex ribonucleoproteins that fit together to constitute the functional ribosomes. (Based on J. A. Lake, *Scientific American* 245 (Aug.):84, 1981.)

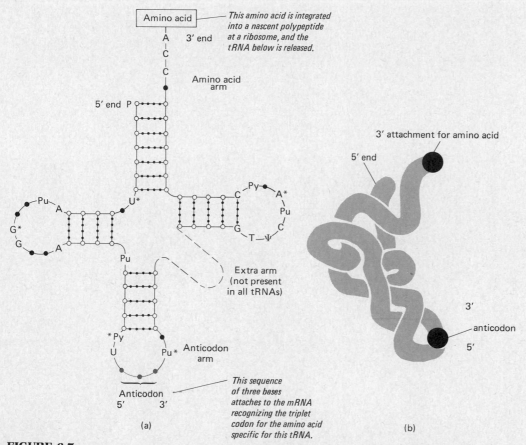

Amino acid — *This amino acid is integrated into a nascent polypeptide at a ribosome, and the tRNA below is released.*

A 3' end
C
C
Amino acid arm
5' end P

Extra arm (not present in all tRNAs)

U*
Pu A
G*
G A
Pu

C Py A*
Pu
G T ψ C

*Py
U Pu* Anticodon arm

Anticodon
5' 3'

This sequence of three bases attaches to the mRNA recognizing the triplet codon for the amino acid specific for this tRNA.

(a)

3' attachment for amino acid
5' end

3'
anticodon
5'

(b)

FIGURE 6.7

A tRNA (transfer RNA) molecule. At least twenty different kinds of tRNA molecules exist in cells, one or more for each of the twenty amino acids used as protein building blocks. All have similar structures.

tRNA

tRNAs carry the amino acids to their proper place in a polypeptide during its synthesis. They have about 76 bases in a single chain that is folded into a three-dimensional structure, as shown in Figure 6.7. A triplet *anticodon* is located on one of the loops as shown in the figure. This triplet is complementary to a codon in the mRNA to which the tRNA attaches during translation, as shown in Figure 6.8. For each of the twenty amino acids present in proteins, there is at least one tRNA present in all cells specific for it. Some amino acids have more than one tRNA. Therefore a cell will generally be capable of producing more than twenty different tRNAs, but they correspond to twenty amino acids.

A tRNA is *charged* with its amino acid by the acid attaching at the 3'-ACC end. All tRNAs have an ACC sequence at their 3' end. The bond between the amino acid and the adenine is formed in the presence of specific enzymes. ATP provides the energy (Fig. 6.8).

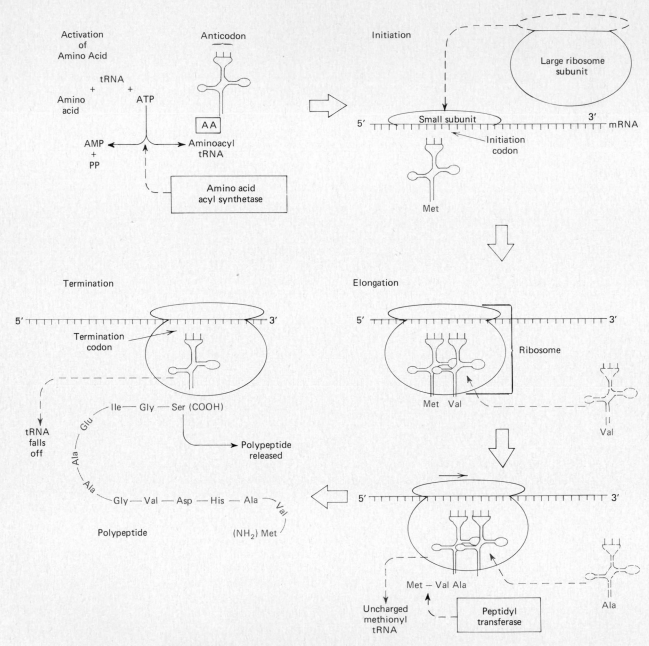

FIGURE 6.8
The process of translation of messenger RNA into a polypeptide. (Redrawn from R. P. Wagner, B. H. Judd, B. G. Sanders, and R. H. Richardson, *Introduction to Modern Genetics*, John Wiley & Sons, 1980.)

mRNA and Structural Genes

mRNA is transcribed from structural genes, but the eukaryotic structural gene has a great deal more DNA than appears in the messenger RNA. In the first place, a structural gene may be flanked by large segments of DNA that may not be transcribed at all but that may play an important role in regulation, including signals for transcription to start. Second, that part of the gene that is transcribed is not always represented fully by complementary RNA sequences in the final or mature mRNA from that gene. The complete transcript from the gene is hnRNA. This is the precursor to the functional mRNA from which the polypeptide will be translated. This is not true for all structural genes, but is seems to be for the majority of them. The histone genes, for example, are one of the exceptions.

The diagram in Figure 6.9 shows what apparently occurs in the cell nucleus,

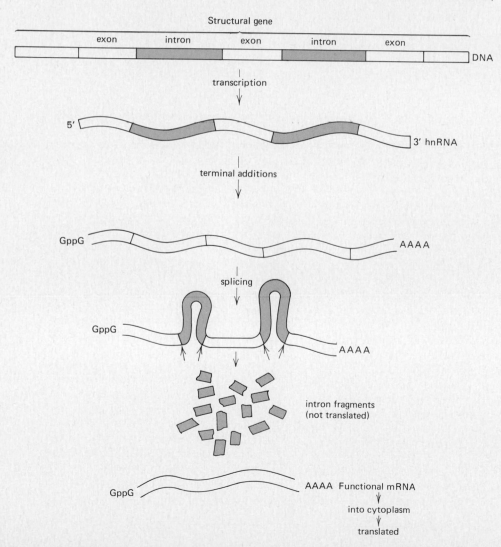

FIGURE 6.9
The transcription of a structural gene and the formation of messenger RNA from hnRNA.

starting with transcription of a structural gene and the formation of hnRNA. The hypothetical gene is represented as a bar at the top of the figure. Within it, three areas are distinguished by differences in color and shading. The entire gene may be as much as 10,000 or more base pairs long, but of this only 2000 or 3000 bases in a sequence may be transcribed. That part which is transcribed is hnRNA, and it in turn may have 10–100 times more bases in it than the mRNA to be derived from it.

The process of mRNA formation from hnRNA is complicated by the complex structure of some structural genes and the modification of the message after transcription. Box 6.1 discribes some of these details. The process of translation is also further described in Box 6.2.

———————————————————————————— Box **6.1**

Structural Genes, Exons, Introns, Caps, and Tails

Many if not most structural genes are partitioned into segments called *exons* and *introns*. Both kinds of DNA segments are transcribed as shown in Figure 6.9, but after transcription the hnRNA segments complementary to the introns are cut out. The RNA strands corresponding to the exons are joined together in a process called *RNA splicing*. The result is functional mRNA after it is capped and a *poly-A tail* is added. Generally, before or simultaneously with splicing, the prospective mRNA has a GpppG sequence called a *cap* added at the 5′ end and a polynucleotide made of 50–100 or more adenine ribonucleotides added to the 3′ end. This is called the poly-A tail, and it and the cap are characteristics of eukaryotic mRNAs. They are not found on the mRNAs of bacteria and other prokaryotes, nor are they found in the products of certain eukaryotic genes such as the histone structural genes. The prokaryotes do not have introns and exons either—still another significant difference between them and eukaryotes.

At present no one really knows what the functions of the introns, cap, and poly-A tails are. Guesses range from the introns serving in some regulatory capacity, or providing for more variability in gene products, to being fossil remnants of genes no longer used. The last possibility seems the least likely, since the introns in different distantly related eukaryotes have been maintained for millions of years of evolution essentially unchanged in their base sequences. This can only mean that they are important in the functioning of the genetic material. The GpppG caps may function as signals for the beginnings of translation, since translation does proceed from the 5′ to the 3′ end.

The number of bases in the intron sequences of a gene is generally greater than that for the exons. For example, the rabbit globin gene coding for the beta-polypeptide of its hemoglobin has about 1295 base pairs, of which approximately 705 are in two introns. More base pairs in introns than in exons seems to be the general rule for structural genes in humans and other eukaryotes.

Posttranslational Modification

After some polypeptides are translated, they may be cleaved at specific points and some sections discarded. This, for example, is true for certain digestive enzymes produced by the pancreas. A cleaved product becomes the active form. In the coagulation of our blood, many of the proteins involved become active in blood clotting only after they are modified by cleavage. Besides cleavage, polypeptides may become phosphorylated, meaning that they have phosphate groups attached to them (Fig. 6.10). These modifications all have a considerable effect on the structure and function of the proteins derived from these polypeptides and may play an important role in the processes of development.

Proteins from Polypeptides

Polypeptide chains do not function as such, as we have already described in Chapter 3. They coil and fold to form compact molecules called proteins. These protein units may then aggregate to form still larger super- or *macromolecules.*

The Interactions Between Protein Subunits

Most proteins that we know anything about consist of more than a single polypeptide and are therefore *multimeric,* in distinction to proteins consisting of a single polypeptide, which are called *monomeric.* A major protein constituent of our muscle is myoglobin, which is monomeric. Most of our other proteins, including nearly all of our enzymes, are multimeric. The forces that hold the

FIGURE 6.10
Some types of posttranslational modification of proteins.

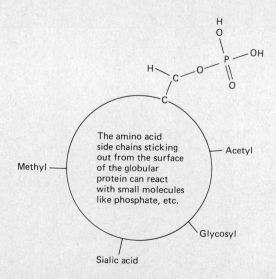

Box **6.2**

Polypeptide Synthesis in the Cytoplasm

So far as we are aware, all polypeptide and hence all protein synthesis occurs in the cytoplasm of cells, most of it in the membranous endoplasmic reticulum (Fig. 4.3). A very small but essential part takes place in the mitochondria. The process begins with an mRNA receiving a ribosome, as shown in the diagram, at the 5′ end where the GpppG cap is located. This may be followed by a stretch that contains other signals and then the codon for methionine, AUG. The ribosome at this point receives the tRNA for methionine (tRNA^{met}) and translation begins.

The ribosome moves to the next codon and the next with successive tRNAs coming into the production line. Simultaneously the amino acids are tied together by enzymes forming peptide bonds. The carboxyl group of the first amino acid to enter, methionine, forms a peptide bond with the amino group of the next to enter, leaving the amino group of the methionine free. The result is that all polypep-

tides start out with a free amino group and end with a free carboxyl group. When the ribosome reaches the stop signal, which will be one of the codons UAA, UAG, or UGA, it will drop off, terminating translation for that polypeptide. The polypeptide may now be modified by having the initial methionine at its amino end removed (Fig. 6.10).

The first ribosome to enter the message will be followed by others, and new chains of amino acids will start forming. The result is that one messenger will be economically used to form many polypeptides with the same sequences until the messenger disintegrates and is replaced by a new one from the nucleus. The whole process is a production line that works with great precision and accuracy to make large numbers of polypeptides with specific sequences of amino acids.

In addition to the polypeptide synthesis associated with the endoplasmic reticulum (ER), synthesis also occurs in the

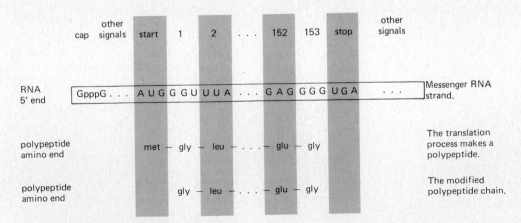

Some details of the translation process.

Box 6.2

(*continued*)

mitochondria. This synthesis is quantitatively much less that that occurring in the ER, but it is vital to the cell's function, since it forms several of the polypeptides that function in the process of respiration in which ATP is generated and O_2 is used.

The mitochondrial DNA is transcribed and the mRNA transcripts translated just as in the ER. Both transcription and translation are carried out in the presence of enzymes, tRNAs, rRNAs, and mRNAs either made in the mitochondria under the direction of mitochondrial genes or transported in from the sites of synthesis outside the mitochondria.

polypeptide subunits together in a multimeric protein are essentially no different from those which hold a polypeptide in its tertiary configuration.

In general, two kinds of multimeric proteins occur: those that are *homomeric*, having only one kind of peptide, and those that are *heteromeric*, which, as the name implies, have two or more different kinds of polypeptides. Figure 6.11 gives a diagrammatic example of each of these kinds of multimers. Perhaps the best known heteromultimeric protein, and certainly one of the most studied, is vertebrate hemoglobin. In Figure 6.11, human hemoglobin A is shown diagrammatically as a *tetramer* with four polypeptide chains, two designated α and two β. Hemoglobin A is therefore denoted $\alpha_2\beta_2$. The polypeptides α and β actually have some similarities in their amino acid sequences, but they are coded by different genes, which, in fact, are located on different chromosomes. With rare exceptions, of which insulin is the best example (Fig. 6.12), heteromeric proteins are the

FIGURE 6.11
Examples of protein multimers. (a) Hemoglobin A is a heteromeric molecule that consists of two each of two kinds of subunits, α- and β-globin chains. It is therefore also a tetramer. (b) Glutamine synthetase consists of twelve identical subunits and is therefore a homomeric molecule.

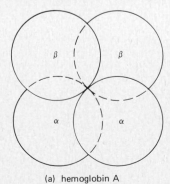

The α and β globin chains are represented as spheres. Actually they are folded polypeptides.

(a) hemoglobin A

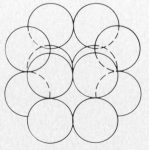

Each folded polypeptide is identical.

(b) glutamine synthetase

113

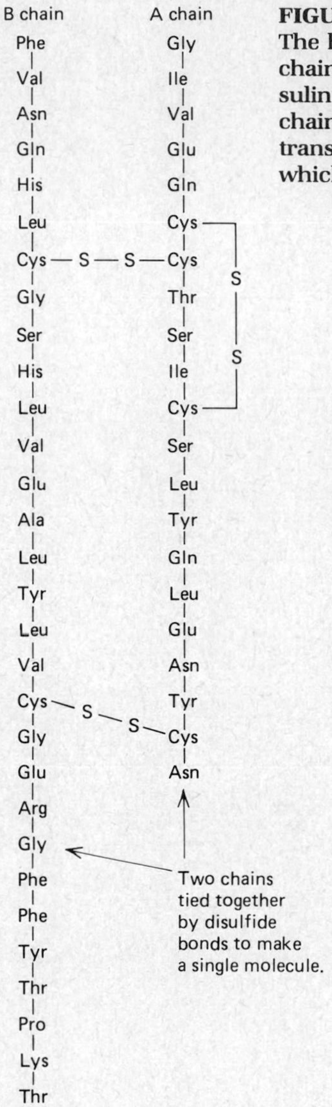

```
 B chain          A chain
  Phe              Gly
   |                |
  Val              Ile
   |                |
  Asn              Val
   |                |
  Gln              Glu
   |                |
  His              Gln
   |                |
  Leu              Cys ─┐
   |                |   │
  Cys ─ S ─ S ─ Cys    │
   |                |   S
  Gly              Thr  │
   |                |   │
  Ser              Ser  │
   |                |   S
  His              Ile  │
   |                |   │
  Leu              Cys ─┘
   |                |
  Val              Ser
   |                |
  Glu              Leu
   |                |
  Ala              Tyr
   |                |
  Leu              Gln
   |                |
  Tyr              Leu
   |                |
  Leu              Glu
   |                |
  Val              Asn
   |                |
  Cys ─ S ─ S ─ Tyr
   |                |
  Gly              Cys
   |                |
  Glu              Asn
   |                ↑
  Arg
   |
  Gly ←─────   Two chains
   |           tied together
  Phe          by disulfide
   |           bonds to make
  Phe          a single molecule.
   |
  Tyr
   |
  Thr
   |
  Pro
   |
  Lys
   |
  Thr
```

FIGURE 6.12
The human insulin molecule. The two polypeptide chains that comprise insulin were formed from proinsulin, a single polypeptide consisting of these two chains joined by a connecting polypeptide. Proinsulin is transcribed and translated from a single gene, after which the intermediate segment is cut out by enzymes.

products of more than one gene. In this way, they are different from homomeric proteins, which are the product of a single gene.

The Role of Proteins

In Chapter 4, we described some of the roles that proteins play in the body as structural elements and as functional elements. Of all the different kinds of molecules within us, they are the most important elements in the formation of our total phenotypes. We have only a vague idea about how a complex individual develops from an egg, but undoubtedly a most important part of the process is the determination of the kinds and amounts of proteins synthesized in the different parts of the developing embryo. These processes are collectively under some sort of regulatory control.

The Regulation of Protein Synthesis and Metabolism

If all the structural genes in our cells were active at all times to the same extent, we could not exist as humans. For one thing, all our cells would be the same. However, it is a clearly established fact that the different kinds of cells in our bodies produce different spectra of proteins. Liver cells, kidney cells, nerve cells, and muscle cells not only look different from one another but are also quite different in their metabolic capacities and protein constitution. The basic reason for these differences is that most of their structural genes are "turned off," that is, are not transcribing and hence not producing the mRNA necessary for the synthesis of the proteins for which they code. Not all the same genes are turned off in the different cell types, however. The result is they differ in many of their proteins even though they may have some of the same proteins in common.

Regulation and Development

During the process of development of an animal or plant from a single egg cell, more than simple cell division occurs. *Differential growth* is the increase in total cell mass in a directed way, and *differentiation* is the formation of the different kinds of cells in our bodies from undifferentiated embryonic cells (Fig. 4.15). Differentiation is the result of the regulation of gene activity, and unfortunately we do not understand the mechanism of this regulation. It involves in part the regulation of transcription; in eukaryotes regulation of translation is also involved.

Proteins can be classified as *general* or *differentiated*. General proteins are those essential for the function of every cell, whatever the specialized functions of the cell may be. Examples would be structural elements necessary to form cell membranes and tubules, mitochondrial enzymes, ribosomal proteins, and so forth. Differentiated proteins are those found only in certain differentiated cell types. They may be absolutely essential for the organism as a whole, but it is not essential for every cell in the body to manufacture them. Hemoglobin is a prime example. A brain cell functions very well without making its own hemoglobin, so long as there is an adequate supply from elsewhere. The result of regulation during development is that genes are turned on and off as the embryo proceeds through its various stages to become a fully developed infant ready to be born. We discuss this process in more detail in Chapter 14.

Regulation associated with the process of differentiation is generally quasi or completely irreversible in the sense that a liver cell having been derived from an embryonic cell makes only more liver cells; its daughter cells do not revert to an embryonic condition. This kind of regulation may well be unique to the eukaryotes, especially the higher animal eukaryotes. It is the kind of regulation that is associated with diploid organisms that develop complex body structures.

Transient Regulation

In addition to the regulatory mechanism(s) involved in development, there are others that are transitory in nature. Mechanisms for these have been successfully analyzed in bacteria, where it has been shown that genes involved in the transcription of messengers for enzymes involved in the synthesis of specific amino acids can be turned on or off reversibly depending on the need for these amino acids. Figure 6.13 gives an example of this for the biosynthetic pathway for the amino acid histidine in the bacterium *Salmonella typhimurium*. Note that when sufficient histidine for the organism's needs is synthesized, the additional synthesis is inhibited.

Transitory changes of this nature also occur in eukaryotes. Our livers, for example, produce many different kinds of enzymes, some of them at a relatively constant rate, and some only on demand. These latter are called *inducible enzymes*. The liver is probably the most metabolically active organ in our bodies, and besides making all sorts of proteins, storing and releasing sugar, etc., it also acts as a policeman. The blood that carries the digested food from our intestines goes first to our liver. If substances that may be injurious are present, liver enzymes

FIGURE 6.13
Transient feedback regulation in the metabolic pathway for histidine synthesis in the bacterium *Salmonella typhimurium*.

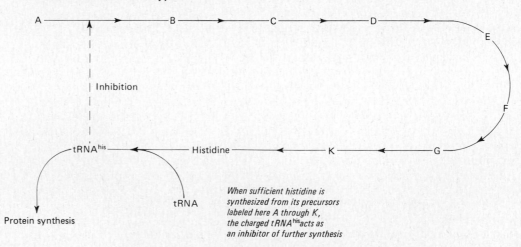

that may have the capacity to degrade these substances to a state where they are no longer injurious will be induced. A whole class of inducible enzymes called *mixed function oxidases* are produced (or induced) on demand in the presence of suitable substrates.

In addition to enzymes induced in the presence of substrate, presumably because the genes involved are activated by the substrate, other regulatory systems of great importance exist. A prime example is the action of the hormone estrogen. This hormone, which is a complex organic compound called a steroid, is a regulator in the sense that it causes certain genes to start transcribing in target cells (cells that respond to estrogen) and producing mRNA for protein important in the maintenance of the cells in the ovary in such a state that ovulation does not take place. Hence, the action of estrogen as a birth control agent depends upon its ability to modulate gene activity.

X-Chromosome Inactivation

In view of the major phenotypic effects caused by having one too many or too few autosomes, the similarity of males and females for genes on the X chromosome is somewhat surprising. For most, the one X chromosome of males produces the same amount of gene product as the two X chromosomes of females. It was long supposed that some type of dosage compensation must occur.

The explanation that proved to be correct was proposed by a British geneticist, Mary Lyon, who works primarily with mice. Her theory was that in any one cell, only one X chromosome is active, regardless of the number of X chromosomes present. Therefore, both XX females and XY males would have the same number of *active* X chromosomes in a cell. Furthermore, in order to explain the mosaic coat color patterns seen in certain mice, she proposed that

1. the X chromosomes that become inactive do so early in embryogenesis although after some number of cell divisions has occurred;
2. in each cell in which inactivation occurs, the choice of which chromosome is to become inactive is purely a matter of chance;
3. once an X chromosome is inactivated, the same X chromosome is inactive in all descendants of that cell.

The Lyon hypothesis, as it is commonly called, has proved to be true for the long arm of the human X chromosome, where nearly all known genes are located. It is not true for some portion, perhaps all, of the short arm, where a few genes are now known to be. These do not show dosage compensation. The inactive X chromosomes replicate later than active X chromosomes or autosomes, but in metaphase spreads the two X chromosomes are identical in appearance in female cells. In interphase, the inactive X chromosome (or the inactive part of it) is highly condensed and is visible as the sex chromatin body (Box 12.1).

As a consequence of X-chromosome inactivation, females are a mosaic of tissues, in some of which one X is active, in some of which the other X is active. This is not ordinarily noticeable. But rarely a woman who is heterozygous for an X-linked trait can be shown to have mosaic expression of the trait.

The Final Phenotype

The phenotype of a multicellular eukaryote, such as a human infant, is the result of the action of many genes and the interactions of their products among themselves and with the intrauterine environment. Probably the one factor of overwhelming importance is the temporal one. By this we mean not just the tempo but the timing of the period of activity of the various genes.

Although the main processes of development of a human are completed at the time of birth, the processes that make us human continue. Our phenotypes change from the time we are born to the time we die, but the temporal changes are slower than when we were *in utero*. It is, however, certain that the activities of our genes change as we grow older, and because of this and the interactions with our environment our phenotypes change.

Review Questions

1. Define the term gene as you presently understand it, and enumerate the several kinds known to exist.
2. Briefly describe the processes of transcription and translation and compare their similarities and dissimilarities.
3. What is hnRNA? How is it formed?
4. Distinguish between exons and introns.
5. What is meant by redundancy in the genetic code?
6. The anticodon is complementary to what?
7. A molecule of human hemoglobin A consists of what four polypeptides?
8. Enumerate the various roles played by proteins in cells.
9. Distinguish between general and differentiated proteins.
10. What are the roles of tRNA, ribosomes, and mRNA in protein translation?

References and Further Reading

Judson, H. F. 1979. *The Eighth Day of Creation.* Simon and Schuster, New York. 686 pp. An account of the revolutionary advances in cell biology, chemistry, and genetics, starting approximately in the 1940s, which have led to our present understanding of human genetics. Very chatty and readable.

Lake, J. A. 1981. The ribosome. *Scientific American* 245: 84–97. Details of the structure and functioning of the ribosomes and their central role in protein synthesis as presently understood.

Stent, G. S., and R. Calendar. 1978. *Molecular Genetics*, 2nd ed. Freeman, San Francisco. 773 pp. A well-written account of the present understanding of the underlying biochemical and molecular aspects of genetics.

Watson, J. D. 1976. *Molecular Biology in the Gene*, 3rd ed. Benjamin/Cummings, Menlo Park, Calif. A very readable account of the elements of molecular genetics.

Up to this point in our discussion we have concerned ourselves not with genetics per se, but with aspects of cellular and molecular biology and biochemistry that are closely allied to, if not inseparable from, genetics. We now turn to a discussion of what most people associate with the word *genetics*: the passage of traits or characteristics from one generation to the next. This part of genetics is ordinarily called inheritance or heredity.

What we or any other organism inherit is DNA. This being so, it should follow that aside from understanding the mechanism of DNA replication little else needs to be understood. But nature has conferred sex upon us organisms and this has complicated inheritance profoundly. As a consequence of sex and related phenomena, each new eukaryotic generation is a result of reshuffling and reassortment of genes during what amounts to a crap game of heroic dimensions. We can best begin to understand this generational reshuffling of the genetic material by analyzing the results of a few simple crossing experiments with mice. This will introduce us to the Mendelian principles of inheritance.

The Mendelian Principles of Inheritance

These principles were first established by Gregor Mendel using the common pea plant, as explained briefly in Chapter 2. Here we will use mice instead of peas. The same principles apply in either case.

The Monohybrid Cross

We start with two separate *lines* or *strains* of mice: gray mice and white mice. Each strain is *inbred*; that is, gray mice have been mated only to other grays and whites only to whites. Each line *breeds true* for its own color; gray parents have only gray offspring and white parents only white offspring.

To start the experiment, a gray parent is crossed to a white parent (Fig. 7.1). *All* progeny are gray without exception, in accordance with Mendel's observations that one form is often dominant to the other. These progeny are designated the F_1 (first filial) generation. Although they are similar to the gray parent in appearance, they are not entirely the same because when these F_1 grays are crossed to one another, their progeny (the F_2 generation) are either gray or white. If enough F_2 progeny are obtained, it can easily be seen that, in accordance with Mendel's law of genetic segregation, the two colors occur in a ratio of about three gray to one white. In other words, if 100 F_2 progeny are obtained from crosses between the F_1 mice, about 75 should be gray and 25 white. Notice that the modifier "about" is used rather than "exactly." The reason for this inexactitude will be explained presently.

If the experimental crosses are continued with the F_2 mice as parents, the following results are obtained.

1. The F_2 white mice, when mated with each other, produce *only* white offspring.
2. The F_2 gray mice, when mated with each other, produce *both* gray and white offspring.

These results can be simply explained by making two reasonable assumptions.

7

The Mechanics of
Inheritance in
Eukaryotes

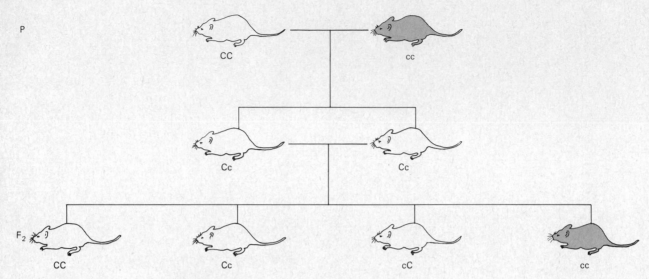

FIGURE 7.1
A monohybrid cross: gray × white mice.

1. There is a gene on one of the mouse chromosomes that controls the formation of pigment in the mouse fur (hair).
2. The gene exists in at least two forms—one form acts *positively* toward the formation of gray pigment, the other second form does not.

Let's call the first form of the gene *C* and the second *c*. To reduce verbiage, we designate *C* an *allele* of *c*. *C* and *c* can each be called genes too, but they are allelic genes, or *different forms of the same gene*.

If we apply these assumptions to the crosses described, we designate the true breeding gray parents *CC* and the true breeding white parents *cc*. Actually, this is a shorthand way of describing them as true breeding. A *CC* individual crossed to another *CC* (*CC* × *CC*) can only produce *CC* (gray) offspring, and a *cc* × *cc* cross can only produce *cc* (white) offspring. Thus, when the cross

$$CC \times cc$$

is made, the gray parents will produce only *C* gametes and the white *cc* only *c* gametes. As a result, all offspring will be *Cc*. Since all offspring are gray even though their cells carry the allele *c*, *C* is said to be *dominant* to *c*, or contrariwise *c* can be called *recessive* to *C*. Although *CC* and *Cc* mice are both gray, they are different in the alleles they carry. We call their allelic designations *CC* and *Cc* their *genotypes*, and their gray color their *phenotypes*. They have different genotypes but similar phenotypes because of the dominance of *C* over *c*.

The F_1 *Cc* offspring are next crossed to one another.

$$Cc \times Cc$$

What do we expect from this cross? First we determine the kinds of gametes expected. Refer back to the discussion of meiosis in Chapter 2 and review it. To add to the definition of allelic genes given in a previous paragraph, we can say that they occupy equivalent positions on homologous chromosomes. Thus, we can write the F_1 parental genotypes as

Furthermore, we know that among the forty chromosomes of the mouse $(n = 20)$ this gene $(C$ and its allele $c)$ is on chromosome 7 about halfway between the centromere and the opposite end. This is the gene's *locus* (position) on the seventh chromosome. How we know this will be described presently. When a cell carrying both alleles prepares to undergo meiosis we can show the seventh homologous pair as

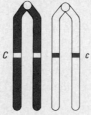

and as a result of this first division we will have

because in the first division of meiosis the homologues separate. Next, the second division will produce gametes with these genotypes:

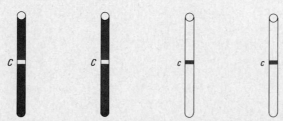

and obviously C gametes and c gametes in equal numbers.

We can then write the cross as

♀ \ ♂	½C	½c
½C	¼CC	¼Cc
½c	¼Cc	¼cc

Thus   ¼ of the offspring will be *CC* and gray
¼ + ¼ of the offspring will be *Cc* and gray
¼ of the offspring will be *cc* and white

or in a ratio of ¼*CC* : ²⁄₄*Cc* : ¼*cc*. We discard the denominators since they are all the same and get 1 *CC* : 2 *Cc* : 1 *cc*. This is called the *genotypic ratio*. The *phenotypic ratio* will be 3 gray : 1 white, since ¼*CC* + ²⁄₄*Cc* is equivalent phenotypically to ¾ gray. Although *Cc* and *CC* are phenotypically similar or identical (gray), they are obviously genotypically different, *and this difference can be demonstrated only by making crosses*: that is, *CC* × *CC* gives only grays, whereas *Cc* × *Cc* gives both gray and white progeny. To identify this genotypic difference we call *CC* animals *homozygous* (both alleles the same) and *Cc* animals *heterozygous* (two different alleles present). The *cc* animals are also homozygous, but homozygous for *c* rather than *C*.

The F_2 ratios 1*CC* : 2*Cc* : 1*cc* and 3 gray : 1 white resulting from the cross, *Cc* × *Cc*, are called Mendelian ratios. Other kinds of Mendelian ratios are also possible, if, for example, the cross

$$Cc \times cc$$

is made. We expect the heterozygous gray parent to produce two kinds of gametes, *C* and *c*, in equal numbers but the homozygous recessive white to produce only one kind, *c*. Thus, the cross will give

	c
½C	½Cc
½c	½cc

or a phenotypic ratio of ½ gray : ½ white. This kind of cross is also called a *test cross*, since you can distinguish gray (*Cc*) mice from gray (*cc*) mice with it. (*CC* × *cc* gives only gray; *Cc* × *cc* gives ½ white.)

By postulating the existence of two alleles, *C* and *c*, and invoking what we know about meiosis, we can explain all the results described in the first part of this section.

Some Elementary Probability I.   Events that may or may not happen to us all have a certain probability of occurrence. For convenience the probability is measured on a scale of 0 to 1. The 0 is absolute certainty that a given event will not occur, and 1 is absolute certainty that it will. We may accept as 0 the probability that we will grow wings tomorrow and become angels, and the probability that we will be older tomorrow than today as 1. In between 0 and 1 are an infinite number of probabilities about other things that may or may not happen to us, such as whether we get such and such a disease, or get into an automobile accident tomorrow, or have a fire in our house next Tuesday.

What we are interested in immediately are the probabilities associated with meiosis and fertilization. Consider the mouse crosses we described above and ask yourself the question: What are the probabilities the *CC* animals will produce only *C* gametes and *cc* animals only *c* gametes? Obviously the answer is 1. (Actually it is slightly less than 1, more like 0.999999. We will explain why in Chapter 9.) What about the heterozygous *Cc* animals, what proportion of *C* and *c* is probable? If we define the probability of an event occurring as being *equal to the number of ways it can occur divided by the total number of ways it can occur or fail*, we can assign a number. Two possible events are equally probable: A gamete can be either *C* or *c*. The *sum* of the probabilities that a gamete will be *either C* or *c* is 1, since these are the only possible outcomes. Each has an equal chance of occurring; therefore the probability for each is ½.

$$\text{Total probability} = \text{probability of } C + \text{probability of } c$$
$$= ½ + ½ = 1$$

We have assumed in our discussion that *C*- and *c*-bearing homologues in a heterozygote will segregate normally and that both *C* and *c* gametes have an equal chance of survival and ability to fertilize or be fertilized. The analogous situation would be the chances of heads or tails with a coin. Tossing a head with a well-balanced coin would be ½, but if the coin is so altered that heads are more likely to come up than tails then the probability will be >0.5 for heads and <0.5 for tails. In any case the probability of heads plus the probability of tails must always equal one, since these are the only two possible events.

On page 124 we have drawn what is called a Punnett square to show the results of the cross *Cc* × *Cc*. Since the probabilities of *C* and *c* gametes occurring in both parents are 0.5, we can show the square as

♀ \ ♂	0.5 *C*	0.5 *c*
0.5 *C*	0.25 *CC*	0.25 *Cc*
0.5 *c*	0.25 *Cc*	0.25 *cc*

The offspring are the result of a sperm meeting an egg. The chance of a *C* sperm fertilizing a *C* egg is 0.5 × 0.5 = 0.25 (or ¼). The event *C* sperm with a probability of 0.5 is *independent* of the event *C* egg also with a probability of 0.5. *The probability that two or more events, independent of one another, will occur together is the product of the probabilities of the separate events.* Hence, we get 0.25 for the *CC* event. In the same way we get 0.25 for each of the other possibilities of fertilization.

Since Cc progeny may occur in two different ways (a C sperm fertilizes a c egg, or a c sperm fertilizes a C egg), we add the two probabilities to get the sum 0.5. In the same way, the probability that a mouse from this cross will be gray is $0.25 + 0.50$ (probability of CC plus probability of Cc) = 0.75.

Consider this question. What is the probability that a gray mouse in the litter from the cross $Cc \times Cc$ is heterozygous? The probability that it is Cc is 0.5 and that it is CC is 0.25. The sum is 0.75. The probability that a given gray mouse is heterozygous is therefore $0.50/0.75 = 0.67$ (or $\frac{1}{2} \div \frac{3}{4} = \frac{2}{3}$). What we are doing here is making 0.75 the total of possibilities. We ignore the possibility of cc occurring.

We now turn to a more complicated cross than gray $\times$ white and use our new knowledge of probability to figure out the results.

The Dihybrid Cross

The monohybrid cross involved a single pair of alleles of the coat color gene C. We now introduce a second pair of alleles of a gene of a different chromosome from C. C is on chromosome 7 of the mouse, and our second gene, W, curly whiskers, is on chromosome 9. Straight whiskers, W, is dominant to its allele, w, curly whiskers. We now cross two mice heterozygous for both these gene loci.

$$Cc\ Ww \times Cc\ Ww$$

These parents are both gray and straight whiskered. What kinds, and in what proportions, of offspring are expected? To answer this, first recognize that what happens to the C and c alleles in meiosis is quite independent of what happens to the segregation of the W and w alleles in meiosis, since they are on different chromosomes. Think of this cross therefore as two independent crosses.

$$(1)\ Cc \times Cc \qquad \text{and} \qquad (2)\ Ww \times Ww$$

(1) will give gametes in the ratio 0.5 C : 0.5 c for both parents, and (2) will give gametes in the ratio 0.5 W : 0.5 w. We now calculate the joint occurrence of the different alleles in the same gametes. The meiotic events are independent so we multiply

$$0.5\ C \times 0.5\ W = 0.25\ C\ W \ (\text{or } \tfrac{1}{2} \times \tfrac{1}{2} = \tfrac{1}{4})$$
$$0.5\ C \times 0.5\ w = 0.25\ C\ w \ (\text{or } \tfrac{1}{2} \times \tfrac{1}{2} = \tfrac{1}{4})$$
$$0.5\ c \times 0.5\ W = 0.25\ c\ W \ (\text{or } \tfrac{1}{2} \times \tfrac{1}{2} = \tfrac{1}{4})$$
$$0.5\ c \times 0.5\ w = 0.25\ c\ w \ (\text{or } \tfrac{1}{2} \times \tfrac{1}{2} = \tfrac{1}{4})$$

The four possible different kinds of gametes occur with equal probabilities and hence equal numbers. We can now make a Punnett square as follows.

Male Gametes

	0.25 C W	0.25 C w	0.25 c W	0.25 c w
0.25 C W	0.0625 CC WW	0.0625 CC Ww	0.0625 Cc WW	0.0625 Cc Ww
0.25 C w	0.0625 CC Ww	0.0625 CC ww	0.0625 Cc Ww	0.0625 Cc ww
0.25 c W	0.0625 cC WW	0.0625 cC Ww	0.0625 cc WW	0.0625 cc Ww
0.25 c w	0.0625 cC wW	0.0625 cC ww	0.0625 cc wW	0.0625 cc ww

Female Gametes (label at left of rows)

Or we can use what we know about probability in a different way and consider the independence of the (1) and (2) crosses. The probabilities of getting CC, Cc, and cc offspring are 0.25, 0.50, and 0.25, respectively, and the same is true for WW, Ww, and ww offspring. Then, since we are dealing with independent events, we can write

$$
\begin{aligned}
0.25\ CC \begin{cases}
0.25\ WW = 0.25 \times 0.25 = 0.0625\ CC\,WW \\
0.50\ Ww = 0.25 \times 0.50 = 0.125\ \ \ CC\,Ww \\
0.25\ ww = 0.25 \times 0.25 = 0.0625\ CC\,ww
\end{cases}
\end{aligned}
$$

$$
\begin{aligned}
0.50\ Cc \begin{cases}
0.25\ WW = 0.50 \times 0.25 = 0.125\ \ \ Cc\,WW \\
0.50\ Ww = 0.50 \times 0.50 = 0.25\ \ \ \ Cc\,Ww \\
0.25\ ww = 0.50 \times 0.25 = 0.125\ \ \ Cc\,ww
\end{cases}
\end{aligned}
$$

$$
\begin{aligned}
0.25\ cc \begin{cases}
0.25\ WW = 0.25 \times 0.25 = 0.0625\ cc\,WW \\
0.50\ Ww = 0.25 \times 0.50 = 0.125\ \ \ cc\,Ww \\
0.25\ ww = 0.25 \times 0.25 = \underline{0.0625}\ cc\,ww
\end{cases}
\end{aligned}
$$

Total   1.0000

Collecting terms for identical genotypes and phenotypes, we get

		Genotype Frequency	Phenotype	Phenotype Frequency
CC	WW	.0625	gray, straight	
CC	Ww	.1250	gray, straight	
Cc	WW	.1250	gray, straight	0.5625
Cc	Ww	.2500	gray, straight	
CC	ww	.0625	gray, curly	
Cc	ww	.1250	gray, curly	0.1875
cc	WW	.0625	white, straight	
cc	Ww	.1250	white, straight	0.1875
cc	ww	.0625	white, curly	0.0625
		1.0000		1.0000

These results are obtained because the two different genes are on different chromosomes (7 and 9). Since their alleles segregate independently, this is called *independent assortment*, sometimes also called Mendel's law of independent assortment. The phenotype ratio of 0.5625 gray, straight : 0.1875 gray, curly : 0.1875 white, straight : 0.625 white, curly may also be read $9/16 : 3/16 : 3/16 : 1/16$ or $9 : 3 : 3 : 1$. We may also calculate it directly since gray is 0.75 and white 0.25, just as straight is 0.75 and curly is 0.25. Then

$$
\begin{aligned}
\text{Gray, straight} &= 0.75 \times 0.75 = 0.5625 = 9/16 \\
\text{Gray, curly} &= 0.75 \times 0.25 = 0.1875 = 3/16 \\
\text{White, straight} &= 0.25 \times 0.75 = 0.1875 = 3/16 \\
\text{White, curly} &= 0.25 \times 0.25 = 0.0625 = 1/16
\end{aligned}
$$

In this way we can get the phenotypic ratios much more easily and quickly than by drawing a Punnett square.

The Variations Among the Progeny of Diploid Eukaryotes

When two heterozygotes such as *Cc* are crossed, *three* different genotypes are expected: *CC*, *Cc*, and *cc*. When a double heterozygote such as *Cc Ww* is inbred, *nine* different genotypes are expected. Nine is equal to 3^2. If we had three gene loci on three different chromosomes and each gene represented by a pair of alleles, we should expect $3^3 = 27$ different genotypes. The mouse has twenty pairs of chromosomes. Let us suppose that each of these twenty pairs has a gene locus represented by a pair of alleles in a certain mouse strain. If a cross is made between two heterozygotes of this type, we should expect 3^{20} *different genotypes*. This is roughly 3.5×10^9 or 3.5 billion different possible genotypes. Of course, under natural conditions, it is highly improbable that two such individuals will mate. However, just consider the number of different kinds of gametes a human female is capable of producing. Assume that she is heterozygous for each of her 23 pairs of chromosomes (which is not at all improbable). How many kinds of eggs can she produce? For one pair of alleles, she can produce two, for two pairs four, for three pairs eight. You can see the pattern. 2^n equals the number of different kinds of gametes possible, where *n* equals the number of pairs of segregating alleles. Our example female, then, is potentially capable of producing 2^{23} different kinds of eggs. Again, we get a number we scarcely can comprehend: about 8.4×10^6 or 8.4 million.

Since probably everyone is heterozygous for at least one gene on each of our 23 pairs of chromosomes, you can readily appreciate that every egg or sperm we produce by meiosis is different from every other with a probability close to 1. Furthermore, and this is most important, no two people are ever identical genotypically with the possible exception of identical twins. Each of us is unique genotypically. The probability is essentially 0 that our identical genotype has existed in another person. That it will ever exist again in a future generation is also a probability of essentially 0. But this is not all. Variability is even more extensive than we have indicated, because of the phenomenon of crossing over.

Allelic Genes, Dominance, and Epistasis

Allelism

We have previously described allelic genes as being different forms of the same gene. This definition almost sounds metaphysical. What it means is that there is a stretch of DNA (a gene) at a particular site (or locus) on a specific chromosome that has a *specific function*. The function, let us say, is to transcribe the necessary mRNA for a specific polypeptide that acts as an enzyme to catalyze a specific metabolic reaction. However, differences in base sequences can and do occur within this stretch of DNA by mutation, as we shall describe in Chapter 9. These constitute the different forms or alleles, as shown in Figure 7.2.

The polypeptides formed by each allele will be different but generally not greatly

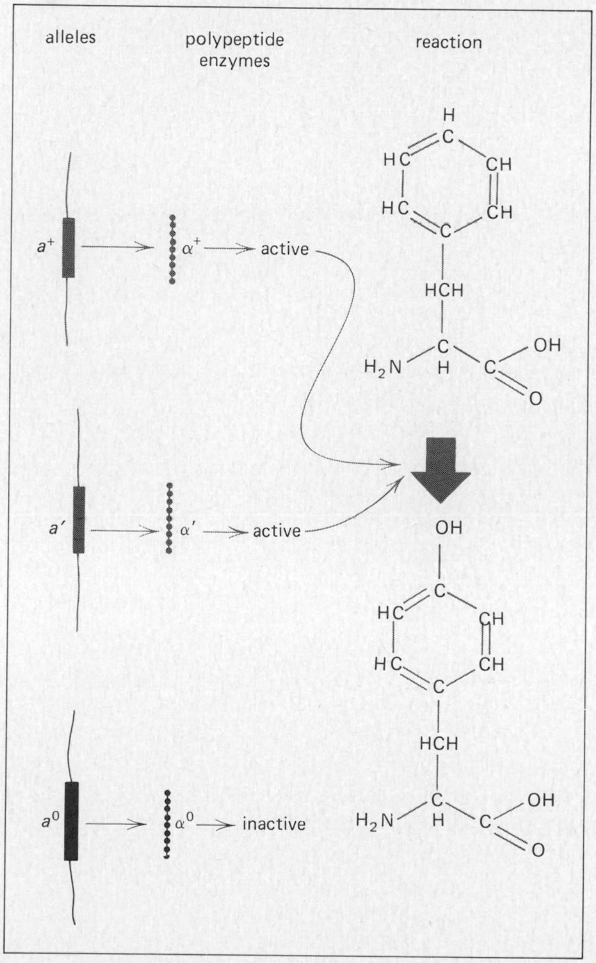

FIGURE 7.2
The gene a is a structural gene that codes for the enzyme phenylalanine hydroxylase which catalyzes the conversion of phenylalanine to tyrosine. Two alleles are known in human populations: a^+ and a^0. When a^+ is present, either in the homozygous state or in heterozygous combination with a^0, there is sufficient hydroxylase activity to give the normal phenotype. But when a^0 is homozygous, all the protein produced is inactive and the phenotype is phenylketonuria. Other alleles such as a' are thought also to exist. These alleles may produce hydroxylases that are less active than that produced by a^+ but more active than the inactive product of a^0.

different. A single amino acid difference out of a hundred or more may be the extent of the difference. Some of the different polypeptides may function effectively enough, but some may not. Indeed, some may not function as active enzymes at all.

In the human population, as in all other animal and plant populations, many gene loci are *polymorphic*. This is simply a way of saying that many of our genes are represented by two or more alleles that are common in the population. The different ABO and Rh blood groups are examples of polymorphic loci, as are the loci that influence eye and hair color, although we know very little about the inheritance of these latter traits. Often the different alleles are recognized by the polypeptides they produce. If the two polypeptide products from two different alleles differ by an amino acid, this difference may be detected if there is a charge difference. Box 7.1 describes the technique of electrophoresis by which such differences can be analyzed and described.

—————————————————————————————————— Box **7.1**

Allelic Variation and Electrophoresis of Proteins

The separation of proteins by electrophoresis provides a powerful means of detecting small differences in proteins because of small differences in their electric charges. A protein or even an amino acid has a characteristic ability to ionize because of the amino and carboxyl groups that are present (Chapter 3). For example, a simple amino acid such as glycine has the potential for one positive and one negative charge.

$$H-N-CH_2-C-OH \rightleftharpoons H-{}^+N-CH_2-C-O^-$$

The carboxyl (—COOH) group on the right end of the molecule can give up a positively charged proton (a hydrogen ion) and becomes negatively charged. The amino (—NH₂) group can bind a proton and becomes positively charged. The ionized form of glycine shown would

be electrically neutral, since there is the same number of positive and negative charges.

A peptide bond cannot ionize. Therefore a polypeptide composed of a chain of glycine residues would still have the ability to form only one negative charge at the free carboxyl end and one positive charge at the free amino end. If a glutamic acid or aspartic acid is inserted, an additional negative charge is possible because of the carboxyl group in the side chain.

$$H-{}^+N-CH_2-C-N-CH-C-N-CH_2-C-O^-$$

glycine   aspartic acid   glycine

Similarly, lysine and arginine both have amino groups in their side chains and

Box **7.1**

(*continued*)

can bind a proton to become positively charged, as can histidine, which has a nitrogen that is not an amino group but that can nevertheless bind protons. Glutamic acid and aspartic acid are *acidic* amino acids and add to the negative charge of a protein. Lysine, arginine, and histidine are *basic* amino acids and add to the positive charge of a protein. All other amino acids are *neutral.* The net charge of a protein is determined by the relative numbers of acidic and basic amino acids.

The actual charge of a protein also depends on the acidity (pH) of solution in which the protein is dissolved. At low pHs (high acidity), the high concentration of hydrogen ions tends to suppress ionization of the carboxyl groups and promote ionization of the amino groups. The net charge of the protein would be positive. At high pHs (low acidity = high alkalinity), the carboxyl groups are ionized, but ionization of the amino groups is suppressed since there are few protons available. The net charge of the protein would be negative.

Electrophoresis is the separation of charged molecules in an electric field. As frequently carried out in the detection of

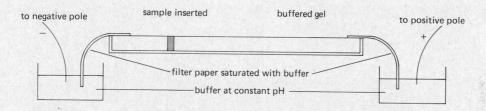

A mixture of proteins is inserted into the buffered gel, and a voltage is applied. Proteins will move toward the electrode of opposite charge. A mixture of human hemoglobins would give the pattern below at alkaline pH. The gel is viewed from above.

Box **7.1**

(*continued*)

genetic variants of proteins, the protein is dissolved in a buffer. A buffer is a solution of a mixture of chemicals that keeps the pH constant. The protein solution is inserted into an electrophoresis system, consisting of filter paper, a porous membrane, or a gel. The system is connected to electrodes, and the charged protein molecules migrate toward the electrode of opposite charge; that is, a protein with a net negative charge would be attracted to the positive electrode (anode). A typical arrangement is shown in the figure.

A change in a codon often results in substitution of one amino acid for another. If the new amino acid has a charge different from the old, the protein will have a different net charge. Examples would be replacing lysine with valine for a loss of one positive charge (equivalent to gain of one negative charge) or replac-

ing valine with glutamic acid for a gain of one negative charge (equivalent to loss to one positive charge). Therefore, one possible effect of gene mutation is to change the charge of its protein product, a change that often can be detected easily by electrophoresis. Most of the known inherited variants of human hemoglobin, of which there are over 200, were detected by electrophoresis, and many other loci are known to have a variety of alleles, as revealed by the separation of the corresponding proteins by electrophoresis. It also follows though that mutations that cause substitutions of amino acids of similar charge cannot be recognized by electrophoresis. It is estimated that about ⅓ of amino acid substitutions cause a change in electric charge.

Dominance and Recessiveness of Alleles

One of the phenomena that puzzled the early geneticists, including Mendel, was that of *dominance*. By this we mean that the phenotype of the homozygote *AA* is similar or identical to the heterozygote *Aa*. The allele *a* is *recessive* because it is expressed phenotypically only when it is homozygous *aa*.

Now that we know something about gene action, the mystery surrounding dominance is removed for the most part. It should be quite obvious that if there exist two or more alleles for a gene responsible for the production of an enzyme, the alleles that produce active enzymes will be dominant to those that do not, provided of course that the active allele produces sufficient enzyme to maintain a normal phenotype when present but once. This is generally the case in humans and other animals.

Dominance and recessiveness then are not properties of the alleles in a strict sense. One could not examine an isolated allele and say that it is dominant or recessive. Rather, *a dominant allele is one whose presence can be detected when there is only one copy*. Otherwise stated, a dominant allele can be detected in heterozygous combination. This definition does *not* require that the phenotypes of homozygotes and heterozygotes be the same. In many instances they are quite different. *A recessive allele is expressed only when two copies are present.* Occasionally an allele may be dominant when combined with one allele and recessive when combined with another. That is, in the combination a_1a_2, the phenotype associated with a_1 is expressed. But in the combination a_1a_3, the phenotype is determined by a_3.

There are now over 100 human inherited diseases known in which the gene involved has inactive alleles that are recessive to the dominant active allele—active in the sense that it produces mRNA that is translated into an active enzyme. A pedigree showing such recessive inheritance for albinism in a Hopi Indian family is given in Figure 7.3. The normal allele at the albino locus, often referred to as the "wild-type" allele, is responsible for production of the enzyme tyrosinase that converts the amino acid tyrosine to melanin, the pigment that colors our skin and the irises of our eyes. The normal allele need be present only once to give a full complement of pigment.

Generally, beneficial inherited traits are dominant. The dominant alleles of structural genes produce active enzymes that carry out essential functions. The absence of active enzymes can cause such mild conditions as albinism, that are only partially disabling, or other more severe conditions ranging from early death to low IQ. Some of these recessive inherited conditions in which the deficiency of a specific enzyme is known are listed in Table 7.1. Other conditions leading to metabolic changes resulting from possible defects of proteins in membranes are described in Table 7.2.

Many of the inherited detrimental traits in the human population are dominant. The pattern of inheritance for one of these, brachydactyly, is shown in Figure 7.4. This is a relatively mild condition in which the fingers are broad and short, and it was the first Mendelian trait demonstrated in humans. Other more extreme effects are caused by other dominant alleles at other loci. One example, familial hypercholesterolemia, is listed in Table 7.2.

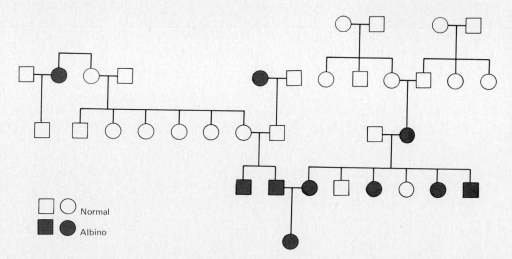

FIGURE 7.3
Pedigree showing autosomal recessive inheritance of albinism in a Hopi Indian family. (Redrawn from C. M. Woolf and R. B. Grant, *American Journal of Human Genetics* 14:391, 1962.)

□ ○ Normal

■ ● Albino

TABLE 7.1   Some Deficiencies of Metabolism in Humans That Are Inherited in a Simple Mendelian Pattern

All these conditions are recessive and are apparently the result of a single gene mutation. This is only a small sample of the inherited metabolic errors known in humans.

Disorder	Enzyme Deficiency	Frequency per 100,000 Births	Syndrome
Phenylketonuria	Phenylalanine hydroxylase	6–7	Mental retardation; accumulation of phenylalanine and its products in the body. May be alleviated by low phenyl-alanine diet.
Severe combined immunodeficiency disease	Adenosine deaminase	4	Lack of ability for immune response. Cannot withstand viral or bacterial infections.
Galactosemia	Galactose-1-phos-phate uridyl transferase	1–2	Mental retardation; accumulation of galactose in body unless galactose is excluded from diet.
Maple syrup urine disease	Branched chain amino acid amino transferase	0.5	Mental retardation; accumulation of valine, isoleucine, and leucine.
Tay–Sachs disease	Hexosaminidase A	28 among Ashkenazi Jews; very rare among others.	Mental retardation; certain lipids accumulate in brain. Early death.
Lesch–Nyhan syndrome	Hypoxanthine phos-phoribosyl transferase		Mental retardation, high blood uric acid, compulsive biting of fingers and lips.

TABLE 7.2
Some Membrane Defects Inherited as Mendelian Dominant Traits

Not all inherited diseases are the result of deficiencies in enzyme activity. Some are the result of alterations in membranes, presumably as the result of changes in the proteins in the membranes, which act as receptors or which regulate the passage of substances through the membranes. Such deficiencies in activity of membrane proteins can have just as serious consequences as deficiencies in enzyme activity.

Disorder	Chemical Basis	Frequency per 100,000 Births	Syndrome
Familial hypercholes-terolemia	Receptors on sur-face of cells do not bind the carrier of cholesterol	100–200; domi-nantly inherited	High blood choles-terol; vulnerability to coronary disease.
Cystinuria	Defect in membrane transport in kidneys and other cells	7–8	High levels of cystine, lysine, ornithine, and arginine in urine. Kidney stones made of cystine.

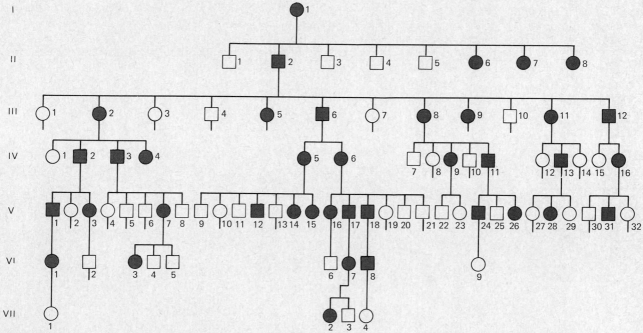

FIGURE 7.4
Pedigree showing dominant inheritance of brachydactyly, a condition in which the fingers are short. This was the first trait shown to follow simple Mendelian transmission in humans (W. C. Farabee, *Papers Peabody Museum Amer. Arch. Ethnol., Harvard Univ.* 3: 65–78, 1905). The present pedigree is the same family reported by Farabee, as brought up to date in the 1960s by V. A. McKusick. (Redrawn from V. A. McKusick, *Human Genetics,* 2nd ed., Prentice-Hall, Englewood Cliffs, N.J., 1969.)

The dominance of structural gene alleles that produce active enzymes over their alleles that produce inactive products is easily understood. However, the dominance of alleles that cause conditions such as brachydactyly, Huntington disease, and so forth, is puzzling. The general consensus is that most of these genes have some control over the course of development. They may be regulatory genes that control the timing of events in ways not presently understood. Some also appear to involve genes whose protein products have a structural role. For example, collagen that includes some abnormal protein subunits in its formation is understandably different in its properties from collagen with only normal subunits. We shall return to a discussion of this in later chapters.

Epistasis

This term is widely used by geneticists to describe interaction between *nonallelic* genes. In contrast, dominance and recessiveness describe *allelic* interaction. We can use albinism in humans as an example of epistatic action.

Humans come in many shades of color, ranging from blond, light-skinned Scandinavians to black-haired, black-skinned central Africans. These pigment differences appear to be controlled by allelic variations at a number of different loci. All human populations carry the allele for complete albinism, and those individuals who are homozygous for it are albinos, who do not form pigment, whether they carry other nonallelic genes for light pigmentation or dark. An albino Norwegian and an albino negroid African have the same amount of pigment— none. For this reason, the albino gene is said to be epistatic to the other color genes that determine the degree of pigmentation.

Crossing Over and Recombination

When two mice that are heterozygous for two genes such as *Cc Ww* are mated, one expects the 9:3:3:1 ratio among the offspring, because, as we have explained, the two loci *C* and *W* are on different nonhomologous chromosomes. Hence, their alleles segregate independently. But this is not always the case. Each chromosome of mouse and human bears many genes. Genes on the same chromosome are said to be *linked* (or *syntenic*), and it is obvious that if two gene pairs are linked, they will *not* segregate independently.

For example, consider in the mouse the gene loci *C* and *F*. Both of these are on the same chromosome, number 7. *F* and its allele *f* affect the condition of the fur coat: *FF* and *Ff* mice have normal fur; *ff* mice have frizzy fur. Hence, the locus is called *frizzy*. Suppose we make the following cross.

$$\frac{C\,F}{c\,f} \times \frac{C\,F}{c\,f}$$

We have indicated that *C* and *F* are coupled on the same chromosome, while their alleles *c* and *f* are coupled to each other on the homologous chromosome 7. According to earlier statements, we should expect two kinds of gametes from each parent: *CF* and *cf*. And only two kinds of offspring would be expected: normal color and fur condition (*C–F–*) and albino frizzled (*ccff*) in a 3:1 ratio. In fact, we get the other combinations too. The reason is that *crossing over* occurs during meiosis in the germ cells of the parents with the result that gametes with all four products, *CF*, *cf*, *Cf*, and *cF*, are produced. However, all four are *not* produced in equal numbers, for reasons that become evident by studying Figure 7.5. In this figure, (a) represents the two homologs of chromosome 7 as sister chromatids, paired to form tetrads in the first prophase of meiosis, (b) is a diagram of what happens while they are paired in this four-strand stage. Nonsister chromatids break and rejoin, not with themselves but with one another to give the result shown in (c). This is crossing over. Obviously, if it occurs between *C* and *F* loci in a heterozygote, the result after the second meiotic division will be four different chromosomes, as shown in (d). Two of them are *crossover* or *recombinant* products, *Cf* and *cF*, and two are of the original parental types or *noncrossover*, *CF* and *cf*.

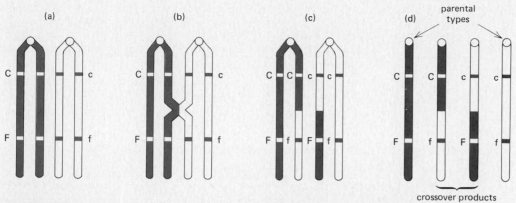

FIGURE 7.5
Crossing over between the chromatids of two homologous chromosomes.

If crossovers occurred between a pair of nonsister chromatids in every tetrad, then we should expect the four types, *CF*, *Cf*, *cF*, and *cf*, in equal numbers, but in fact this only occurs if they are far apart on the chromosome. Generally the parental types will outnumber the crossover types if the two loci are fairly close together. Crossing over outside the *C-F* segment obviously will not produce recombinant gametes. In fact, experimental results show that the nonparental gametes, *cF* and *Cf*, represent only 22% of the gametes produced. The other 78% are the *CF* and *cf* parental types.

Linkage of two loci therefore does not preclude recombination or the formation of nonparental type gametes. But it does reduce recombination under many circumstances. We discuss this important phenomenon and its consequences further in Chapter 8.

Sex Determination and Sex Linkage

In most animals, including the human variety, two kinds of chromosomes exist: *autosomes* and *sex chromosomes*. Autosomes are the ordinary type of chromosome, such as chromosome 7 in the mouse. The two homologous chromosomes 7 are identical in appearance, and each carries the same complement of genes (ignoring allelic variations). But one pair of chromosomes is heteromorphic in human and mouse males (and in other mammals). One of this pair is the X chromosome. It is typically one of the larger chromosomes and bears many genes. The other is the Y chromosome, which is much shorter, is highly heterochromatic, and bears only one clearly identified gene, the locus responsible for the H-Y antigen. In contrast to the XY chromosome constitution of males, females have two X chromosomes. The X and Y chromosomes are the *sex chromosomes*, as contrasted to the autosomes.

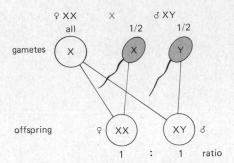

FIGURE 7.6
Sex determination of the XY type found in humans.

The X and Y chromosomes do not pair in meiosis like true homologs. Rather, they pair end-to-end, as if only a small segment were homologous. During normal meiosis, males produce two kinds of sperm with respect to these chromosomes, the Y-bearing and X-bearing sperm, in equal numbers, whereas females produce only one kind of egg, the X-bearing egg. Figure 7.6 shows the results. The male:female sex ratio approximates 1:1, but in fact it nearly always shows a slight preponderance of males over females at birth in many species of animals including humans. In addition, the ratio varies slightly from year to year, from about 104 males for each 100 females to about 106 males for each 100 females, for reasons that are not known.

Genes on the X chromosome are described as *X-linked* or *sex-linked*. The inheritance pattern for sex-linked genes is somewhat different than that for autosomal genes. Figure 2.6 gives an example of the inheritance of hemophilia among the descendants of Queen Victoria of England. Note that (1) the males have the hemophilia and (2) they inherit it through their mothers, who are not affected, and never from their fathers. The reasons for this are (1) males have but one X chromosome, and if that X bears the gene for hemophilia, males express the disease. Females have two X chromosomes and must be homozygous for the recessive hemophilia gene for the disease to be expressed. (2) Males receive their X chromosomes only from their mothers.

Mitochondria and Cytoplasmic Inheritance

Mitochondrial Characteristics

In Chapter 4 it was pointed out that the eukaryotic cell contains a number of different kinds of highly organized organelles such as the endoplasmic reticulum, Golgi apparatus, and mitochondria, in addition to a nucleus. The mitochondria deserve special mention because, unlike the other organelles, they contain DNA. This makes them partly independent of the nucleus but by no means completely so.

The mitochondrial DNA is in the form of rings of DNA imbedded in the mitochondrial matrix interior (Fig. 7.7A). The circumference of these rings is about

FIGURE 7.7
Mitochondrial DNA. (a)
Photograph of mitochon-
drial DNA from disrupted
human mitochondria.
The DNA has been coated
with platinum to make it
visible with an electron
microscope. (b) Diagram
of a circle of human mi-
tochondrial DNA with the
loci of the known genes
indicated. (The photo-
graph was supplied by
Dr. Donald L. Robberson,
The University of Texas
System Cancer Center,
Houston.)

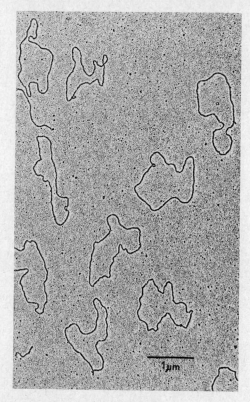

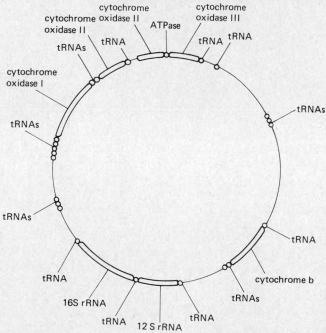

5.5 μm. They each contain about 16,500 nucleotide base pairs. Compared to the DNA of a haploid set of human nuclear chromosomes, which has *in toto* about 800,000 μm of DNA and 2.4×10^9 nucleotide base pairs, mitochondrial DNA would appear to be relatively insignificant. But in organisms, small can be beautiful and very important. The human mitochondrial genome has the capacity to transcribe all of the tRNAs and rRNAs necessary to translate several messenger RNAs also transcribed in the mitochondria.

Figure 7.7B shows a diagram of a single DNA circle and the regions of the DNA that code for the transfer and ribosomal RNAs as well as the messengers for the polypeptides that become important components of at least three key enzymes in the electron transport system: cytochrome b, cytochrome oxidase, and ATPase. If any one of these enzymes were to malfunction, instant death would occur because ATP synthesis and respiration would cease.

The proteins mentioned make up only about 5% of the total mitochondrial protein. The other 95% is made in the cytoplasmic endoplasmic reticulum with mRNA made in the nucleus and is transported to the mitochondria. Hence the mitochondria are by no means independent of the nuclear DNA, but their role is vital nonetheless.

Mitochondria are about the size of bacteria and, like bacteria, they reproduce by simple fission. This means that a mitochondrion, upon reaching a certain size and state, divides to make two mitochondria. As might be expected, as a cell divides, each daughter cell receives a quota of mitochondria that themselves divide in preparation for the next cell division.

In addition to having the capacity to synthesize proteins, the mitochondria also contain the enzymes necessary for the replication of their DNA. These polymerases and accessory enzymes are made outside the mitochondria, however, as are the enzymes necessary for translating the mitochondrial messengers into the aforementioned vital polypeptides.

All in all, mitochondria fit the role of obligate symbionts in the sense that their host cells cannot do without the mitochondria and the mitochondria cannot do without the host cells. It has been suggested that the mitochondria are descendants of ancient bacteria that invaded and established themselves a billion or so years ago in the primordial proto-eukaryotic cells. These cells were unable to use oxygen and hence led an *anaerobic* existence. When, however, combined with the *aerobic* (oxygen-using) bacteria, they became the ancestor of all present-day eukaryotes. The advantage of an aerobic, oxygen-using existence is, as was pointed out in Chapter 4, the capacity to generate many times more ATP molecules from ADP and energy-containing foods than would be the case for an anaerobic existence. Organisms would never have arisen from the primordial slime had they not become capable of generating large quantities of ATP for their energy needs.

Cytoplasmic Inheritance

Since our mitochondria have DNA, we inherit certain of our characteristics through them. Obviously not all of what we inherit from our parents is contained within the nuclei of our mother's eggs and father's sperm. But here an interesting factor comes into the picture. The unfertilized human egg, like animal and plant eggs in general, is heavily endowed with mitochondria. But the sperm is not. Generally, animal sperm have only a few mitochondria (Fig. 7.8) located in the *midpiece*. These mitochondria generate the ATP that provides the energy that drives the tail that provides the motility that enables the sperm to approach the egg. The best evidence we now have is that a sperm that penetrates an egg contributes only its nucleus. The paternal mitochondria either do not enter the egg, or if they do, they disintegrate and do not contribute to the genetic makeup of the mitochondrial population of the individual that develops from that egg. This is *maternal* or *cytoplasmic inheritance*.

By means of techniques using enzymes that break the phosphate bonds in the mitochondrial DNA at sites where there are specific sequences of bases, it is possible to demonstrate that the mitochondria in the human population are genetically variable, or polymorphic, just as is the nuclear genetic material. Figure 7.9 shows a pedigree of three generations of a family in which a mitochondrial variation has been detected. Note that the "atypical" mitochondrial DNA is passed on only by the mother. This kind of cytoplasmic inheritance has also been found in rats, mice, and donkeys. It may be a general phenomenon among plants and animals.

The criterion for judging whether a trait is passed through the cytoplasmic rather than the nuclear genetic material is to determine if its inheritance is only through the mother. All children will be like their mother and not like their father, if he is different from the mother. And of course we do not expect Mendelian ratios among the progeny because there is no segregation. These patterns will be apparent if one compares the pedigree in Figure 7.9 with those in Figure 2.6 (hemophilia, an X-linked recessive), Figure 7.3 (albinism, an autosomal recessive), and Figure 7.4 (brachydactyly, an autosomal dominant).

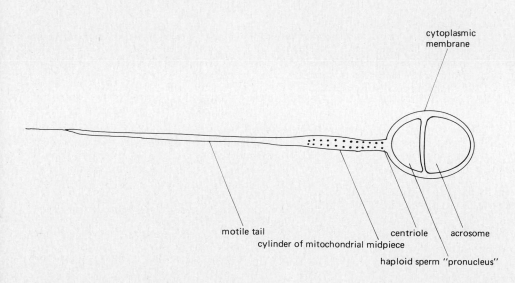

FIGURE 7.8
The anatomy of a human spermatozoon.

cytoplasmic membrane

motile tail

cylinder of mitochondrial midpiece

centriole

acrosome

haploid sperm "pronucleus"

FIGURE 7.9
Pedigree of a family showing cytoplasmic (maternal) inheritance. The trait is a mitochondrial variant and is transmitted only through the mother to all her offspring. (Redrawn from R. E. Giles, H. Blanc, H. M. Cann, and D. C. Wallace, *Proceedings of the National Academy of Sciences* 77:6715, 1980.)

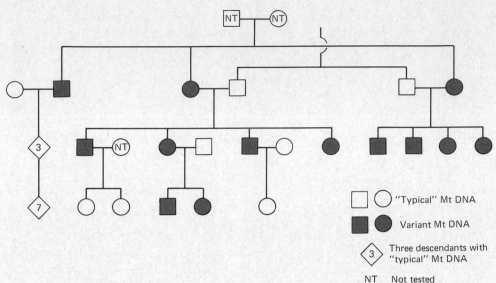

Review Questions

1. The dominant gene *A* controls pigment formation in humans. Its recessive allele *a* prevents pigment formation when homozygous *aa*. Homozygotes are albinos with no pigment in skin, hair, or eyes.
   (a) An albino man has children by a woman with normal pigmentation but whose mother was an albino. (1) What is the genotype of the woman? (2) What will be the phenotypes of the skin, hair, and eyes of their children and in what ratio?
   (b) Assume that both parents of the woman in (a) were normally pigmented and that the albino man and his wife have six children, all normally pigmented. (1) What is the probable genotype of the woman? (2) How would your answer change if one of the children were albino?

2. Galactosemia is an inherited metabolic disease in which the affected person cannot use the sugar galactose. It is an autosomal recessive trait, the gene being on chromosome 9. A man with galactosemia who has been raised on a diet free of galactose marries a woman whose father has hemophilia, an X-linked recessive trait. The woman's family has no history of galactosemia.
   (a) What are the chances that they will have normal daughters?
   (b) What are the chances that they will have normal sons?
   (c) What are the chances that they will have daughters heterozygous for galactosemia? Heterozygous for hemophilia? Heterozygous for both galactosemia and hemophilia?

3. It is said that sons are more closely related to their mothers than to their fathers. Why?

4. What is the probability of a male child inheriting the Y chromosome of his father's grandfather? We all have two grandfathers; which of the father's grandfathers does the child get his Y chromosome from?

5. Huntington disease is a rare autosomal dominant disease characterized by degeneration of the central nervous system beginning usually at ages 35 to 45.
   (a) What are the chances that a son or daughter of a person who develops Huntington disease will have the gene for the disease?
   (b) What are the chances that a grandchild of the affected person will later develop the disease?

6. A red-green colorblind man has children by a red-green colorblind woman. What are the chances that their children will be colorblind?

7. A colorblind man has children by a woman who has normal color vision. Her father and mother also had normal color vision, but both her grandfathers were colorblind. What is the probability that the couple will have (a) colorblind sons? (b) colorblind daughters?

8. A man known to be heterozygous for Tay–Sachs disease marries a woman who is heterozygous for galactosemia. The loci for these genes are on chromosomes 5 and 9, respectively.
   (a) What are the chances of their having children who are phenotypically normal?
   (b) What are the chances that any particular child would carry neither gene, i.e., would be homozygous normal at both loci?
   (c) What are the chances that any particular child would be a heterozygous carrier for both diseases?

9. What is the difference between crossing over and recombination?

10. A man possesses a certain characteristic that also occurs in his mother but not his father. Examination of the family histories of both parents reveals that the trait has been transmitted only by the females of his mother's line. None of this man's seven children have the trait. Give a plausible explanation for this kind of inheritance.

11. What is meant when two genes are described as allelic?

12. Sometimes a genotypic ratio is the same as a phenotypic ratio. Sometimes it is not. Why?

13. Distinguish epistasis, crossing over, and dominance.

References and Further Reading

Garrod, A. E. 1909. *Inborn Errors of Metabolism*. (Reprinted with a supplement by H. Harris, 1963.) Oxford Univ. Press, London. 207 pp. A great classic in which Garrod reviews his hypothesis that defects in metabolism are inherited according to Mendelian principles.

Harris, H. 1980. *The Principles of Human Biochemical Genetics*, 3rd ed. Elsevier/North Holland. Amsterdam. An authoritative treatise on the biochemical basis of inherited human disorders.

Novitsky, E. 1982. *Human Genetics*, 2nd ed. Macmillan, New York. 487 pp. An introductory text for persons with a limited background in genetics.

Sutton, H. E. 1980. *An Introduction to Human Genetics*, 3rd ed. Saunders College, Philadelphia. 592 pp. An advanced text, surveying the major aspects of human genetics.

Wagner, R. P., B. H. Judd, B. G. Sanders, and R. H. Richardson. 1980. *Introduction to Modern Genetics*. Wiley, New York. A general textbook of genetics that includes many human applications.

Whitehouse, H. L. K. 1973. *Towards an Understanding of the Mechanisms of Heredity*, 3rd ed. Edward Arnold, London. A very readable, literate explanation of genetic principles with references to their historical background.

One of the principles established by Mendel in his classic paper is that different inherited characteristics are transmitted independently of each other. When a haploid gamete is formed in a diploid organism, whether the gamete receives *A* or *a* from one locus does not influence whether it receives *B* or *b* at a second locus. This is sometimes referred to as Mendel's *law of independent assortment*.

We learned in Chapter 7 that when two heterozygotes such as *AaBb* are mated, a 9 : 3 : 3 : 1 ratio among the progeny is expected. But this is true only if the two gene loci are on different nonhomologous chromosomes, and we also learned that this is not always true. Genes may be linked, and if linked they may be separated by crossing over, even though on the same chromosome. In this chapter we extend our discussion of this very important area of genetics.

Demonstration of Genetic Linkage

The number of genes in higher organisms is not known but is on the order of tens of thousands. Yet the number of chromosomes on which these genes are located is typically less than fifty pairs. It is obvious then that there must be many different genes on each chromosome. Since chromosomes are thought to remain intact throughout the cell cycle, even during interphase when they are highly extended and cannot be readily observed, it follows that genes on the same chromosome should not segregate independently of each other.

The first demonstration that independent assortment does not occur for genes on the same chromosome was by A. H. Sturtevant in 1913 (Fig. 8.1). Sturtevant analyzed the segregation at two loci on the X chromosome of *Drosophila*, one that has an abnormal allele *r* causing rudimentary wings, the other with an abnormal allele *v* causing vermilion eyes rather than the normal red. The wild-type alleles at these loci are symbolized by *R* and *V*, respectively. Both loci were known to be on the X chromosome because of the characteristic pattern of transmission in experimental crosses. The wild-type alleles are dominant in females, but the abnormal alleles are expressed in hemizygous males.

Sturtevant crossed female flies of genotype *RRvv* with males of genotype *rV*. The F_1 offspring were either *RrVv* females or *Rv* males (Fig. 8.1). But the female offspring should have had *R* and *v* together on the maternally derived X chromosome and *r* and *V* on the paternal X, since these were the combinations in the parental gametes. We will therefore write their genotypes as *Rv/rV* to show how the alleles are coupled.

When these F_1 females were crossed with the *Rv* males (which we will write *Rv/Y*), it was possible to recognize the allele combinations in the gametes of the females by the phenotypes of the F_2 male offspring. As anticipated, the majority of the F_2 received either the *Rv* combination from their mother or the *rV* combination. But 27% received *RV* or *rv*. Thus, recombination of genes on the same chromosome is possible, but it does not necessarily occur at the frequencies expected if the genes assort freely.

The explanation, of course, lies in crossing over between homologous chromo-

8

Linkage and Recombination

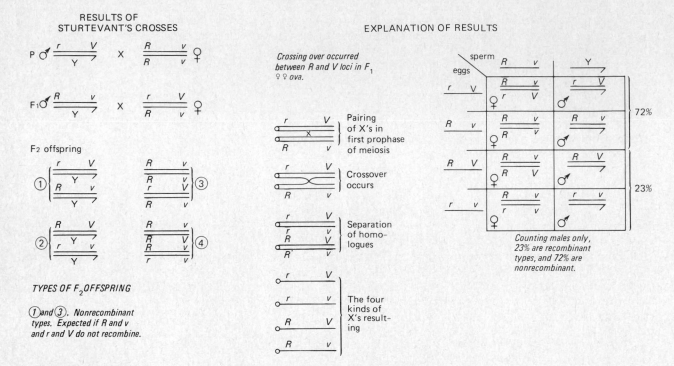

FIGURE 8.1

Diagram of Sturtevant's experiment showing nonrandom assortment of the X-linked genes *r* (rudimentary wings) and *v* (vermilion eyes) in *Drosophila*. In the F_1 females, *R* is coupled with *v* and *r* with *V*. These are the nonrecombinant arrays. If crossing over occurs between the *R* and *V* loci, the recombinant ova *RV* and *rv* are produced. In the F_2 male offspring, the four combinations can be distinguished.

somes during meiosis, as we described in Chapter 7. The chiasmata observed microscopically in meiotic prophase reflect the exchange of chromatid segments between homologous chromosomes. If a crossover occurs between two loci, the parental combinations of alleles will no longer exist. The new chromosome combinations are described as *recombinant* and the parental combinations as *nonrecombinant*. Loci that are close enough on the same chromosome to show reduced recombination are said to be *linked*.

If two loci are very close together on a chromosome, the likelihood that a crossover will occur between them is less than if they are far apart. Therefore, the frequency of recombination between loci is a measure of the distance between them. Two loci that show only 1% recombination are quite close to each other compared to the 27% recombination observed by Sturtevant for the *R* and *V* loci.

The frequency of recombination is therefore useful to measure the distance between linked loci, and a *linkage map* can be constructed by comparing the frequency of recombinants from various pairs of loci. By convention among geneticists, the total percentage of recombinants among the total number of progeny is called the *map distance* between the two loci being followed in the cross. The percentage values are called *map units* or *centimorgans.*

Not every pair of loci on a chromosome show reduced recombination. If the loci are far apart, two or more crossovers may occur between them, leading to completely random association of alleles in the gametes. If this occurs, the recombinant and nonrecombinant gametes will be equal in number. Therefore, 50% is the upper limit of recombinant gametes that can be recovered, and loci that show 50% recombination are said to assort independently. They may be far apart on the same chromosome or they may be on different chromosomes. One can demonstrate linkage directly only if the loci are close enough to show less than 50% recombination.

If three loci on the same chromosome are used in an experimental cross (a *three-point cross*), one can test whether linkage map distances are additive and one can determine the order of genes if they are linked. For example, consider a diploid organism heterozygous at three loci *A*, *B*, and *C* (Fig. 8.2). We know from the parental genotypes that the parental gamete combinations are *ABC* and *abc*.

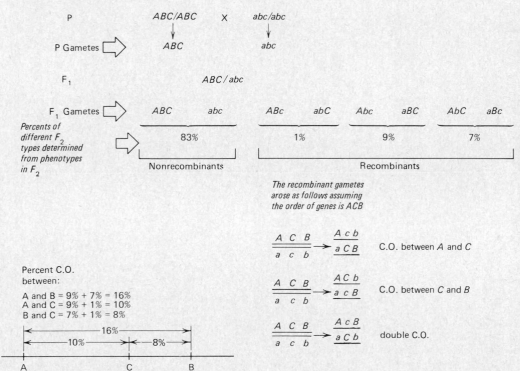

FIGURE 8.2
Diagram of a hypothetical three-point cross showing the additivity of recombination fractions. The gametes produced by the F_1 generation are scored as recombinant (R) or nonrecombinant (NR) for each pair of markers. The only sequence of markers that is consistent with the recombination fractions is *A–C–B*. The gametes *ABc* and *abC* must therefore involve double crossovers, causing them to be scored as nonrecombinant for the *A–B* pair, when in fact they should be scored as two crossovers for each gamete recovered.

Therefore the genotype of the F_1 heterozygote can be written ABC/abc. But we do not know whether these loci are linked or, if they are linked, in what order they may occur on the chromosome.

If the F_1 is backcrossed to one of the parental lines, the various gametic combinations can be identified and the map distances between pairs of loci measured. Let us assume that the results of this particular cross showed A and C to be 10 map units (centimorgans) apart. (One map unit = 1 centimorgan = 1% recombination.) A and B are found to be 16 units apart, and B and C are 8 units apart. These results suggest that A and B are farther apart than are A and C or B and C. Therefore the order must be ACB (or BCA, which is equivalent since we could start with either end first).

If we compare the distances between these loci, we find that 10 units between A and C plus 8 units between C and B is 18 units, which is approximately equal to the 16 units measured directly. The sum of the component distances would exactly equal the distance between outside loci if double recombinants were counted as two rather than as zero. Short distances are more accurate than longer distances because of double crossovers, and linkage maps are constructed by adding the distances between successive pairs of loci. In this way it is possible also to build up map distances greater than 50%. A typical chromosome may be 100–200 units long, but such a map would have to be assembled by adding together the much smaller measurable distances.

Loci known to be on the same chromosome are described as *syntenic*. Often syntenic loci show independent assortment because they are far apart. For example, all X-linked loci obviously are syntenic, but, of the more than 100 X-linked genes known in humans, only a few have been shown to be linked to other X-linked genes by reduced recombination. Generally when genes are described as being in the same linkage group, they have been placed with respect to each other by recombination techniques. At the present time, two distinct linkage groups have been identified on the human X chromosome, although both, as well as all other X-linked genes, are in the same syntenic group. Presumably someday the placing of genes between the two linkage groups will permit their connection into a single large linkage group.

When two loci are linked, the alleles that are found together on a particular chromosome are said to be *coupled* or to be in the *cis* configuration. For example, in Figure 8.1, R and v are coupled in the F_1, as are r and V. Similarly, R and V are said to be in *repulsion* or in the *trans* configuration, since these particular alleles are on opposite chromosomes.

Since crossing over occurs between all loci—indeed, it appears to occur at any point in the DNA and thus within genes as well as between—all combinations of particular alleles can be produced. In the population as a whole, combinations of alleles of linked genes should occur in the same proportions as if the loci assorted independently. Thus, AB, Ab, aB, and ab will occur in the same frequencies whether the A and B loci are linked or on different chromosomes. There should be no association between the presence of an A allele and a B allele considering the population as a whole. If the loci are linked, however, there will be associations *within* families. A particular combination aB will be transmitted as a unit until crossing over occurs to produce some other allelic combination. Consistent associations between two genetic traits in a large population cannot be explained on the basis of linkage. The term *linkage* is occasionally misused by persons not trained in genetics to mean association. Blond hair and fair skin are often associated in persons, presumably because they are influenced by some of the same genes. But these two traits are not linked in the genetic sense. If they were, one would expect blond hair to be found with dark skin as often as with fair skin.

Measurement of Linkage

Measures of Recombination

The traditional methods of measuring linkage are based on the same approach as that used by Sturtevant. In brief, experimental organisms or persons heterozygous at two loci of interest are crossed with other individuals of appropriate genotypes so that recombination can be detected in the offspring. (In human beings, of course, the crosses are not made experimentally. Rather, families are located that include persons with informative genotypes.) It is essential that one member of the mating be a double heterozygote, otherwise the mating will not be informative. Crossing over may occur between the *A* and *B* loci in a *AB/aB* person, but the crossover products cannot be distinguished from the nonrecombinant gametes because the person is homozygous at the *B* locus.

The simplest examples in human beings are the X-linked genes (Fig. 8.3). This is

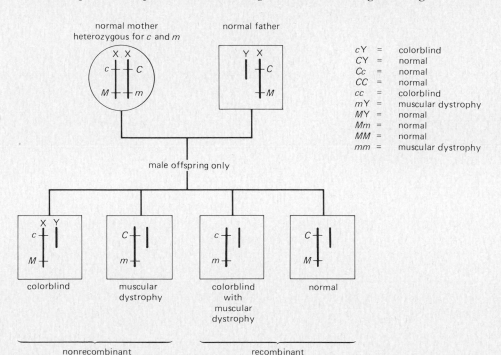

*c*Y	=	colorblind
*C*Y	=	normal
Cc	=	normal
CC	=	normal
cc	=	colorblind
*m*Y	=	muscular dystrophy
*M*Y	=	normal
Mm	=	normal
MM	=	normal
mm	=	muscular dystrophy

FIGURE 8.3
Recombination between the two X-linked genes for colorblindness and muscular dystrophy in humans. The woman is doubly heterozygous for these two genes, which are *trans*, having come from different parents. Her sons, being hemizygous, would express whatever combination of genes is received from her and can readily be classified as recombinant or nonrecombinant. Her daughters would all be normal and hence noninformative unless their father also had both of these traits.

because X-linked recessive alleles are expressed in sons. Furthermore, since the father contributes a Y chromosome to his sons rather than an X, the father's genotype (and identity) is irrelevant. If a woman is heterozygous for two X-linked genes, muscular dystrophy and colorblindness, for example, her sons may inherit both, one, or neither. Which combinations are recombinant and which nonrecombinant will depend on the coupling phase in her cells. For example, if her father was colorblind but did not have muscular dystrophy, he would have transmitted the colorbind allele c coupled with the normal allele at the muscular dystrophy locus M, since he only has one X chromosome. Therefore, the woman's genotype must be cM/Cm, with the Cm chromosome having come from her mother. If the two loci were closely linked, the majority of her ova would have the nonrecombinant chromosomes, cM or Cm, and the majority of her sons would have either colorblindness or muscular dystrophy. Few would have both or neither. If the loci were not closely linked, the recombinant chromosomes CM and cm should occur with high frequency.

Somatic Cell Hybrids

A very powerful new technique for demonstrating synteny of human genes is provided by somatic cell hybrids. The principles are discussed in more detail in Chapter 10. Briefly, if one has a hybrid cell that contains a full complement of mouse chromosomes but only one or a few human chromosomes, then any human genes expressed in the hybrid cell must be located on those particular human chromosomes.

Several hundred human genes have now been assigned to chromosomes based on somatic cell hybrid results. In order for this method to be used, the trait must be expressed in cultured cells. Therefore, many human traits cannot be studied this way. On the other hand, a large number of enzymes are expressed in cultured cells and have been mapped to specific chromosomes.

Evolution of Linkage Groups

The amino acid sequences—and the nucleic acid sequences—are very similar for homologous proteins in different species. The more closely related the species, the more similar are their genes. Similarly, closely related species have chromosomes that are much alike in structure. It follows therefore that the arrangements of genes on the chromosomes of closely related species should also be very similar.

This is borne out by observation. It is most easily demonstrated in X-linked genes. The mammalian X chromosomes seem to have very much the same genes in all species. This is sometimes called Ohno's law. As more information becomes available on autosomes, the same situation is seen. Much less is known about nonhuman primate chromosomes than about human chromosomes. But the information available corresponds exactly to that expected from cytogenetic similarities (Figure 8.4). Even in more distantly related species, where homologies

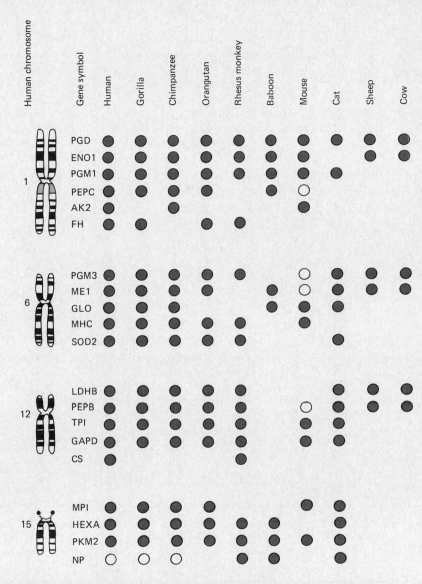

FIGURE 8.4
Comparison of the linkage relationships of several loci that have been studied in various species of mammals. Solid circles that are grouped to correspond to one of the human chromosomes all belong to the same linkage group, that is, they are on the same chromosome. Open circles are genes that have been shown not to be part of that linkage group. Blank spaces indicate the information has not yet been obtained. In most instances, the position of a gene on the chromosome is not known, and in some species the groups of linked genes have not yet been assigned to particular chromosomes. Much more information is available on the X chromosome (not shown), which seems to have changed little during evolution of mammals.

with human chromosomes have not been well established, the grouping of certain genes seems to have survived millions of years of separate evolution.

The Human Linkage Map

Present knowledge of the chromosomal location of some of the better known human genes is given in Table 8.1. In some instances, the location of a gene is

TABLE 8.1
A Partial List of Human Gene Loci That Have Been Assigned to Specific Autosomes[a]

Chromosome	Trait
1	Rh blood group
	Duffy blood group
2	Antibody subunit (kappa light chain)
3	Plasma transferrin (the iron-transporting protein)
4	MN blood group
	Plasma albumin
6	Major histocompatibility complex (MHC)
	Congenital adrenal hyperplasia (21-hydroxylase deficiency)
7	Histone genes (H1, H2A, H2B, H3, H4)
9	Galactosemia
	ABO blood groups
	Nail-patella syndrome
	Leukocyte interferon gene family
11	Insulin
	Wilms tumor
	Aniridia (absence of iris of eye)
	Albinism (tyrosinase-negative)
	Hemoglobin beta-gene complex (beta, gamma, delta, epsilon)
	Acute intermittent porphyria
13	Ribosomal RNA
	Retinoblastoma (cancer of the eye)
14	Ribosomal RNA
	Severe immune deficiency (nucleoside phosphorylase)
	Antibody subunit (heavy chain)
15	Ribosomal RNA
	Tay-Sachs disease (hexosaminidase A)
16	Hemoglobin alpha chains
17	Thymidine kinase-1 (soluble)
	Growth hormone gene family
19	Polio virus sensitivity
20	Severe immune deficiency (adenosine deaminase)
21	Ribosomal RNA
22	Chronic myeloid leukemia
	Antibody subunit (lambda light chain)

[a]A complete list would have over 300 entries.

known with some precision. Most often, it is known only to be somewhere on the chromosome.

As yet, knowledge of human linkage has contributed little to the solution of genetic problems of humans. This will change in time, as more and more is learned of the exact location of human genes. For example, it should be possible eventually to determine exactly which segments of which chromosomes are inherited by a child. Although many genes will not be expressed until later in life or only under certain environmental circumstances, one might know with certainty that a particular allele is present or absent.

As an example, the PKU (phenylketonuria) gene cannot be detected directly. Furthermore, the normal allele is active only in liver and activity is low before birth. Heterozygotes are difficult to recognize except by having been parents of a child with PKU. The position of the PKU locus is not known. For purposes of illustration, let us assume that it is close to the Rh blood group locus. There are a number of alleles at the *Rh* locus, of which three are common in most populations. If, in a particular family, a PKU allele is shown to be coupled to an R^2 allele, it should be possible to assess the PKU status of persons in that family by testing the red blood cells for Rh type to see whether they have the R^2 allele. Such tests could be done on fetal or newborn blood cells as well as cells from older persons.

In another family, the PKU allele might be coupled to a R^1 allele. Each family would have to be considered individually, and crossing over between the *Rh* and *PKU* loci would introduce error. As the positions of more loci are established, it will be possible to test for several closely linked loci, making this application much more reliable. An important part of genetic counseling in the future will be the analysis of the transmission of linked complexes of genes, only one of which may be of concern because of its detrimental effect on health.

Review Questions

1. A particular combination of alleles on the same chromosome may be separated at meiosis by what process?
2. Two loci, *A* and *B*, on a chromosome are said to be 10 map units apart. How often would crossing over occur between them during meiosis?
3. What is the difference between a syntenic group and a linkage group?
4. Why do loci that are far apart on the same chromosome show random recombination?
5. The X-linked traits G6PD deficiency and red–green colorblindness are 5 centimorgans apart. A woman is heterozygous for both traits, having received the genes for them from her father. What is the likelihood that her son will have both traits? How would your answer differ for her daughters?
6. How have mouse–human hybrids been of value in the study of linkage?
7. A rare mouse is found that appears to have a form of G6PD deficiency similar to that in humans. On what chromosome would you expect this mutation to have occurred? Why?
8. What is the difference between linkage of two traits and association of the traits?
9. It is expected that linkage will become important in genetic counseling. What information will linkage provide? Why is it not often used now?
10. How many human linkage groups are there?

References and Further Reading

McKusick, V. A. 1983. *Mendelian Inheritance in Man*, 6th ed. Johns Hopkins Univ. Press, Baltimore. 1378 pp. In addition to listing all known genetic loci in humans, this compendium also has tables of linkage and syntenic groups of mice and men.

Sturtevant, A. H. 1913. The linear arrangement of six sex-linked factors in Drosophila, as shown by their mode of association. *J. Exp. Zool.* 14:43–59. This is the seminal paper in which the demonstration of linkage was first reported.

Inherited alterations in the DNA constitution of an organism that are *not* the result of independent assortment or crossing over as described in Chapters 7 and 8 are called *mutations.* Mutations occur in germ cells and in somatic cells. If they occur in germ cells they will be passed on to succeeding generations provided they are not *dominant lethals.* A gamete carrying a dominant lethal will result after fertilization in a zygote that will die immediately or produce an embryo that will die before completing development. When a nonlethal mutation occurs in a somatic cell it will be passed on to the daughter cells and succeeding cell generations. A mutant *clone* may thus be established within and confined to an individual, and the mutation will not be passed on to that individual's progeny. Figure 9.1 shows the difference between germinal and somatic mutations and explains what is meant by a clone.

The geneticist distinguishes two general kinds of mutations: *gene mutations* and *chromosomal mutations* or *chromosomal aberrations.* We shall consider both kinds in this chapter.

Gene Mutations

A gene mutation is a change in the DNA structure of a chromosome that is detected by a change in phenotype of the individual carrying it, but it produces no easily detectable change in chromosomal structure. Those mutations that do produce detectable chromosomal changes are classified as aberrations and are considered in the next section. Gene mutations result in alleles that may be dominant, codominant, semidominant, neutral, or recessive. The effects of these various types are delineated in Box 9.1.

Types of Gene Mutations

The two known basic types of gene mutations are *missense* mutations and *frameshift* mutations. Both are the result of alteration of the nucleotide base sequence in DNA with the result that the reading of the sequence of codons is altered during transcription.

   Missense Mutations.   A missense mutation is simply a *substitution* of one kind of pyrimidine or purine base for another. Figure 9.2 shows the results of a missense mutation. If such a substitution transforms a codon reading for one amino acid into one coding for a different one, it is obvious that the polypeptide resulting after transcription and translation will have a different amino acid in the position corresponding to the altered codon.

   In Figure 9.2, we show a DNA strand that has the sequence of codons expected in the gene that codes for the β-globin chain of human hemoglobin A. This polypeptide has 146 amino acids, but only the first ten starting at the amino end are shown in the figure. In normal β chains, the sixth amino acid is glutamate (Glu). In some persons, a β chain is present that has a valine (Val) in the sixth position. This *mutant*

9

Alterations in the Genome

FIGURE 9.1
The germ line, the somatic cell line, and mutations that occur within them. A clone is a group or family of cells derived by mitosis from a single mother cell. Therefore, it can be said that all the cells of our body are of a single clone. However, we are also constituted of subclones that arise through mutation, either in the somatic cells or germ cells. Those mutations that occur in the germ line may pass on to succeeding generations.

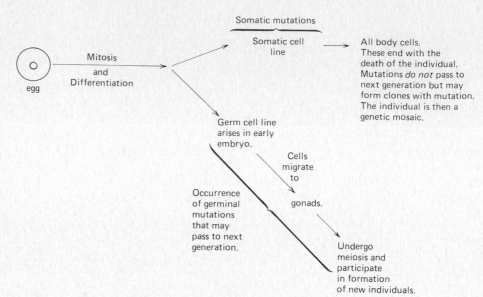

Somatic mutations

egg — Mitosis and Differentiation →

Somatic cell line → All body cells. These end with the death of the individual. Mutations *do not* pass to next generation but may form clones with mutation. The individual is then a genetic mosaic.

Germ cell line arises in early embryo.

Cells migrate to gonads.

Occurrence of germinal mutations that may pass to next generation.

Undergo meiosis and participate in formation of new individuals.

Box **9.1**

Various Types of Alleles That Arise from Mutation

Every gene has the potential of mutating to many different kinds of alleles. Some are easily distinguished from one another by their effects on the phenotype. Others may have little or no effect. Whether or not an allele is expressed in the phenotype depends on several factors. The corresponding allele at the homologous gene locus is most important. The genotype at other loci and the environment may be important. The methods of examining the phenotype often determine whether a mutant allele is detected.

The following terms are commonly used in genetics to describe the influence of particular alleles on the phenotype. By itself, an allele is neither dominant nor recessive. Such terms are meaningful only in the context of an organism, where the allele functions (or fails to function) in the presence of other alleles and other loci. The terms are very useful, however, in describing patterns of transmission of traits.

Dominant allele   An allele *A* is dominant if its corresponding phenotype can be observed when a single copy of *A* is present in the genotype. Very often, the combinations *AA* and *Aa* are identical. In that case, the allele *a* would be recessive to *A*.

Box 9.1

(continued)

Codominant alleles	Two alleles, A_1 and A_2, are codominant with respect to each other if, in the heterozygous combination A_1A_2, the characteristic phenotype of each is expressed. For example, in the ABO blood groups, the presence of both an *A* allele and a *B* allele produces the blood type AB.
Semidominant allele	An allele is semidominant if the phenotype of the heterozygote is intermediate to the phenotypes of the two homozygotes. Symbolically, *AA* > *Aa* > *aa*. For example, if *AA* is black and *aa* is white, *Aa* might be gray.
Recessive allele	An allele is recessive if it is expressed only when in homozygous combination. The *O* allele of the ABO blood groups is recessive, since type O blood is associated only with the genotype *OO*. An allele may be both recessive and dominant, depending on the combination with other alleles. The A_2 allele of the ABO blood groups is dominant to the *O* allele, since A_2O persons have type A_2 blood. But A_2 is recessive to A_1, since A_1A_2 persons have A_1 blood.
Isoalleles	Isoalleles are alleles that have identical effects on the phenotype, at least under certain conditions. Symbolically, $A_1A_1 = A_1A_2 = A_2A_2$. The phenotypes may differ if the genetic background or the environment is changed. Some can only be distinguished by chemical analysis of the protein product of the gene or by the DNA sequence of the gene.
Lethal gene	Many mutant alleles are lethal, causing death of the zygote, the embryo or fetus, the juvenile, or even the adult who has not reproduced. The term is quite relative, since even an allele that causes sterility in an otherwise healthy person is lethal insofar as genetic considerations are concerned. Lethal alleles may be dominant or recessive, or the lethal effect may be associated primarily with heterozygous combinations. An example of the last is found in the Rh blood groups, in which heterozygous *Rr* offspring are lost from *rr* (Rh-negative) mothers.

β is the result of a mutation in the DNA codon in which a thymine deoxyribonucleotide has been substituted by adenine deoxyribonucleotide. Reference to the RNA code on page 105 will show that, if the *DNA* codon is converted to CAT, the corresponding RNA codon will go from GAA to GUA. While the RNA codon GUA codes for valine, the codon GAA codes for glutamate. Thus, by one nucleotide change an amino acid is changed.

This is a missense mutation that we can be certain has occurred in the human population. Furthermore, in populations that have been screened for β-chain differences, at least two other substitutions have been found at position six; in one case a lysine (Lys) replaced the glutamate and in the other it is in alanine (Ala). A second look at the codons for Glu shows that a single substitution will give a codon for either Lys (AAA from GAA) or Ala (GCA from GAA).

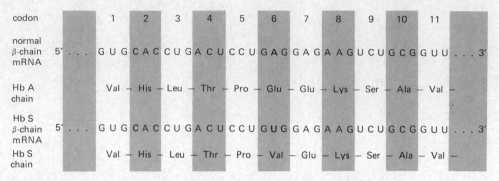

FIGURE 9.2

A missense mutation involving the β chain of hemoglobin A. The normal allele for the β chain (β^A) codes for a sequence of 146 amino acids, beginning with the sequence shown. The nucleotide sequence of the corresponding messenger RNA is shown. For convenience, the nucleotides are grouped into codons. Substitution of U for A in the second position of the sixth codon of the mRNA produces the altered β chain that results in sickle cell hemoglobin.

Many missense mutations have been spotted in the human population (as well as in all plants and animals that have been analyzed) by the use of protein electrophoresis. Hundreds of amino acid substitutions have been identified in both the α and β chains of human hemoglobin A. In the β chain alone, more than sixty of the amino acid positions show substitutions. In most of these the amino acid substitution has resulted in a change in charge. A charge change is expected when a valine is replaced by a glutamate. But what is more important than a charge change is the change that may occur in the functioning of the protein.

Missense mutations in general cause only a single amino acid change, but under certain conditions they may have a drastic effect. To see why, refer again to the RNA code. You will note that three codons are labeled *stop*. These codons, UAA, UGA, and UAG, are labeled *stop* or *termination* codons because cells do not ordinarily have tRNAs capable of recognizing them, as described in Figure 6.7. If they occur in the codon sequence prior to the proper termination point, an incomplete polypeptide will be formed. This is because a codon that specifies an amino acid can be converted by a missense mutation (sometimes called a *nonsense* mutation in this instance) into a termination codon. For example, UCA codes for serine. A change to UAA or UGA would terminate polypeptide synthesis at that point. As a consequence, this type of missense mutation results in a drastic change in the ability of the gene product to function. Termination codons can also be converted to readable codons by missense mutations, causing the polypeptide chain to be longer than normal.

Frameshift Mutations.   Frameshift mutations are the result of the deletion or duplication of bases. Figure 9.3 illustrates the two types that in either case result in the change in the *reading frame* and therefore a change in message reflected in the sequence of amino acids. As might be expected, a frameshift mutation will generally be completely nonfunctional, and such a mutation will result in an allele that is lethal when homozygous or even when heterozygous. Termination codons

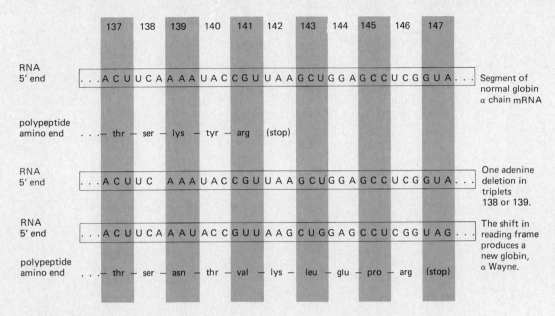

FIGURE 9.3

Example of a frameshift mutation. The α chain of hemoglobin A normally has 141 amino acids. The α-chain variant, hemoglobin Wayne, has an additional five amino acids on the carboxyl-terminal end caused by deletion of one of the A's in codons 138 or 139. Since the messenger RNA is read three nucleotides at a time, this causes the nucleotides to the right of the deletion to shift to the left one position with respect to the reading frame. An entirely new set of codons is thus generated. The normal α chain is terminated by the UAA codon at position 142. In the new reading frame, this termination codon no longer exists, and translation continues until the UAG codon is encountered at position 147.

are formed frequently as a result of frameshift mutations. The result is that a shorter polypeptide is formed than would otherwise be the case.

The Origin of New Alleles

New alleles arise from old alleles by mutation. Presumably most of them are the result of missense mutations—at least those that survive for geneticists to examine. We have no way of knowing how many mutant alleles don't survive.

Every population has more than one allele segregating for a great many of its genes. Estimates for the human populations vary, but a figure of about 30% is generally accepted. This means that 30% of the genetic loci are *polymorphic* (have more than one common allele). The definition of "common" is to be present in greater than 2% of the individuals. The true figure may be higher or lower, depending on various biases.

The point that we wish to make here is that new alleles are constantly arising by mutation in the human population at a rate of about one mutation at each locus per million gametes. Some loci are known to mutate at a higher frequency; others may be lower. Probably a large number of these mutations simply vanish because the gamete carrying it did not participate in fertilization or the zygote that received it died for other reasons. We discuss this important matter in later chapters. Meanwhile, feel confident that, because of polymorphism in the human populations, you are bound to be heterozygous at a number of your loci.

The Effect of Gene Mutation on the Phenotype

We are by no means aware of all the differences in proteins resulting from missense mutations in the human or any other population. But we do know that many of our proteins are polymorphic. Consider as a specific example the hemoglobin β chain with the substitution at position 6 that we discussed in the previous section. This mutant form of β is usually referred to as β^S to distinguish it from the "normal" form, β^A. The frequency of the allele that produces β^S is essentially zero in the white population in the United States. But in the black population the frequency of this allele is about 4%. In other words, in the American black population, 4% of the alleles at the β locus are β^S, while 96% are β^A. This is not exceptional for proteins other than the hemoglobins. But let's concentrate on the β polypeptide for this discussion of the effect of gene mutations on the phenotype. We go into the interesting matter of the distribution of polymorphic loci in Chapters 17 and 20.

The hemoglobin molecule consists of four protein subunits, called globins (Fig. 6.11). The principal hemoglobin in normal adults is Hb A, which has two α and two β globin subunits. Hence, its formula is generally written $\alpha_2^A \beta_2^A$. The α subunits are the products of a separate α locus. Individuals who carry the allele determining the β^S globin may be either $Hb_\alpha{}^A Hb_\alpha{}^A / Hb_\beta{}^A Hb_\beta{}^S$ or $Hb_\alpha{}^A Hb_\alpha{}^A / Hb_\beta{}^S Hb_\beta{}^S$, depending on whether they are heterozygous or homozygous for the mutant allele.

One might expect that since β^S differs from β^A in only 1 amino acid out of 146, there should be essentially no difference in the properties of the hemoglobins that have β^S compared to those with β^A. But such is not the case. Both heterozygotes and homozygotes (AS and SS) have hemoglobins that differ from the standard $\alpha_2^A \beta_2^A$, because hemoglobin polypeptide β^S has different physical and chemical properties than β^A. Red blood cells containing only β^S in their hemoglobin react differently from those with only β^A in the presence of low levels of oxygen, the condition found in the venous system. The arterial blood gives up its oxygen and gains carbon dioxide as it passes through the capillaries. Under these circumstances, the red blood cells with only β^S "sickle," that is, they assume various odd shapes because their hemoglobin tends to crystallize (Fig. 9.4). Heterozygotes (AS), with both β^A and β^S globins in their hemoglobin, do not have this problem, but SS homozygotes have *sickle cell disease*, which manifests itself as a severe anemia accompanied by intermittent severe pain and eventually death at an early age.

However, the effect of the mutation is not all bad. Those individuals who are heterozygous and therefore have hemoglobin with both β^S and β^A do not have a debilitating anemia, but they do have greater resistance to malaria caused by the protozoan *Plasmodium falciparum* than do homozygous normal individuals. The

FIGURE 9.4
Red blood cells from a person with sickle cell anemia. The cells are viewed with a scanning electron microcope. The cell on the left has the normal biconcave disk shape. The one on the right has become distorted into the typical "sickled" shape, from which the disease gets its name. (From B. F. Cameron, R. Zucker, and D. R. Harkness of the Miami Comprehensive Sickle Cell Center. Reproduced from *Annals of the New York Academy of Sciences* 244:60, 1975.)

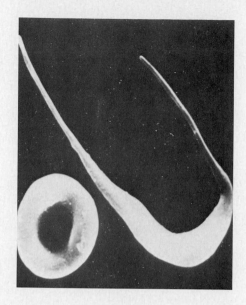

homozygous *SS* individuals presumably have this resistance too, but it should do little good in areas where malaria is prevalent since the sickle cell disease is much more serious than the malaria.

Many of the proteins synthesized in the human may differ from each other by one or two amino acids only. Hence, their alleles would differ only by one or two codons. Such small differences are not always easy to detect, and many of what are considered to be single alleles may in fact be multiple alleles that have not been distinguished. The phenotypic effects of a single amino acid substitution may range from lethality to neutrality. (See Box 9.1.) That is, a mutant allele may be lethal when homozygous, or its effect may be identical to the normal allele. In between these two extremes lie a whole range of possibilities resulting from missense mutations causing amino acid substitutions. Recognize that in a polypeptide like the β chain of hemoglobin there are at least 146 amino acids and hence at least $3 \times 146 = 438$ base pairs in the DNA coding for it. If we assume that any one of these base pairs is subject to change by missense mutation, then it is evident that the number of possible alleles is very large for the β gene. And, of course, the same applies to all other genes that code for polypeptides.

The reasons why some amino acid substitutions have little effect on a protein's function while others have a drastic effect are simple enough. Some of the amino acids in a polypeptide chain are more important than others in their role of maintaining a certain three-dimensional tertiary structure for the functional unit. A single change can cause a drastic alteration in the active site of an enzyme, for example, that may make it impossible to accept the substrate. Hence, although the protein is synthesized, it is inactive. It is also apparent that substitutions in the lipophilic amino acids that are folded into the interior of the tertiary structure are particularly disruptive to the protein's function, particularly if the mutation results in the substitution of a lipophilic amino acid such as valine, isoleucine, or leucine by a hydrophilic one.

Mutations that confer inactivity on a protein or even just reduce its activity below a certain level can cause severe to relatively mild disorders depending on the enzyme activity affected. Such mutations are generally recessive and are easily recognized only in homozygotes. It is convenient to call their effects *metabolic blockage* caused by *genetic blocks*. Obviously, when an enzyme necessary for the catalysis of a particular reaction is inactive the reaction cannot take place or it proceeds so slowly (the more likely case) that no effective amounts of product are formed.

To take a relatively frequent and well-studied inherited condition in humans, we turn to the disease *phenylketonuria*. About 1 in 10,000 newborns have the disease, and since they are invariably mentally retarded unless treated early, the condition can be considered to have a significant impact on the general population.

The classical form of the disease is the result of a mutation of a gene that results in a deficiency in the activity of the enzyme *phenylalanine hydroxylase* that catalyzes the reaction converting phenylalanine to tyrosine (Fig. 9.5). both phenyl-alanine and tyrosine are essential amino acids and the average diet contains more than sufficient amounts of both for protein synthesis. Why, then, is the enzyme so important? The reason is quite simple though not immediately obvious. More phenylalanine than can be used in protein synthesis is generally taken into the body from food. If this excess is not converted to tyrosine, excess phenylalanine accumulates or is converted to the deaminated form, phenylpyruvic acid, and other phenyl compounds. These compounds are excreted by the kidneys, but even so their level in the blood of phenylketonurics is much higher than in normal persons. An average of about 1 mg of free phenylalanine is found in 100 ml of blood from an unaffected person, whereas a phenylketonuric may have a 50 times greater concentration. This high level of phenylalanine has a number of deleterious effects, the most significant of which is mental retardation. Some phenylketonurics have an IQ in the normal range, but most are severely retarded, confined to institutions for the retarded.

Another metabolic block in the phenylalanine–tyrosine pathways in our bodies was recognized about 80 years ago by the British physician A. E. Garrod. This block results in failure of conversion of a product of tyrosine catabolism, homogentisic acid (Fig. 9.5), with the result that this compound is excreted through the kidneys in large amounts. The condition is called *alkaptonuria*, and it causes no major problems, unlike phenylketonuria. When first passed the urine is normal in appearance, but it becomes black on standing because of the oxidation in air of the homogentisic acid to a red to black pigment. Garrod recognized that this inability to convert homogentisic acid to the normal end products, CO_2 and H_2O, was inherited according to a Mendelian pattern. He recognized this only a year after Mendelism was rediscovered and called this disease and others that he recognized *chemical sports* or *inborn errors of metabolism*. He is now referred to as the father of human biochemical genetics.

Well over 100 different kinds of genetic blocks or inborn errors of metabolism caused by mutations in structural genes for enzymes important in human metabo-lism are now known. A sample of them is given in Table 7.1. It is evident that practically all of them cause serious pathological symptoms.

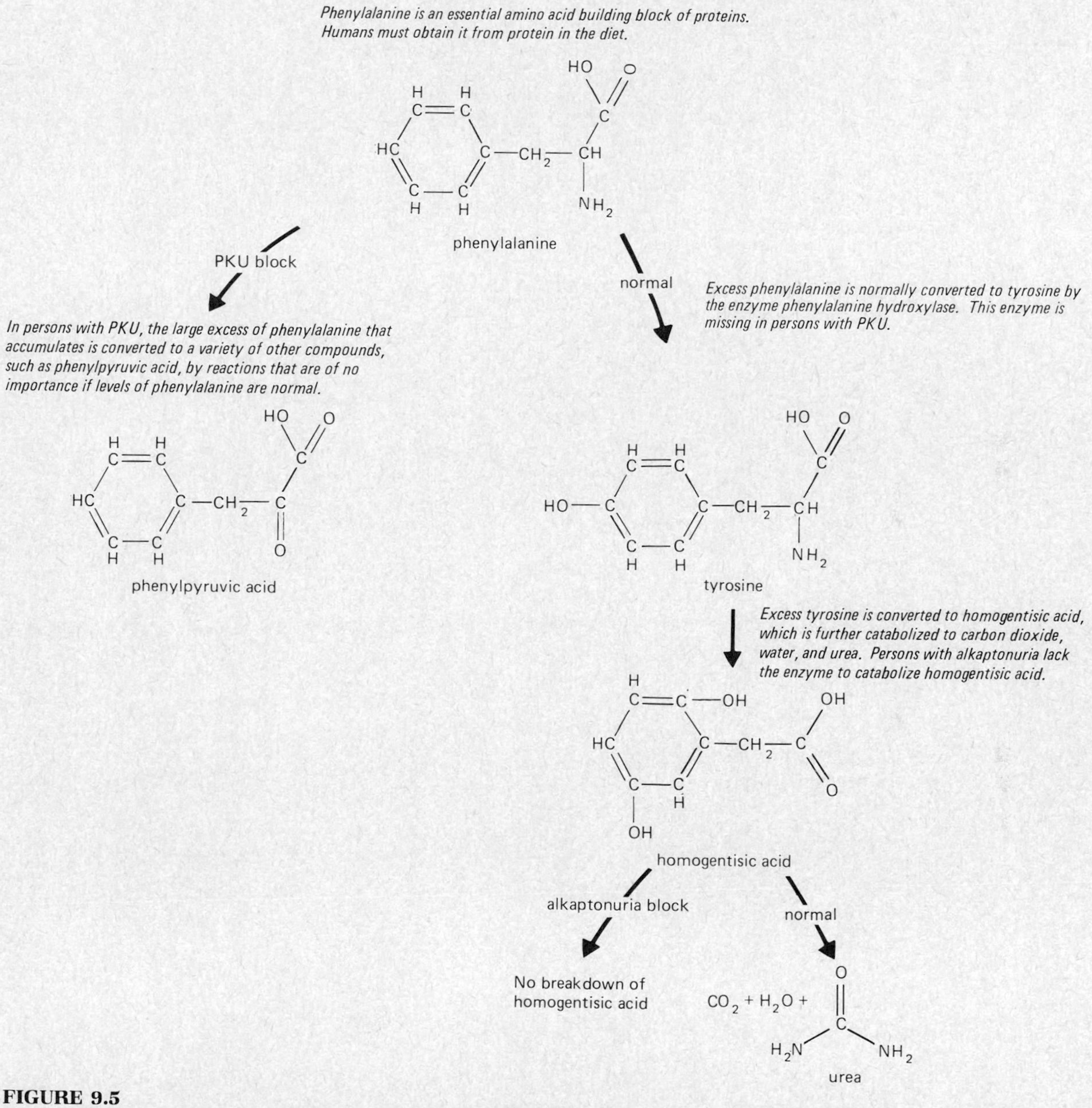

Phenylalanine is an essential amino acid building block of proteins.
Humans must obtain it from protein in the diet.

phenylalanine

PKU block

normal

In persons with PKU, the large excess of phenylalanine that
accumulates is converted to a variety of other compounds,
such as phenylpyruvic acid, by reactions that are of no
importance if levels of phenylalanine are normal.

Excess phenylalanine is normally converted to tyrosine by
the enzyme phenylalanine hydroxylase. This enzyme is
missing in persons with PKU.

phenylpyruvic acid

tyrosine

Excess tyrosine is converted to homogentisic acid,
which is further catabolized to carbon dioxide,
water, and urea. Persons with alkaptonuria lack
the enzyme to catabolize homogentisic acid.

homogentisic acid

alkaptonuria block

normal

No breakdown of
homogentisic acid

$CO_2 + H_2O +$

urea

FIGURE 9.5

The metabolism of phenylalanine and tyrosine. The accumulation of excess phenylalanine and its deaminated derivatives in a normal person is prevented by its conversion to tyrosine, which, in turn, does not accumulate because it is catabolized to CO_2,
H_2O, and urea excreted from the body through the lungs, kidneys, and skin. In phenylketonurics, phenylalanine conversion is blocked. In alkaptonurics, homogentisic acid
breakdown is blocked. Its accumulation does not cause a serious problem, however,
because it is excreted through the kidneys and the level in the body remains low.

Chromosome Aberrations

No precise division can be made between gene mutations and chromosome aberrations other than, if you can't see it, it's a gene mutation, and if you can, it's an aberration. The reasons for this will become clear as we discuss aberrations.

Four general types of chromosomal aberrations are identifiable in eukaryotes with chromosomes large enough to detect alterations in chromosome structure. These are described as

1. deletions
2. duplications
3. inversions
4. translocations

All four types have been found in humans. Here we consider each type in turn using examples drawn from humans.

Deletions

Deletions (or deficiencies) are what might be expected from the name: the result of loss of a piece of a chromosome, either from an end (terminal) or, more usually, from within its length (interstitial) (Fig. 9.6). Most deletions have drastic phenotypic effects, even when heterozygous. Either they are lethal at an early stage of development, resulting in a spontaneous abortion, or (unfortunately) the child is born and is defective. Generally there are anatomical malformations and mental retardation.

An example of a deletion that occurs in relatively high frequency (1 in 50,000 births) is "cri du chat," so called because afflicted infants have a cry almost identical to the mewing of a kitten (Fig. 9.7). Additionally, they have marked cranial and facial malformations and severe mental retardation. Generally their IQ is in the range of 20 or lower. These children may attain adult age, but they remain below

FIGURE 9.6
Some of the types of chromosomal aberrations found in humans.

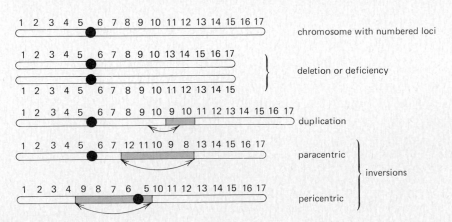

FIGURE 9.7
A child with cri-du-chat syndrome. (From A. de Capoa, D. Warburton, W. R. Breg, D. A. Miller, and O. J. Miller, *Amer. J. Human Genet.* 19:586, 1967.)

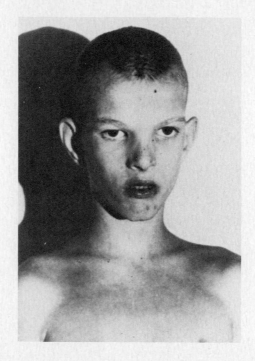

normal weight and size and do not learn to speak more than a few words. The deletion has been identified as loss of a small segment in the short arm of chromosome 5. Several of the other chromosomes of the human complement have also been found to have deletions with severe phenotypic effects.

Duplications

These aberrations are the opposite of deletions—a segment of a chromosome is repeated (Fig. 9.6). The phenotypic effects are rarely as drastic as in the case of deletions, and none has been identified with pathological effects in humans except ones also associated with deletions. The reasons for this association are given later in this chapter. Duplications are generally considered to be important in the evolution of organisms by providing for the origin of new genetic material. Duplicate genes provide the possibility of one of them mutating to a new function while the other continues to carry on the original function.

Inversions

Inversions are, as the name implies, inverted segments of chromosomes, as illustrated in Figure 9.6. They are relatively common in the human population and generally have no detectable phenotypic effect in persons that carry them. They

FIGURE 9.8
The results of a crossover within an inversion loop in a germ cell heterozygous for an inversion chromosome and a normal chromosome. The gametes bearing the crossover products will have a deficiency or duplication of genetic material that will produce genetically unbalanced and usually inviable zygotes after fertilization. Homozygous inversions will produce "normal" gametes, with or without crossovers.

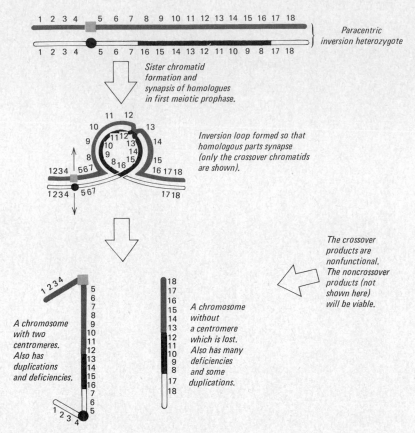

can, however, cause trouble for the offspring of the persons who are heterozygous for inversions.

Two kinds of inversions are distinguished by geneticists: *paracentric* and *pericentric.* The former involve a chromosomal segment outside the centromere and the latter include the centromere, as shown in Figure 9.6. Both kinds are found in humans as well as in many plants and animals whose chromosomes are large enough and can be banded or have detectable natural bands. We can safely assume that inversions occur in all eukaryotes and have probably been important in evolution. When one compares the chromosomes of human origin to those of the chimpanzee, one finds not only an amazing resemblance, as shown in Figure 19.9, but also some interesting differences. You can see clearly that chromosomes such as 1, 4, 5, 9, 12, 15, 17, and 18 have pericentric inversions that place the centromeric regions in different positions relative to the arm lengths.

When an individual is heterozygous for an inversion, a configuration such as shown in Figure 9.8 may be formed in the meiotic prophase. If a crossover occurs *within* the inversion loop shown in the figure, some of the gametes formed will have an abnormal content of genetic material. They may be deficient in some of it and have an excess of others of it. As a result they will produce zygotes with an

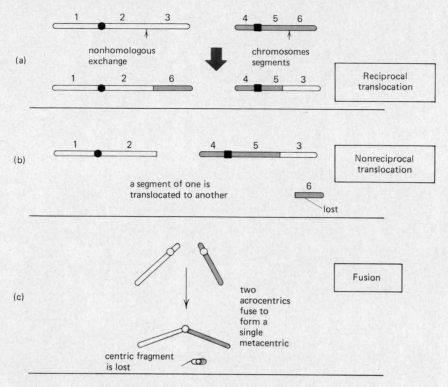

FIGURE 9.9
Types of translocations
found in humans. (a) Re-
ciprocal translocation. (b)
Nonreciprocal transloca-
tion. (c) Centric fusion or
Robertsonian fusion.

abnormal complement of genes. These may die before many cell divisions or produce abnormal offspring.

Translocations

When two different nonhomologous chromosomes break, the broken ends may rejoin with no evident change in the structure of the two chromosomes. On the other hand, the broken ends of the two chromosomes may be exchanged (Fig. 9.9). The result is called a translocation.

Translocations occur regularly in humans. Figure 9.10 illustrates one that has been recently discovered in a family. Translocations of this nature may have no particular phenotypic effect or they may have drastic consequences. Figure 9.11 shows the effect of one that results in a condition known as *Down syndrome*.

Euploidy and Aneuploidy

Ploidy and Euploidy

The 46 chromosome complement of normal female human diploid cells consists of 23 pairs of homologues, and the cells are said to be *euploid* (*eu* = true ploid refers

FIGURE 9.10
Karyotype of a person with a balanced translocation between chromosomes 5 and 8, indicated by arrows. Since all of the chromosomal regions have been preserved in the translocation, the complement of genes is normal and the phenotype is therefore normal. Many normal persons have been identified with balanced translocations of this type. They have a high risk of abnormal offspring, however, because of the problems of pairing and segregation of chromosomes during meiosis. (Provided by Dr. Patricia N. Howard-Peebles, The University of Texas Health Science Center at Dallas.)

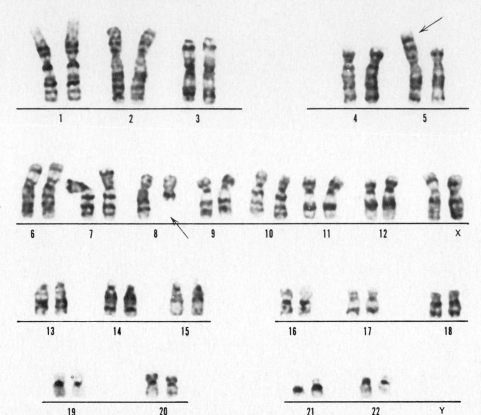

to a chromosome set, 23 in humans). The number of sets may be single and thus *haploid* (*haplo* = single), double or *diploid* (*diplo* = twofold), or *polyploid* (*poly-* = many). Polyploidy is any condition in which there are more than two sets of chromosomes. Triploids (3n) have three sets, tetraploids (4n) have four sets, and so on. All are euploids if each set is complete. Polyploidy is quite rare in animals except among amphibians, but it is common among the higher plants. Many of our important food plants are polyploids. For example, the cultivated wheat grown in the western world is hexaploid (6n), described further in Chapter 21.

Triploid fetuses invariably abort unless they are mosaic, that is, unless they contain some diploid cells and tissues as well as the triploid. Pure triploidy appears to be incompatible with complete development of a human fetus. However, triploids constitute 7% of the spontaneously aborted fetuses whose karyotypes have been determined (Table 9.1). Therefore, their nonoccurrence among live births does not mean that triploidy is rare in human zygotes.

Aneuploidy

Aneuploidy is a condition in which cells are *not* euploid. In other words, they do not have all chromosome sets complete. For example, some humans are born with three copies of chromosomes 8, 9, 13, 18, 21, or 22. They are said to be *trisomic* for those chromosomes and are therefore *aneuploids*. Other chromosomal trisomies also occur in zygotes, but virtually all, as well as many of the trisomies listed above, spontaneously abort. Trisomics that do survive to birth are invariably mentally retarded and have multiple developmental abnormalities. Figure 9.12 gives some examples of these.

Down Syndrome. One of the commonest aneuploidies in humans is trisomy 21, a condition that causes Down syndrome. Many of these trisomies abort as fetuses, but approximately 1 in 600 live births is a child with Down syndrome. These children are mentally retarded and have a characteristic facial appearance (Fig. 9.12). Many other parts of the body are also affected. They age prematurely, although some live into their eighties. They have a high risk of respiratory infections and leukemia. Approximately 50% die before their fifth year. Of special interest is the occurrence of a baby chimpanzee with Down syndrome due to trisomy for the chromosome homologous to the human 21.

Monosomy. In addition to excess chromosomes, there may also be a deficiency. This aneuploid condition is called *monosomy*. An example is given in Figure 9.7. Children who survive until and after birth with monosomy have only partial monosomy. That is, only part of a chromosome is missing. This is therefore a deletion. Often also the person is a genetic mosaic, with a mixture of cells, some with normal chromosome complements and some with the monosomy. The normal cells appear to compensate for the deficiencies of the aneuploid cells. It is probable that a complete nonmosaic monosomy for an autosome would not be viable.

Aneuploidy of the sex chromosomes is relatively frequent because most aneuploids are viable. These are considered in detail in Chapter 12.

Table 9.1 gives a partial list of some of the chromosomal aberrations that have been found in human abortuses. Every chromosome in the human complement has been found to be involved in one way or another. These conditions are not

Although most of the abnormalities are due to nondisjunction, twenty were structural rearrangements that required chromosome breakage.

**TABLE 9.1
Frequency of Normal and Abnormal Karyotypes Observed in 1000 Spontaneous Abortions**

Karyotype	Number of Cases
Normal 46,XX	307
Normal 46,XY	230
Sex chromosome monosomy (45,X0)	112
Autosomal monosomy (45,XY, − 21)	1
Sex chromosome trisomy (47,XXX; 47,XXY)	2
Autosomal trisomy	204
Double trisomy	9
Mosaic trisomy	12
Triploidy	70
Tetraploidy	33
Structural rearrangements	20

Data from T. Hassold et al., A cytogenetic study of 1000 spontaneous abortions, *Annals of Human Genetics* 44: 151–178, 1980.

FIGURE 9.11
Karyotypes of patients
with Down syndrome.
This rather common con-
dition is caused by having
more than the diploid
amount of chromosome
21. The unbalanced
genotype may arise by
two different mecha-
nisms. (a) A fertilized egg
may start out with a com-
plete extra 21, called
trisomy 21. (b) Less often
it may contain extra ma-
terial from chromosome
21 in the form of a translocation. In the karyotype
shown, there is a 14/21
translocation, indicated
by the arrow, in addition
to the normal comple-
ment of the two chromo-
somes 21. (c) If a parent
is heterozygous for a
translocation between
chromosome 21 and
another chromosome
(usually 13, 14, 15, or 22),
the parent may be quite
normal but may produce
unbalanced gametes that
lead to offspring with
Down syndrome. (Photo-
graphs provided by Dr.
Patricia N. Howard-
Peebles, The University of
Texas Health Science Cen-
ter at Dallas.)

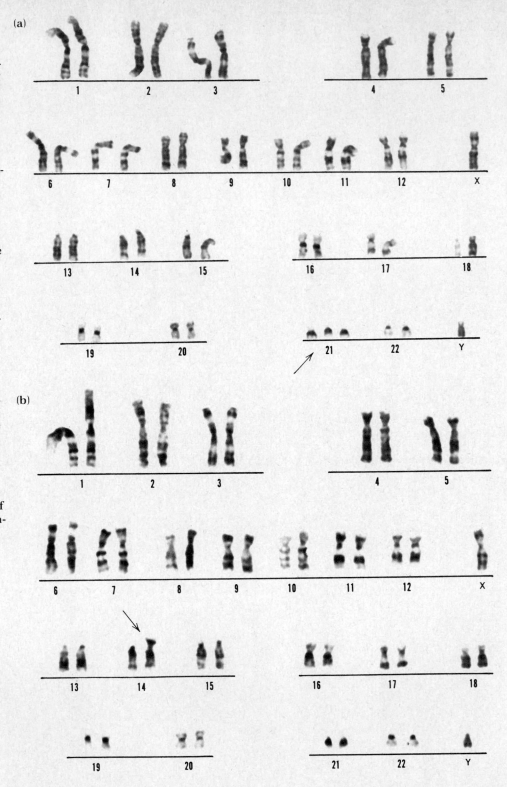

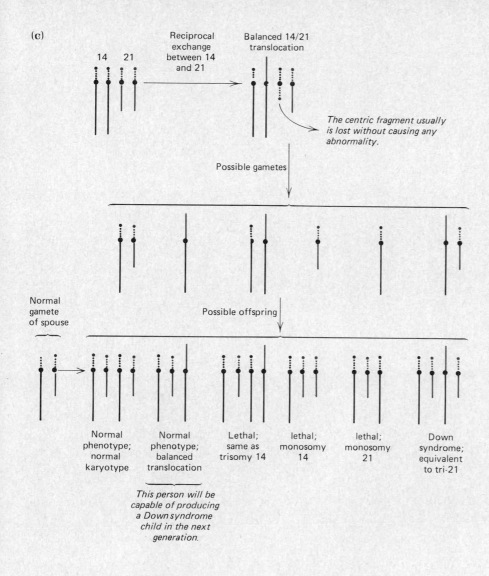

(c)

14   21

Reciprocal exchange between 14 and 21

Balanced 14/21 translocation

The centric fragment usually is lost without causing any abnormality.

Possible gametes

Normal gamete of spouse

Possible offspring

Normal phenotype; normal karyotype

Normal phenotype; balanced translocation

Lethal; same as trisomy 14

lethal; monosomy 14

lethal; monosomy 21

Down syndrome; equivalent to tri-21

This person will be capable of producing a Down syndrome child in the next generation.

FIGURE 9.11
(continued)

uncommon. Fifty percent of the fetuses that abort spontaneously have abnormal karyotypes detectable by current cytogenetic techniques, which are not extremely sensitive. Unfortunately, many of these do not abort and survive to be born with gross malformations.

The Causes of Aneuploidy. The reason for the relatively high frequency of chromosomal aberrations in humans (and other eukaryotes as well) is that

1. Chromosomes break and do not always reconstitute or restitute themselves as they were before the break.
2. Mitosis and meiosis, although remarkably precise and accurate, are not invariably so. Accidents do happen, and the result is some kind of aneuploidy.

Many agents cause chromosomes to break, including ionizing radiation and chemicals (discussed in more detail in Chapter 16). Some occur spontaneously without the help of any recognized agent. When a chromosome breaks, it may rejoin properly so that no permanent damage occurs. Or errors may occur in the rejoining, producing some of the aneuploid conditions described. All would involve structural changes in the chromosomes.

Changes in numbers of chromosomes, such as trisomies and monosomies, are due to *nondisjunction.* At the time of cell division, chromosomes (or sister chromatids) *disjoin* or separate and go to opposite poles, as shown in Figures 5.5 and 5.6. When *disjunction* is normal, normal gametes and daughter cells are formed. If nondisjunction occurs in meiosis, as shown in Figure 9.13, abnormal gametes are formed. If it occurs in mitosis, a clone of aneuploid cells is formed, as shown in Figure 9.14.

Nondisjunction occurs with much higher frequencies in meiosis in older mothers. This means that the risk of having a child with Down syndrome, which is about 1 per 1000 if the mother is young, increases to about 1 in 20 if the mother is over 40 years old. Nondisjunction occurs in the germ cells of both mothers and fathers, but the fathers do not show the strong increase with age.

(a)

(b)

(c)

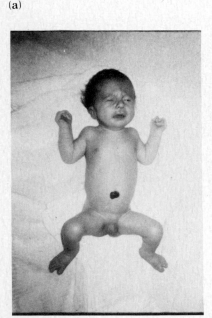
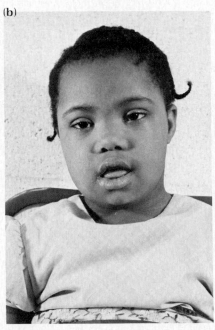
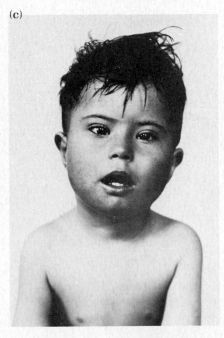

FIGURE 9.12
Some examples of children with trisomy. (a) A 3-week-old infant with trisomy 18. The hands are flexed in a manner characteristic of this syndrome. (b) and (c) Children with trisomy 21 (Down syndrome). (The Photograph in (a) was supplied by Dr. J. Friedman, The University of Texas Health Science Center at Dallas; (b) and (c) were supplied by The March of Dimes Birth Defects Foundation.)

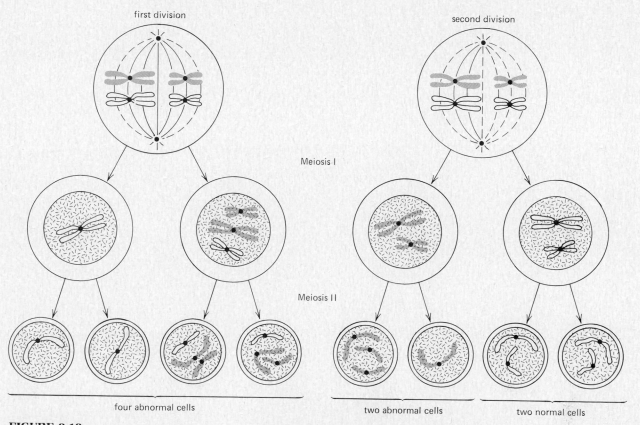

FIGURE 9.13
Nondisjunction in meiosis. (Redrawn from R. P. Wagner, B. H. Judd, B. G. Sanders, and R. H. Richardson, *Introduction to Modern Genetics*, John Wiley & Sons, Inc., 1980)

Factors that Influence the Rate of Mutation

Under "ordinary and natural" conditions, mutations, whether genic or chromosomal, are relatively rare events. It is generally considered that human genes mutate to a condition that can be recognized in a phenotypic change at a rate of about 1 in 10^5 to 1 in 10^6 gametes for each locus. But even at this low rate, a large number of gametes carry mutant genes. If we assume that we have about 10,000 genes per haploid set (a very conservative guess), then we should expect about 1 in 10 to 1 in 100 of our gametes to carry at least one mutation that occurred within our bodies.

The "natural" rate of mutation is generally referred to as the *spontaneous* rate by

FIGURE 9.14
Nondisjunction in mitosis.

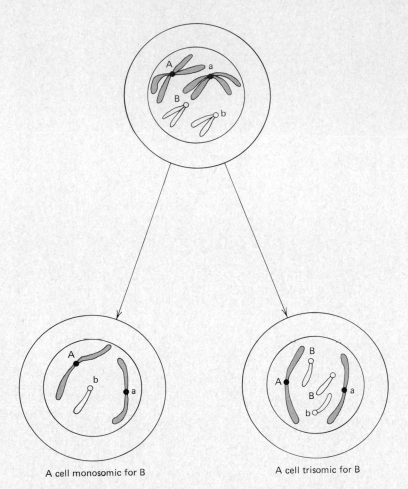

A cell monosomic for B   A cell trisomic for B

geneticists. However, this rate can be increased by environmental factors, such as ionizing radiations and many different kinds of chemicals in the food we eat, the water we drink, and the air we breathe. As our environment becomes more and more polluted by our technological activities, the problem of mutation in the human population becomes more and more important as a health hazard. We devote Chapter 16 to this topic.

The Repair of DNA

The DNA in our cells is not a highly stable chemical compound. It is continually subjected to breaks and even alterations of base sequences, which can and frequently do have deleterious consequences, as discussed in previous sections.

These disturbances of DNA structure may be spontaneous or induced by external sources of radiation or chemicals.

Despite these continued insults to our DNA, genetic continuity of generations of healthy humans is maintained by systems of enzymes that repair the damaged DNA. Without these enzymes, we would soon disappear as a species. Indeed it is doubtful that any kind of life as we know it could exist without mechanisms for the maintenance of the integrity of DNA.

Repair of Missing and Mismatched Bases

Any modification of the DNA of the nucleus or mitochondria that alters its coding properties or its normal functions in replication and transcription can generally be repaired. For example, if a nucleotide is missing in one strand of a DNA double helix, it can be replaced by reinsertion, using the intact complementary strand as a template (Fig. 9.15). In the same way, a large segment of a strand can be replaced accurately by matching the complementary strand.

Occasionally an incorrect base is inserted during replication of DNA. To correct these errors, a mismatch repair system functions by traveling along the DNA seeking out mismatches and repairing them. The wrong base is removed and replaced with one that matches the complementary strand. One presumes also that occasionally the "correct" base is removed and replaced with one that matches the incorrect one, producing a mutation.

Repair of Altered Bases

Radiations and chemicals may not only cause breaks and mismatches in DNA, but they may also alter the chemical nature of the bases themselves. If not repaired, this alteration would cause a mutation. A good example of this in bacteria and in cultured human cells is the effect of ultraviolet light (uv) on DNA structure. Ultraviolet light in the range of 260 nm wavelength causes the formation of covalent bonds between adjacent pyrimidines such as thymines, forming "dimers," as shown in Figure 9.16. These must be removed by excision and then repaired. Many other kinds of base alterations are made by radiations and chemicals, and repair

FIGURE 9.15
The repair of DNA by replacing or adding nucleotides.

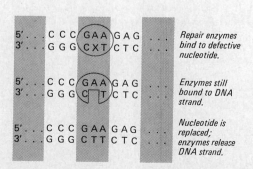

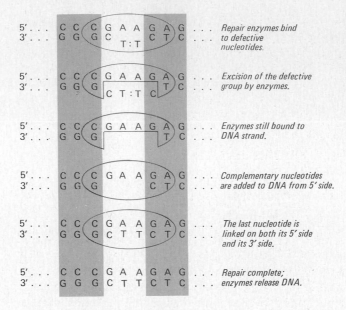

FIGURE 9.16
The excision of damaged DNA and its repair. In the example, adjacent thymines have been altered structurally because of irradiation with ultraviolet light. They are removed (as are several adjacent but undamaged nucleotides) and replaced with correct nucleotides.

systems similar to the thymine dimer excision–repair system described are present in normal cells to act on them.

Inherited Repair Deficiencies in Humans

The absence of parts of the repair system, caused by mutations in the structural genes that code for the repair enzymes, can have serious effects. Table 9.2 lists a number of conditions involving chromosome instability or radiation sensitivity (or both) that are thought to be defects in repair systems. Most are associated with a high frequency of cancer, believed to be mutational in origin. Premature aging and early death also are common.

The most striking example of premature aging is the condition known as progeria. An affected child at age 13 may look, act, and have the physiological characteristics of a person of 90. These children generally die in their teens of heart disease. One theory about the basis of normal aging is that it is caused to a large extent by unrepaired alterations in the DNA of our cells. Over a period of years, these somatic mutations accumulate and are the chief factors in the aging process. We have no proof of this at present, but the theory is credible. Certainly, the better our repair processes, the healthier we will be, other things being equal.

Review Questions

1. Give a general definition of mutation in genetic terms.
2. Distinguish between gene mutations and chromosomal mutations or aberrations.

TABLE 9.2
Hereditary Deficiencies
in DNA Repair in
Humans

*These diseases involve chromosome instability and radiation sensitivity, known to be
associated in some instances with repair deficiencies. The specific nature of the repair
defect has not been established in most of them.*

Ataxia telangiectasia	Neurological deterioration, hypersensitivity to ionizing radiations, high frequency of chromosome aberrations. Autosomal recessive.
Bloom syndrome	Retarded growth, highly sensitive to uv, high frequency of chromosome breaks and aberrations. Autosomal recessive.
Cockayne syndrome	Dwarfism, mental retardation, premature aging, hypersensitivity to uv. Autosomal recessive.
Dyskeratosis congenita	Hyperpigmentation of skin, anemia, and other symptoms associated with skin and its derivatives. X-linked recessive.
Fanconi anemia	Retarded growth, high frequency of chromosome aberrations. Autosomal recessive.
Progeria	Very early senility, coronary artery disease. Probably autosomal recessive.
Xeroderma pigmentosum	Hypersensitivity to uv, skin cancers on exposed areas. Autosomal recessive.

3. A missense mutation may result in a fully formed but altered polypeptide or an incomplete polypeptide, i.e., one that has fewer than the normal complement of amino acids. Explain.
4. How does a frameshift mutation differ from a missense mutation?
5. What is meant by a polymorphic genetic locus?
6. State the probable cause or causes underlying an inherited metabolic block.
7. What is meant by the term lethal mutation? Does it differ in a basic way from any other kind of mutation?
8. By what mechanisms does a polymorphic locus arise?
9. Give a concise, precise definition of the term allele.
10. What kind of mutation results in the production of sickle cell hemoglobin? What is the phenotype of this hemoglobin at (a) the molecular level? (b) the physiological level?
11. A man with a "normal" phenotype with regard to intelligence and physical appearance is found to have an abnormal karyotype. He is heterozygous for a translocation in which most of a chromosome 21 is fused to one of his chromosomes 15. He also has a normal copy each of chromosomes 15 and 21. He is married to a woman with a normal karyotype. His wife at age 21 has a child with Down syndrome. Their next child born when the mother is in her thirties is normal. Explain.
12. What is meant by the phrase "inborn error of metabolism?"
13. Distinguish between euploidy and aneuploidy.
14. A human male known to be heterozygous for a large inversion in his second chromosome has very low fertility, but his father who was homozygous for that same inversion had thirteen children. What genetic hypothesis would explain the differences in fertility?
15. Why is the repair of DNA in our cells essential even after we become adults?

References and Further Reading

Carlson, E. A. 1981. *Genes, Radiation, and Society: The Life and Work of H. J. Muller.* Cornell Univ. Press, Ithaca and London. 496 pp. A very readable biography about the interesting, often controversial person who discovered the production of mutations by X rays.

Drake, J. W. 1970. *The Molecular Basis of Mutation.* Holden-Day, San Francisco. 273 pp. A basic treatise.

Grouchy, J. de, and C. Turleau. 1977 *Clinical Atlas of Human Chromosomes.* John Wiley, New York. 319 pp. Authoritative treatment of the many different kinds of chromosomal aberrations that appear in humans and their phenotypic effects.

Hamerton, J. L. 1971. *Human Cytogenetics.* Vol. 2, *Clinical Cytogenetics.* Academic Press, New York. 545 pp. An extensive treatment of human cytogenetic abnormalities, including the mechanisms by which they arise.

Howard-Flanders, P. 1981. Inducible repair of DNA. *Scientific American* 245:72–103. A clear account of DNA repair with excellent illustrations.

Muller, H. J. 1927. Artificial transmutation of the gene. *Science* 66:84–87. This short paper was the first report that an environmental factor can cause mutations.

Vogel. F., and Rathenberg, R. 1975. Spontaneous mutation in man. *Adv. Human Genet.* 5:223–318. This long and somewhat technical review discusses the methodology of measuring human mutation rates and summarizes the results of many studies.

The cells of our bodies that are not germ cells are called *somatic cells.* Collectively they constitute the *soma.* Somatic cells arise, as do our germ cells, from the fertilized egg from which we developed by a series of mitoses and the process of differentiation. The germ cells are set aside in our gonads for the procreation of offspring by first undergoing meiosis to produce gametes that, as sperm and eggs, unite by fertilization and thus beget new individuals. The somatic cells, on the other hand, beget by mitosis more cells within the same individual. Thus, the germ cells represent a *germ line* that is potentially "immortal," whereas the somatic line is mortal since the cells reside in a mortal individual (Fig. 9.1). The genetics of somatic cells we call *somatogenetics* to distinguish it from the genetics associated with the germ cells.

Somatogenetics

We have already described mitosis as a process whereby a mother cell produces two daughter cells that possess the same number and kind of chromosomes as the mother cell. It follows from this that all the cells of the adult human body should be identical in genotype. However, this is only approximately true. Changes can and do occur. Figure 10.1 outlines some of them. It is to be noted from this figure that all of the things that may happen to germ cells may also happen to somatic cells. But there is a significant caveat to be imposed. Gene mutations and chromosome aberrations that occur in germ cells may be transmitted from generation to generation, whereas those that occur in the somatic cells simply produce lines or clones with an altered genotype. Genetically, the person within whom they occur thus becomes a *mosaic*. That is, he or she does if the mutation occurs in a cell that divides following the mutational event. In the late fetal stages of life and in the period following birth, some tissue cells stop dividing and will not divide again during the remainder of the individual's life. A notable example of this is nervous tissue. Neurons (nerve cells) are supposed never to divide after birth, which explains why a severed spinal cord cannot repair itself. On the other hand, liver cells and kidney cells can divide, but generally at a very low rate. Cutting out a piece of liver from an adult, however, stimulates mitosis, and the piece removed will soon be replaced with new cells. The cells that undergo the most active mitosis are those that line our gut and other surfaces. These linings, called epithelia, and the germinal epithelium of our skin (Fig. 10.2) have cells in constant division to replace the old cells worn away by wear and tear. You are certainly made aware of this when you wipe your back with a towel and old skin sloughs off.

Stem Cells

Among the most mitotically active cells of our bodies are the *stem cells* of the bone marrow, found in the long bones, the sternum, and the cranium. These bone

10

The Soma and the Culture of Its Cells

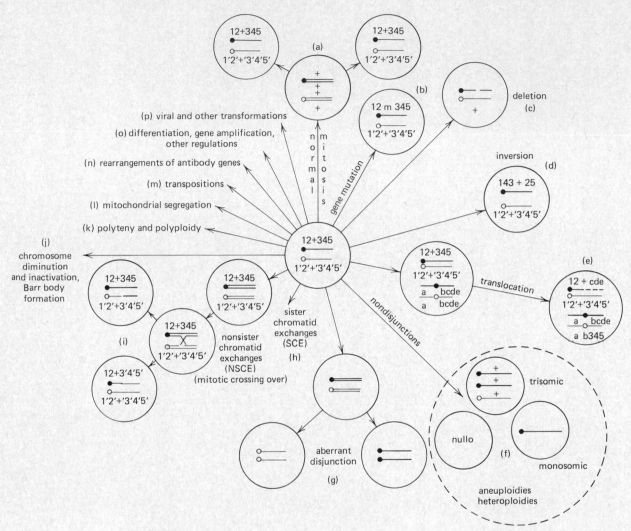

FIGURE 10.1

Genetic and epigenetic changes that may occur in somatic cells. Many of the types of aberrations that occur in germ cells as described in Chapter 9 may also occur in somatic cells. Additionally, other changes may occur in somatic cells during differentiation. These changes may be called *epigenetic,* since they do not involve actual genotypic changes. (a) Represents simple mitosis and no genetic change; (b–i) represent genetic changes in chromosome structure in the classical sense; (j–p) represent changes that range from genetic to epigenetic.

marrow regions in the hollow spaces of our bones are known as the *hemotopoietic* center. In these centers blood cells arise by the constant division of stem cells (Fig. 10.3). Stem cells divide more or less continuously and give rise to differentiated red and white blood corpuscles as well as to more stem cells. Our red blood cells arise initially from stem cells as erythroblasts that, upon losing their nuclei and forming

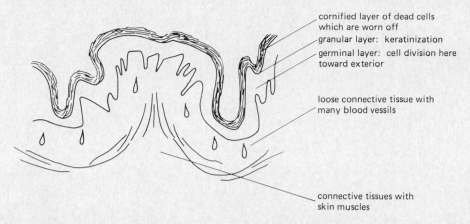

cornified layer of dead cells
which are worn off
granular layer: keratinization
germinal layer: cell division here
toward exterior

loose connective tissue with
many blood vessils

connective tissues with
skin muscles

FIGURE 10.2
The germinal epithelium
of the skin. This layer of
cells shows high mitotic
activity, replacing the
cells that constantly die
and slough off our skin.

large quantities of hemoglobin within their membranes to the exclusion of nearly everything else, become red blood corpuscles or erythrocytes. The corpuscles have an average life span of about 120 days. This means that we must replace those in circulation in a like time span, which requires that we form 2.5 million of them *every second* by cell division in our stem cells! Our white blood cells, which include a number of different kinds, including the lymphocytes, also have limited life spans that require a constant supply of replacements.

This all means that our stem cells in the blood cell-forming centers must be prodigiously active, and that any genetic alteration that occurs in a single stem cell may be multiplied many times in the cells derived from it even though cells such as red blood corpuscles are not capable of further division.

Diseases Associated with Somatic Mutations

Lymphocytes in the circulating blood normally do not undergo mitosis, but they may be induced to do so and then stopped at metaphase, when the chromosomes are most easily observed (Chapter 4). If lymphocyte cultures from a large number of persons are examined, aberrant chromosomes are found to be fairly common. Most of us show, at least in a few of our cells, broken chromosomes, for example, especially if we have or recently have had a viral infection. Why this is so isn't known, and no ill effects are known to result from such broken chromosomes.

The stem cells (shaded) are the progenitors of all cells of that particular tissue.

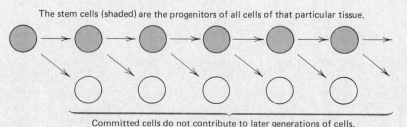

Committed cells do not contribute to later generations of cells.

FIGURE 10.3
Diagram showing the re-
lationship between stem
cells and committed cells.
(See also Box 15.1.) Each
division of a stem cell results in one committed
cell (erythrocyte to be, for
example) and one stem
cell.

Since lymphocytes do not normally divide once they are mature, broken chromosomes in them are probably of no importance.

The frequency of chromosome aberrations is abnormally high in individuals who have inherited certain diseases associated with a predisposition toward the development of cancer. Among these diseases are *Bloom syndrome*, *ataxia telangiectasia*, and *Fanconi anemia*. (See Table 9.2.) All of these are inherited as recessive traits, and fortunately all are rare. But a study of the few known afflicted individuals has provided some valuable insights into that diseased state we call cancer.

Bloom Syndrome.   Bloom syndrome patients are homozygous for an autosomal gene *bl*. They are mentally retarded, short in stature, and have minor anatomical defects. They rarely live past the age of thirty. Death generally occurs from cancer, which may be either a leukemia, carcinoma, or sarcoma. These are cancers that have their origins, respectively, in the lymphocytes of the blood, presumably through the stem cells that produce them, the epithelial cells of the skin and gut lining (carcinomas), and internal organs such as the liver and kidney (sarcomas). The chromosomes in the cells of these individuals are generally abnormal in appearance, showing aberrations ranging from breaks to rearrangements of the type discussed in Chapter 9 (see Fig. 10.4).

Fanconi Anemia.   Anatomical abnormalities and mental retardation have a high incidence among patients with Fanconi anemia, but the most prominent abnormality is found in the bone marrow, which produces fewer platelets, erythrocytes, and white blood cells than normal. This condition is also rare and inherited as an autosomal recessive. Patients usually die at an early age from deficient activity of the blood-forming centers or from leukemia. As in Bloom syndrome, there is a high incidence of chromosomal instability detected as breaks and exchanges.

Ataxia telangiectasia.   Ataxia telangiectasia (AT), also called the Louis–Bar syndrome, is a rare condition like the previous two and is also inherited as an autosomal recessive. Patients start to show neurological deterioration at an early age accompanied by mental deterioration. In most, the immune system does not function well and death frequently occurs from infectious diseases, which the patients cannot fight because of low resistance. The incidence of cancer is much higher in these patients than in the general population, and, as in the case of Bloom syndrome and Fanconi anemia, there is a high incidence of chromosomal aberrations in the somatic cells as detected by karyotypic analysis of lymphocytes. A further peculiarity of this disease is that it is associated with a high sensitivity to ionizing radiations such as X rays. However, this sensitivity does not lead to a higher mutation rate. There is no increased sensitivity to ultraviolet light.

Xeroderma pigmentosum.   A fourth condition inherited as an autosomal recessive and with a predisposition to cancer is *xeroderma pigmentosum* (XP). XP differs from the first three we have discussed because patients with it do not show an abnormal incidence of chromosome aberrations even though they invariably develop cancers, especially of the skin. These cancers are generally the result of exposure to sunlight. XP individuals are extremely sensitive to ultraviolet light, which apparently induces skin cancers in them at a very high rate compared to normal persons. The induction of the cancer is apparently related to the inability of XP persons to repair certain kinds of DNA damage caused by ultraviolet light because of a deficiency of certain DNA repair enzymes that we all ordinarily

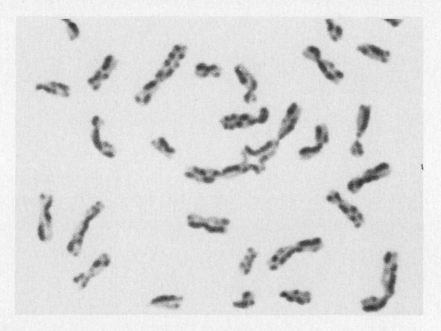

FIGURE 10.4

Chromosome spread from a patient with Bloom syndrome. The white blood cells were cultured through two divisions in the presence of bromodeoxyuridine (BrdU), a chemical analog of thymine that is readily incorporated into DNA in place of thymine. After one cycle of replication, every DNA double helix consists of one old strand without BrdU and one new strand with BrdU. After a second cycle of replication, the double helices would consist of (a) one strand without BrdU and one strand with or (b) both strands with BrdU. For any one chromosome, one chromatid would be of type (a) and the sister chromatid of type (b). If the chromosomes are stained with a fluorescent dye, the sister chromatids can be distinguished because the greater amount of BrdU in type (b) chromatids quenches the fluorescence to a greater extent. One can therefore identify the points at which sister-chromatid exchange (SCE) has occurred by the switching of the dark and light regions from one chromatid to the other. SCE is different from crossing over, which occurs between chromatids of homologous chromosomes during meiosis.

Lymphocytes from normal persons typically have fewer than ten SCE's in an entire chromosome spread. Lymphocytes from persons with Bloom syndrome have many dozens, as shown in this figure, which includes only part of a full complement of chromosomes. Another feature that is rare in normal persons but common in Bloom syndrome is the "quadriradial" figure in the center, in which two homologous chromosomes have exchanged chromatids. The chromosome "instability" of Bloom syndrome is thought to be responsbile for the apparent high mutation rate in somatic cells, as reflected in the high cancer risk, but the molecular basis is unknown. (Figure provided by Dr. James L. German, III, The New York Blood Center.)

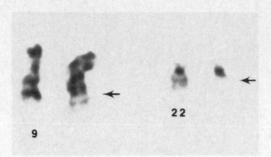

FIGURE 10.5
Translocations found in the somatic cells of persons with chronic granulocytic leukemia. In this instance, the somatic cells are white blood cells. The translocation is
found only in the affected white blood cells; all other somatic cells are normal. The
translocation involves part of the long arm of a number 22 chromosome moved to the
long arm of a number 9. The points of the breaks are indicated by arrows. The deleted
chromosome 22 is often called a *Philadelphia chromosome* after the city where it was
first discovered. (Furnished by Dr. Patricia Howard-Peebles, The University of Texas
Health Science Center at Dallas.)

synthesize in our bodies. The unrepaired DNA damage presumably leads to the
cancer, but how is not at all clear.

The four examples of genetic diseases tell us some things about chromosomes
and cancer that are valuable to our deeper understanding of disease. First of all,
cancers are sometimes associated with chromosomal aberrations, but not always.
Cells from tumors often have normal karyotypes. Second, mutant genes may affect
the stability of chromosomes and cause them to break and form translocations and
inversions. A third and very important lesson we learn here is that, although a
person may be born with a normal array of chromosomes, changes may occur
during later life in somatic cells giving rise to clones of cells with gene and
chromosomal mutations. These may or may not cause pathologies such as cancer.

Since it has become possible to identify unambiguously all the human chromosomes and even their arms and parts of their arms with current banding techniques, the karyotypes of individuals, whether healthy or not, can be examined in
great detail. It has been found that those who develop some of the leukemias may
have chromosomal abnormalities of specific types. The most consistent finding is
in chronic granulocytic leukemia, in which there is a reciprocal translocation of
part of the long arm of number 22 to another chromosome, usually number 9.
These aberrations are found only in lymphocytes and not in other somatic cells
(Fig. 10.5). Hence the persons were not born with them but acquired them. In this
example, the translocation presumably occurs in a single stem cell, causing it no
longer to be regulated and permitting it to outgrow normal stem cells.

The Culture and Genetics of Somatic Cells

The culture of tissues outside the body in media containing the necessary
nutrients to keep the tissue cells alive and reproducing has been practiced for

many years, going back to the early part of this century. It was not until the 1950s, however, that the techniques were fully developed for the culture of a variety of different cell types. It is now possible to culture many human cells as *pure lines*, much like bacteria are handled in the laboratory; that is, all the cells in a culture are a single clone, being presumably of the same genotype since they descended from a single ancestral cell by a series of mitotic divisions. These cells, whether they be of human or other animal or plant origin, can be cultured indefinitely. In addition, the cultured cells may be frozen and stored in small vials in liquid nitrogen at a temperature of about −196° C for periods of many years. To return them to active growth, it is only necessary to bring them back to 37° C and feed them the required nutrients. In a sense, cultured cell lines are immortal, so long as we mortals exist to care for them.

These cultured cell are *somatic* cells because they derive from the soma or nongerm cells. Cells may originate from many of the body organs and frequently maintain the same chromosome constitution as the original cells from which they were derived. Figure 10.6 shows the chromosomes from a line of cells derived from the lung tissue of a Chinese hamster in 1960. This preparation was made in 1980 and shows the same number and kind of chromosomes possessed by the ancestral cells some 20 years earlier. A human cell line called HeLa has been carried in laboratories throughout the world for about 30 years. In contrast to the Chinese hamster line, various HeLa cultures have diverged substantially from the original.

Senescence and Transformation

Although most somatic cell explants yield cells that will grow in culture, such cultures typically do not form immortal cell lines. Rather, they reproduce mitotic-

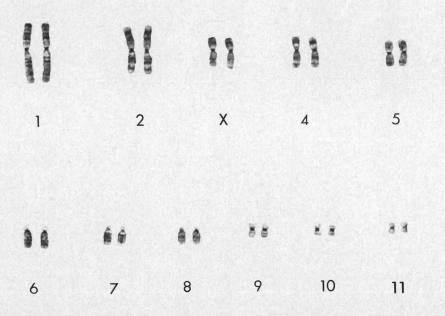

FIGURE 10.6
The karyotype of a Chinese hamster cell from a cell culture that has undergone 20 cell divisions since starting from a cell taken from a live hamster. Its chromosomes are morphologically indistinguishable from those in cells in the living hamster. Some of the cells in such a culture will show some aberrations, such as aneuploidy, however. (Furnished by F. A. Ray, Los Alamos National Laboratory.)

ally for about fifty cell divisions and then become senescent and die. The reasons for this are not known. Various culture conditions influence the onset of senescence, but none has been found to prevent it. Cells taken from an embryo can be cultured through more divisions before senescence then cells from a mature person or animal, suggesting that all our cells have some sort of memory of age. This inability to divide beyond approximately fifty generations is not too serious a problem, however. One cell dividing fifty times would yield 2^{50} cells or 10^{15} cells, approximately the number of cells in a human being.

Rarely, a cell will escape the pathway of senescence and acquire the ability to grow perpetually. Such a cell is said to be *transformed*, and the descendants of that cell are described as a cell *line* rather than a culture. The immortal lines referred to earlier are of this type. The nature of transformation is not understood. In some ways, it is comparable to the transformation of normal tissue into malignant tissue, and transformation of normal cells in the body may well be a step in carcinogenesis. Just as mutation has been suggested as an important step in transforming a normal cell into a cancer cell, so also mutation may be the key event in transforming cells in culture.

Not only do transformed cells have the ability to grow indefinitely; they also are changed in other properties, such as contact inhibition. Normal cells in culture typically grow as a single layer on the surface of the culture vessel. When the surface is filled, growth ceases. Transformed cells may continue to grow, forming mounds of cells and, in some instances, growing suspended in the culture medium.

Of what use are such cell lines and cultures? They have many uses, but here we shall confine our discussion to their usefulness in the study of genetics, especially human genetics. In later chapters we shall show that they may be exploited in a variety of other ways.

Fusion of Cells and the Formation of Hybrids

Cells carried in culture will sometimes fuse and form multinucleate cells, as illustrated in Figure 10.7. If the fusion cell is the product of two different kinds of cells fusing, the result is called a *heterokaryon*, since it has nuclei of different genotypes.

After a period of time the nuclei in fusion heterokaryons may themselves fuse and form a true *hybrid* cell (Fig. 10.7). In the case of the mouse/human hybrid, the fusion nuclei start off with $40 + 46 = 86$ chromosomes. This condition does not last through an indefinite number of mitotic divisions, however. Over a period of time, the nuclei begin to lose human chromosomes until eventually clones may arise that have few or no human chromosomes left. This may appear to be a rather messy result, but actually it provides a powerful tool for determining the linkage relationships of known human genes.

The loss of human chromosomes in the hybrids is entirely random, although if the hybrid cells depend on a human chromosome to supply some particular gene product, then any hybrid that loses that chromosome will not survive. If the hybrid cultures are cloned, some clones will lose one group of human chromosomes, other clones will lose other groups. In this way, a series of hybrid cultures can be

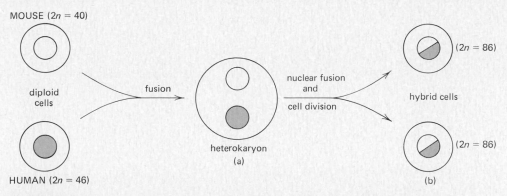

FIGURE 10.7
The fusion of two diploid cells in culture. A mouse cell fuses with a human cell to give a heterokaryon with a nucleus with 40 mouse chromosomes, and one with 46 human chromosomes. When the two different nuclei fuse (which they generally do after a period of time), a hybrid cell with a single nucleus of 86 chromosomes is formed.

derived that have a full complement of mouse chromosomes but only a few human chromosomes, each set different from the next culture.

This series of cultures permits study of the association between a particular chromosome and various phenotypic traits, usually the presence of a human enzyme. In the first application of this technique, the human gene for thymidine kinase (TK) was shown to be associated with the presence of human chromosome 17 (Fig. 10.8). In this particular example, the mouse cells used were TK^-; that is, they lacked a functional TK^+ gene and could not grow in medium containing hypoxanthine, aminopterin, and thymidine (HAT medium). Therefore all clones that could survive in HAT medium retained the human TK^+ gene. It was found that they also retained human chromosome 17, suggesting that the TK gene is located on that chromosome, a suggestion that has subsequently been verified in other experiments.

The use of selective medium such as HAT is not essential. So long as presence of a human trait is always associated with presence of a particular chromosome and is never found without the chromosome, one may conclude that the gene producing the trait is located on that chromosome.

Several hundred human genes have now been assigned to chromosomes based on somatic cell hybrid results. In order for this method to be used, the trait must be expressed in the cultured hybrid cells. Therefore, many human traits cannot be studied this way. On the other hand, a large number of enzymes are expressed in hybrid cells and have been mapped to specific chromosomes.

Cultured Cells and Gene Action

Just as tissues express certain genes and not others, cultured cells usually express only those genes active in the tissue of origin. Liver cells in culture typically synthesize and secrete albumin and other plasma proteins just as the liver does in the body. Kidney cells, on the other hand, do not manufacture albumin either in

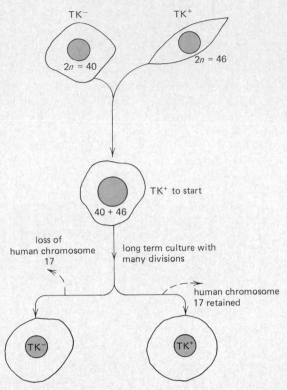

FIGURE 10.8
Illustration of the human/mouse hybrid in the assignment of human genes to specific chromosomes. Any human gene can be assigned to a chromosome by this approach so long as the gene product can be detected in the hybrid. In recent extensions of the technique, the DNA of a particular gene can be detected in the hybrid if that chromosome has not been lost, even if the gene is not active.

the body or in culture. In liver cells, the albumin gene is "turned on," in kidney cells it is "turned off"; in liver cells the albumin structural genes transcribe mRNA for albumin, in kidney cells they do not. The regulatory factors that allow transcription of albumin genes in one but not the other are not understood, but they are obviously very important.

Among the most useful cells for culture are the fibroblasts, spindle-shaped cells commonly found in connective tissue throughout the body. They grow more rapidly than most other cell types, and they express a great variety of genes. Fibroblast cultures can be started from small bits of skin, making it relatively easy to set up cultures from persons without the risks that would be associated with more invasive surgical procedures.

The expression of differentiated functions in cultured cells has proved very useful, both as a tool for studying the properties of the differentiated state of cells and for better understanding the mechanisms of differentiation. As an example of the former, it is possible to do experiments on cultured cells that would be quite impossible on a whole organism, much less a human being. One can expose the cells to metabolic poisons or to highly radioactive materials that would be lethal to a human being. One can control the environment of the cultured cells rigorously.

A further advantage is the ability to study the genetic basis for a rare disease without having to have access to the patient. Fibroblast cultures have been set up from thousands of patients, many with extremely rare inherited diseases. A scientist in St. Louis wishing to study a particular disease may obtain cultures of cells from patients in New York, Rome, and Tokyo. And, in many instances, the patients may have been dead for years.

One example will suffice to show how cultured fibroblasts can be used to pinpoint gene defects. There are a series of related diseases known as *mucopoly-saccharidoses*. As the name indicates, they result from inability to break down mucopolysaccharides, large polymers built up of various sugars that are attached to protein and that are necessary for function of certain nerve cells as well as other cells. Ordinarily these compounds are built up and broken down in a regular cycle. In the case of the hereditary mucopolysaccharidoses, they are built up but not broken down. Hence, they gradually accumulate, interfering with cell function and, in some instances, causing death. *Hurler* and *Hunter syndromes* are the best known examples of the mucopolysaccharidoses. Hurler syndrome is inherited as a simple autosomal recessive trait, whereas Hunter syndrome is an X-linked recessive. Thus the primary gene defects are clearly different, even though the phenotypes are rather similar.

The accumulation of mucopolysaccharides that cannot be broken down occurs in fibroblast cultures from patients with these diseases. Each can be shown to be due to deficiency of a different enzyme. In theory, this could also be shown by direct assay of tissues from the patients, but it is much more acceptable to do the assay on cultured cells, and the supply of such cells is virtually unlimited.

Complementation

A complementation test provides information on whether or not two mutations affect the same or different functions. It was originally devised for use in experimental genetic crosses as shown in Figure 10.9. Consider the case in which an organism is heterozygous for two possibly different mutant recessive genes. If the mutations are in the same functional gene, the consequences should differ depending on whether the mutant genes are *cis* or *trans* with respect to each other. If both mutations were on the same chromosome (*cis*), then the homologous chromosome would have a normally functioning gene and the phenotype would be normal. If the two mutations were on different chromosomes (*trans*), then both

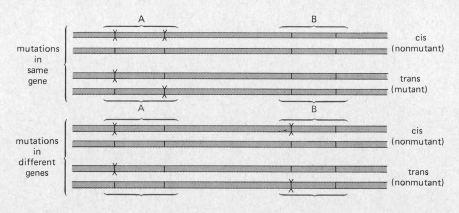

FIGURE 10.9
The cis-trans test for gene complementation. A and B represent different genes with different functions. χ = mutant site.

copies of the gene would be defective, producing a corresponding recessive phenotype. In contrast, if the two mutations were in different functional genes, then the phenotype should be normal, whether the mutant genes are *cis* or *trans*. In the latter case, the mutations are said to *complement* each other, since each has a normal allele also present.

Complementation tests can also be carried out on cultured cells. If cells from Hurler and Hunter patients are grown together, no accumulation of mucopolysaccharides occurs. In other words, each corrects for the other's deficiency. Since the gene defects are different, clearly so in this case because one is autosomal and the other X-linked, each also has functional genes corresponding to the defective locus of the other. Taken together, there is a complete set of functional genes, and the mixed cell culture behaves as if the cells were normal. A positive complementation test thus indicates that the gene defects involve different loci. In the case of the mucopolysaccharidoses, at least 10 complementation groups have been identified by this technique. Each is associated with deficiency of a different enzyme.

Occasionally two diseases that are distinct clinically belong to the same complementation group. For example, *Scheie* syndrome is a mucopolysaccharidosis that belongs to the same complementation group as Hurler syndrome, and the same enzyme has been shown to be defective. The explanation seems to be that the enzyme defect is severe in Hurler syndrome, but there is sufficient activity of the enzyme in Sheie syndrome to avoid the severe clinical problems of Hurler syndrome.

Complementation in mixed cultures depends on the diffusion of the enzyme out of the cell, where it can still function. Most enzymes do not diffuse out of cells but can be tested with cell hybrids. For example, fibroblasts from persons with xeroderma pigmentosum, mentioned earlier in this chapter, are defective in repair of certain types of DNA damage, just as the patients themselves are. Hybrids formed from fibroblasts from two different patients sometimes are still repair deficient. Therefore, they are noncomplementary, which indicates that the defects are in the same gene. Or the hybrids may have normal repair; that is, the defects complement each other and are in different genes. Seven different complementation groups in xeroderma pigmentosum have been established by such pairwise comparisons, an indication that defects in at least seven different loci can lead to this disease.

Prenatal Diagnosis

One of the more important applications of cell culture is in the prenatal diagnosis of many inherited disorders. Briefly, cells obtained from amniotic fluid surrounding a fetus can be cultured. They express the biochemical and chromosomal markers of many disorders. This topic will be discussed in much greater detail in Chapter 22.

Review Questions

1. What is the difference between somatic cells and germ cells?
2. The term *clone* is used rather indiscriminately by journalists and other laymen. What is the more precise biological definition?

3. Is it correct to designate identical twins as a clone?

4. After fertilization, a human egg divides to the point where it is comprised of several hundred cells. A mutation may occur in one of these cells, which may continue to divide. The descendants of the mutant cell may form a vital part of the soma of the individual, who otherwise is quite normal at birth. What term is used by the geneticist to describe such an individual? Can you think of mutant traits that might be easy to identify if such an event occurred?

5. What is meant by the *transformation* of a cell in culture?

6. A Chinese hamster cell in culture has a deficiency in the enzyme adenine phosphoribosyl transferase (APRT). Because it does not have this enzyme, it can grow on a medium containing the toxic substance 8-azaadenine. The hamster cells are fused with human cells to form heterokaryons, which soon become hybrid by fusion of the hamster and human nuclei. The hybrid cells are allowed to grow and divide. Various clones are isolated, and these are then tested for growth on 8-azaadenine. All of the cells die from these clones except for one clone, the cells of which grow on the toxic chemical. A karyotype analysis of these cells show that they have lost both human chromosomes 16. What explanation is suggested by these results?

7. Cells from a patient A with xeroderma pigmentosum (XP) were cultured and shown to have a deficiency in DNA repair (excision repair as described on page 175). Another patient B with XP had his cells cultured and they also showed a DNA repair deficiency. When the cells from A were hybridized with those from B, the descendants of the hybrids did not show a DNA repair deficiency. What is the probably explanation? What is this phenomenon called?

References and Further Reading

Chu, E. H. Y., and S. S. Powell. 1976. Selective systems in somatic cell genetics. *Adv. Human Genet.* 7:189–258. An excellent review of how to find variant ("mutant") cells in culture.

Harris, M. 1964. *Cell Culture and Somatic Variation.* Holt, Rinehart & Winston, New York. 547 pp. A technical summary of the information gained from several decades of cell culture.

Littlefield, J. W. 1964. Selection of hybrids from matings of fibroblasts in vitro and their presumed recombinants. *Science* 145:709–710. The first report of a selective medium that permits hybrid cells to grow but not their parents. Historically, an important paper.

Pollack, R. (ed.). 1973. *Readings in Mammalian Cell Culture.* Cold Spring Harbor Laboratory, Cold Spring Harbor, N.Y. 735 pp. A compendium of important papers that have been published over the past decades, beginning with a short paper by Ross G. Harrison (1907) and a very interesting one by Alexis Carrel (1912).

"It runs in the family." We have all heard that terminology used by others and probably have used it ourselves. When we use that phrase, we are of course making the observation that some trait like easy-going temperament, musical or mathematical ability, or some kind of disease is being passed from generation to generation. In many cases the pattern of inheritance is clear, as in the case of hemophilia (Fig. 2.6), albinism (Fig. 7.3), or brachydactyly (Fig. 7.4). In each of these examples, the inheritance is through an identifiable *single gene*, and the geneticist can, with the family pedigree in hand, make rather precise statements about the odds of a child being born with the disease that runs in the family.

However, many characteristics that seem to run in families cannot be reduced to simple Mendelian patterns of transmission. Physical traits such as hair and skin color or body size and form, family resemblances in facial features, and such traits as temperament or mannerisms in movement and speech are but a few examples of those that do not show a simple pattern of inheritance. The fact that members of the same family have more traits in common than do unrelated persons is an observation that goes back to prehistoric times. It is a fact, but the scientific analysis is far from being in an advanced state.

We can be sure that genic inheritance plays a role in transmitting traits such as these, particularly from the results of breeding animals. Dogs are good examples that come immediately to mind. Body form and shape and differences in temperament, such as irascibility in terriers, placidity in spaniels, and "sharpness" in Dobermans and German shepherds, run in lines or families in these and other dogs.

We cannot breed humans as we do domestic animals, however, and must resort to other techniques to study what we call *complex inheritance* of traits in humans. Much of modern statistics, whether used in business or biology, was invented by persons such as the English scientists Francis Galton (1822–1911) and Karl Pearson (1857–1936) for the express purpose of analyzing the inheritance of complex human traits. Beginning in the 1860s, Galton set out to develop methods for the scientific analysis of human heredity, considering traits such as stature, genius, and fingerprints. Some of his efforts seem a bit misguided today, but it must be remembered that he was a true pioneer. Like many pioneers, some of his pursuits turned out to be more productive than others. He was the first to recognize the individuality of human fingerprints, espousing their use in human identification. He was also the first to recognize the contribution that twins can make in the study of human heredity. He accomplished much without knowledge of Mendelian heredity. Indeed, when Mendelism was rediscovered in his later years, he disdained it as being of no great importance!

The Nature of Complex Traits

It has often been said that Mendel succeeded where his predecessors (and some followers) failed because he was clever enough to study the proper traits. Each characteristic he studied occurred in clearly alternate forms—tall or short, round

11

Inheritance of Complex Traits

or wrinkled, green or yellow. Had he chosen numbers of peas per pod, a trait of some economic interest, he might well have remained as obscure as he was in his own time.

Traits may be "complex" for any of several reasons. We will first discuss what makes them complex. In the following sections, we will then consider some of the ways in which complex traits can be studied.

Single Gene Systems and Complex Traits

The examples of inheritance in previous chapters have been *simple Mendelian traits.* They also may be described as *single gene systems* or *monogenic* traits. These various designations mean that the phenotype is distinctively influenced by the genotype at a single locus. For a dominant allele, one copy of that allele produces a phenotype that signals the presence of the allele, regardless of what alleles may be present at other loci. For a recessive allele, homozygotes are recognized as such whenever the combination occurs. These traits are *simple* because one can ignore variations at other loci, and the different alleles at the locus under consideration produce distinctive phenotypes.

A complex trait may be influenced by variations at more than one genetic locus. Furthermore, the environment may help obscure the genetic differences. In short, one may be able to describe or measure the phenotype, but one cannot attribute it to a specific genotype.

Continuous Versus Discontinuous Variation

Traits can be described as *continuous* or *discontinuous.* A discontinuous trait is one for which distinct phenotypes exist. Mendel's peas were either wrinkled or round with no intermediate appearances. One examines a pea and classifies it, as was true for all the traits he analyzed. The same is true for a child with Tay–Sachs disease. The child either has it or doesn't.

A continuous trait is one that is measured. A wide range of values is possible, and often there is no limit to the values that may occur within a given range. Height is a continuous trait in humans (Fig. 11.1). Between the shortest and tallest persons, there are persons of all sizes. Most are near the middle, but there is no value between the extremes that cannot occur. We may say that someone is tall rather than describe him as 16.3 cm above the mean, but the population is not divided into natural categories of tall, short, or medium.

A trait may be both continuous and discontinuous if some individuals clearly fall outside the usual population distribution. Mendel's short pea plants were not all precisely the same height, nor were his tall plants. Had he measured them, he would have found two continuous distributions separated by a region into which no (or very few) plants fell. Skin pigmentation in humans is a continuous trait and can be measured quantitatively in various ways. Yet albinos, who lack any pigment, fall outside the normal population range and can be classified readily by inspection. Albinism is a simple Mendelian trait, but normal variations in skin pigmenta-

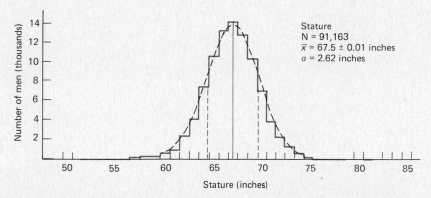

FIGURE 11.1
Distribution of height in a population of English males. Such a distribution is called a *Gaussian* distribution or a *normal* distribution (which does not imply that other distributions are abnormal). The most likely height is in the middle of the range, with fewer and fewer persons as one goes to higher and lower values. There are no "natural" divisions into tall and short. (From G. A. Harrison, J. S. Weiner, J. M. Tanner, and N. A. Barnicot, *Human Biology*, Oxford University Press, 1964.)

tion have a complex inheritance. Facial form appears generally to be a complex trait, although occasionally a particular feature seems to be transmitted as if strongly influenced by a single gene. One of the better documented examples is the Hapsburg lip (Fig. 11.2), transmitted over several centuries through one of the major royal families of Europe and recorded by numerous artists.

Discontinuous traits need not be simple, however. The simplest model for a discontinuous complex trait involves a threshold that must be exceeded for some trait to appear (Fig. 11.3). Consider a trait such as cleft lip (harelip) (Fig. 11.4). It is either present or not. If present it may differ in severity among affected persons. The defect occurs when the lips and palate fail to fuse properly during embryonic development (see Chapter 13). A number of studies indicate that heredity is important, but no single gene has been identified. Rather, it seems that several developmental processes must occur on a well-defined schedule. If any are delayed beyond a certain point, the "threshold" for cleft lip will be exceeded, and fusion will not occur properly. In spite of the relative ease of classifying persons as being affected or not, the underlying genetic factors may be very complex, and so the inheritance is complex.

Multiple Alleles and Multiple Loci

Two of the factors that contribute to complexity in inheritance are (1) the influence on a trait of multiple alleles at one particular locus and (2) the influence on a trait of allelic variation at any of several different loci. In experimental populations, selective breeding enables one to limit the genetic variation influencing a particular trait to a single locus and a single pair of alleles. Natural populations, such as

FIGURE 11.2

The Hapsburg lip, transmitted as an apparent dominant trait over many generations. Even though facial features are generally complex in their inheritance and cannot be classified as simple Mendelian dominant and recessive traits, a particular trait may rarely occur that does appear to be transmitted in a Mendelian pattern. Left: Charles V (1519–1558), Emperor of the Holy Roman Empire. Right: Charles II (1661–1700), King of Spain. Charles II was the great-great-grandson of Charles V, who also ruled as Charles I of Spain. (Photographs of busts in the Kunsthistorisches Museum, Vienna.)

Homo sapiens, may have many alleles at any one locus, and any particular trait may be influenced by variations at many loci.

It is instructive to consider a model showing the effect of multiple alleles on a trait. Let us assume that variation in skin pigmentation is due to different alleles at one locus, perhaps a locus that influences the activity of a key enzyme such as tyrosinase. Figure 11.5 shows the expected levels of pigment if there are two, three, and four alleles, each contributing its characteristic enzyme activity. Although the three phenotypes produced by two alleles would very likely be recognizable, those produced by three alleles and certainly those produced by four alleles would blend so closely as to appear continuous. Thus, what is in fact a simple Mendelian system becomes complex if there are no ways to identify individual alleles except through effects on a continuous scale.

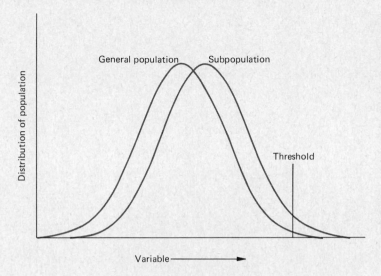

FIGURE 11.3
Hypothetical threshold for a discontinuous trait based on an underlying continuous variable. The general population may have only a few persons whose values exceed the threshold and express the trait. A particular subpopulation—persons with a certain genotype or environmental exposure —may have higher values of the variable and hence more persons who exceed the threshold.

The same results are obtained if two or more loci affect a single variable. Suppose that instead of four alleles there were two loci, each with two alleles. There would be nine possible genotypes compared to ten with one locus and four alleles. Again the effects on a single variable would make it impossible to identify genotypes because of the small differences between neighboring phenotypes.

FIGURE 11.4
A child born with cleft lip and palate. Proper fusion of the embryonic tissues that form the midline of the face did not occur. (Supplied by Dr. Robert J. Gorlin, Univeristy of Minnesota School of Dentistry.)

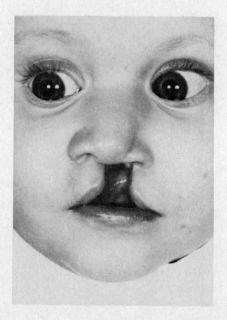

FIGURE 11.5
Diagram showing the expected pigment classes for two, three, and four alleles corresponding to various homozygous and heterozygous combinations. In each case, one allele, a^0, is considered to have no activity. Homozygous a^0a^0 would be "albino." The allele with greatest activity, a^{100}, would produce very dark skin when homozygous. The $a^{100}a^0$ heterozygote would produce half the activity of the a^{100} homozygote. The activity of alleles with intermediate activity would produce corresponding intermediate levels of pigment. Even with only four alleles, it would be virtually impossible to recognize different genotypes by observing the phenotype.

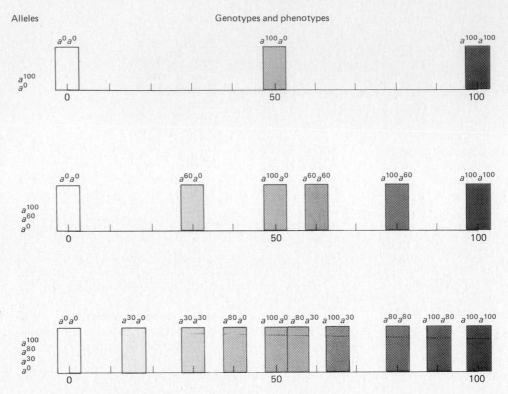

Methods of Studying Complex Traits

In a Mendelian analysis of heredity, specific ratios of affected and nonaffected are predicted from well-defined crosses. If the prediction is correct, the results are said to be consistent with simple dominant, recessive, or X-linked inheritance. If the prediction does not match the actual results, the hypothesis being tested is rejected. A new hypothesis must be formulated or an explanation of why the test was invalid must be advanced. For example, a single well-documented father-to-son transmission of a trait will rule out X-linked inheritance, since fathers do not transmit an X chromosome to their sons.

With complex traits, the predictions are much less explicit. *If a trait is caused by certain combinations of genes, then closely related persons are more likely to have those combinations than are unrelated persons.* Therefore, the prediction against which observations are tested is that the closer the genetic relationship, the more likely a relative of an "affected" person is also to have the trait. Unfortunately the same prediction exists with traits that are entirely environmental. The more closely related two persons are, the more likely they are to have similar environments. The study of complex traits therefore requires environmental as well as genetic considerations. The question becomes: Is a trait the result of genetic factors, environmental factors, or both? If both, then to what degree is each responsible? Answering these questions is most difficult and in many cases borders on the impossible.

Correlations and Associations Within Families

The simplest comparison to make is between close relatives compared to unrelated persons. The choice of relatives will depend somewhat on the trait to be studied. One does not study patterns of baldness in teenagers or breast cancer in males. But generally comparisons are made between first-degree relatives—parent/offspring or sib/sib. First-degree relatives have half their genes in common. More distant relatives may also be included although the proportion of genes in common is less.

Continuous Traits. Two kinds of comparisons are made, depending on whether the trait is continuous or discontinuous. For a continuous trait, the correlation of measurements between family members is calculated. Here we use the terms correlation and correlation coefficient as they are defined technically (Box 11.1). A positive correlation coefficient indicates that family members tend to resemble each other more than they do unrelated persons. This does not mean that the reason is necessarily because they share more genes in common. It is up to the investigator to rule out other explanations if a credible case is to be made for genetic explanations.

Some examples will illustrate how correlations within families are used to obtain evidence in favor of genetic variation. In 1889, Galton published the results of some of his studies of British families in the book *Natural Inheritance*. Stature was one of the traits studied, and he was able to obtain information on 205 sets of parents and 930 of their adult children. Since the average heights of males and females differ, and he was not interested in investigating that particular genetic effect, he multiplied the female heights by 1.08, a factor he found empirically to remove the male–female difference. The results of all persons could then be pooled. Also, he compared the heights of the offspring to a hypothetical *midparent*, that is, the average of the values of the two real parents.

Table 11.1 shows the results obtained when the distribution of offspring heights is compared to the midparent heights. Inspection of the table is sufficient to show that children do indeed resemble their parents in stature. This is even more obvious if the average height of the offspring is plotted against the different midparent heights (Fig. 11.6). The correlation coefficient calculated from these data is +0.46. This is compared to the expected value of +0.71 if the variation in stature were entirely genetic and the various alleles and loci each add or subtract a small increment. Many subsequent studies of families collected with less bias than those of Galton have given an even closer fit of the observed data to the theoretical expectation for 100% heredity.

But does the resemblance between parent and offspring require that we accept genetic variation as the explanation for the similarities? The answer depends on whether we can think of other plausible explanations. One that comes to mind is nutrition. Many studies show that malnutrition causes slower growth and shorter stature. Families would tend to share their nutritional status, and parents who

— Box **11.1**

Correlation: The Similarity of Measurements

Biologists and others often need to express in quantitative terms how similar two measures are in a population. Examples of questions might be: "How closely related is body height to length of forearm?" and "How similar are the heights of sibs?" In the first question, measures of two different features on the same person are being compared. In the second question, a single feature is being compared in pairs of persons who have some special relationship to each other. The same approach can be used to answer questions on the scores students make on an examination versus the time spent studying for it, or the scores people make on a biology test compared to their scores on a history test. In all these examples, a comparison is being made between two series of measurements to see to what extent the measures are related to each other.

A simple way to study such relation-ships is to plot one value against its paired value, as in graphs *a–e*. Inspection of such plots will often tell us what we need to know. If *x* is the hours studied for a test and *y* the score on the test, the outcome in *e* would encourage us to study more, whereas the outcome in *a* would tell us to study less. *c* would suggest that the amount of time spent studying is not important. The most likely results would be as in *d*. Studying improves our score, but other things are important also.

Such subjective conclusions may be adequate for many purposes, but they are not helpful if one wishes to compare the strength of an effect for correlations that fall between the situation in *c* and the perfect correlations in *a* and *e*. The common method of expressing relation-ships between measures is the *correlation coefficient*. If one has a series of measures, symbolized by $x_1, x_2, \ldots, x_n$,

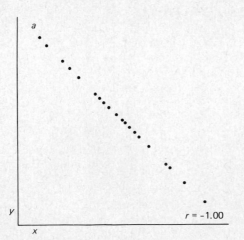

$r = -1.00$

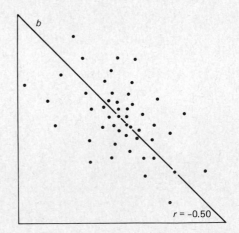

$r = -0.50$

Box 11.1

(*continued*)

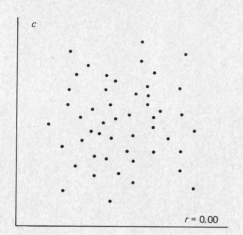

$r = 0.00$

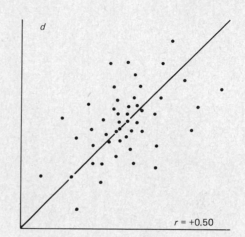

$r = +0.50$

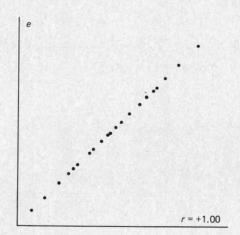

$r = +1.00$

paired with a series of measures y_1, $y_2, \ldots, y_n$, the correlation coefficient r can be calculated by the formula

$$r = \frac{\Sigma(x - \bar{x})(y - \bar{y})}{\sqrt{\Sigma(x - \bar{x})^2(y - \bar{y})^2}}$$

where x and y are the series of measurements and $\bar{x}$ and $\bar{y}$ are the respective means. The derivation and further explanation of this formula can be found in textbooks of statistics. We shall be concerned here only with understanding the meaning of various values of r.

With this formula, the values of r may vary from -1.0 to $+1.0$. In either of these extreme cases, the two measures show perfect correlation; that is, any variation in one measure is matched exactly by variation in the other. For $r = +1.0$, both measures increase or decrease together, as illustrated in e. For $r = -1.0$, one

Box **11.1**

(*continued*)

measure increases as the other decreases, as illustrated in *a*. If the two measures are completely independent of each other, as illustrated in *c*, $r = 0$. This would be the case, for example, if pairs of unrelated persons were drawn at random and compared for some measure such as height.

If two measures tend to be related but not perfectly, the plot of x versus y is scattered as in *b* and *d*. In these situations, a positive correlation greater than zero (or a negative r less than zero) tells us that the two measures share some variation in common but also that each varies independently of the other. The closer the value of r is to -1.0 or $+1.0$, the greater the common variation. The closer the value of r is to zero, the less the common variation.

The expected correlation coefficients between pairs of relatives can be calculated if one assumes that variation in the trait is entirely genetic. Monozygotic twins are genetically identical, and r should be $+1.0$ for any such trait. Other sibs share half their genes on the average, and the expected r is $+0.50$. These theoretical values are rarely obtained in actual observations, but the difference in the observed value and the theoretical value is a guide to the extent to which heredity contributes to variation in a trait.

TABLE 11.1
Distribution of Heights of Offspring Versus Midparent Height in 205 British Families

The trend for taller parents to have taller children can readily be seen.

Height of Adult Children (in.)	Height of Midparent (in.)										
	64	64–65	65–66	66–67	67–68	68–69	69–70	70–71	71–72	72–73	73
74	—	—	—	—	—	—	5	3	2	4	—
73–74	—	—	—	—	—	3	4	3	2	2	3
72–73	—	—	1	—	4	4	11	4	9	7	1
71–72	—	—	2	—	11	18	20	7	4	2	—
70–71	—	—	5	4	19	21	25	14	10	1	—
69–70	1	2	7	13	38	48	33	18	5	2	—
68–69	1	—	7	14	28	34	21	12	3	1	—
67–68	2	5	11	17	38	31	27	3	4	—	—
66–67	2	5	11	17	36	25	17	1	3	—	—
65–66	1	1	7	2	15	16	4	1	1	—	—
64–65	4	4	5	5	14	11	16	—	—	—	—
63–64	2	4	9	3	5	7	1	1	—	—	—
62–63	—	1	—	3	3	—	—	—	—	—	—
62	1	1	1	—	—	1	—	1	—	—	—

Data from Francis Galton, *Natural Inheritance*, Macmillan, London, 1889.

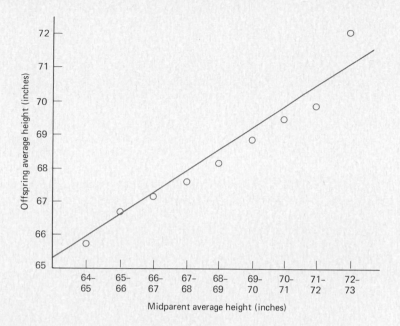

FIGURE 11.6

Average height of offspring plotted against midparent height. The straight line is the calculated relationship based on these data. The average offspring height tends to be closer to the overall population average than are their parents. That is, if both parents are tall, their tall children are not quite as tall on the average as their parents. Similarly the children of short parents tend to be a little taller than their parents. These tendencies toward the mean are offset by children of average parents, whose heights tend to be more deviant.

The value for the tallest category in this graph is out of line, perhaps due to the small number of persons on which this value is based. It may also be that tall parents with tall children were more likely to respond to Galton's request for information. (Data from Francis Galton, *Natural Inheritance*, Macmillan, London, 1889.)

were malnourished as children might well have children who were also malnourished because of the persistence of a low socioeconomic level. It must also be noted, though, that the severe malnutrition that causes stunting of growth is not widespread in western European countries. It was probably less likely to have been present in Galton's sample, because he advertised for his subjects, who responded in writing. They were therefore literate, and probably most were not in the lowest socioeconomic level. Certainly they were not a random sample of the British population. If one must weigh the relative contributions of heredity and malnutrition to variation in stature, the heavier weight would surely go to heredity for any population with reasonably adequate nutrition. There is no evidence that exceptionally good nutrition may add centimeters to one's stature beyond the level for which one is programmed genetically.

No other plausible explanations have been advanced to explain family resemblances in stature. Therefore, such resemblances may reasonably be attributed to heredity. There are a couple of observations that give pause, however. The children of immigrants to the United States are typically taller on the average than their parents. Also, based on suits of armor from the Middle Ages, present-day descendants are taller than their ancestors of several centuries ago. No genetic theory would predict such changes. The frequency of different alleles would not change rapidly enough to account for these reported differences in height. Perhaps nutrition is important in these instances. Whatever the correct explanation, environmental variables can influence stature in some circumstances.

These studies also point out an important aspect of analyzing a trait for complex inheritance. The variation in a population may be due in part to environmental differences among families and in part to genetic differences. If the environment were constant for a population, only genetic variation would remain. In that case, all individual differences would be hereditary. On the other hand, if the population were genetically uniform for loci influencing the particular trait, all variation would be environmental in origin. These two extreme situations probably rarely if ever occur in human populations. But populations do differ from each other in their heredity and in their environment. There is little doubt that skin and hair color are primarily inherited, even though both can be influenced to some extent by environment. (Intentionally applied chemicals would be considered an environmental influence.) Different populations have different frequencies of alleles for skin and hair color, and the variation attributed to heredity will change accordingly. It is no coincidence that most studies of hair color have been done in populations of European descent. It would be a most unpromising research topic to pursue in Japan, where everyone's hair is black. The wide range of hair color in Europeans indicates possible wide genetic variability. The essentially uniform phenotype in Orientals holds no promise of genetic variability. The message from these comparisons is that *heritability* of a trait measured in a particular population may be very different from results obtained in a different population. One could not take the results of Galton's study in England and assume that they would apply to Japanese, Eskimos, or Pygmies.

Heritability.   We have used the term *heritability* without defining it precisely. Whenever we talk about a trait being hereditary or environmental, what we really are talking about is the *variation* in that trait in a specific population. We surely are programmed genetically to have exactly one head. Any exceptions that may occur do not survive. But we cannot speak meaningfully of the genetic control of head number because there is no variation in head number. Similarly, nothing happens unless there is an environment. When we speak of genetic versus environmental control, what we really are talking about is the portion of the *variation* that is attributable to environmental *variation* and the portion that is attributable to genetic *variation*. The latter portion is sometimes referred to as *heritability*. Expressed symbolically,

$$\text{Heritability} = V_G/(V_G + V_E)$$

where V_G is the genetic variation and V_E the environmental variation. The total variation is $V_G + V_E$. The calculations of heritability are somewhat complex, but the concept is simple so long as one remembers that the subject of analysis is the *variation* in the population.

Discontinuous Traits. Associations within families also are useful to estimate the genetic contribution to discontinuous traits. In a typical situation, persons (designated *index cases*, *probands*, or *propositi*) who have a particular characteristic are identified. Their relatives are then studied to see if they are more likely to have the same characteristic than an unrelated person. An example would be a disease such as cancer. If the frequency is elevated in close relatives, and if no environmental explanation is obvious, then one may accept these observations as supporting a genetic contribution to the risk. Figure 11.7 summarizes the results of one such study. When the index case has breast cancer, the incidence of breast cancer among their mothers and sisters is increased several fold over mothers and sisters of women who do not have breast cancer. Similar results have been obtained in other studies of breast cancer as well as other types of cancer. Furthermore, the risk to a woman whose mother and sister both had breast cancer is approximately 30%, a very large increase over the risk if there is no history of cancer in the family.

As with continuous traits, one cannot automatically assume that the increased risk is due to shared genes. But the weight of evidence favors heredity. First, some cancers are inherited as simple Mendelian traits: retinoblastoma, Wilm's tumor, multiple polyposis of the colon, xeroderma pigmentosum. (The cancer *per se* is not inherited. Rather a susceptibility is inherited that almost always results in

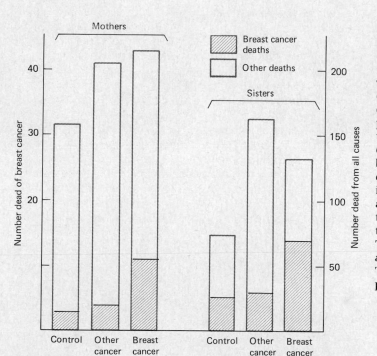

FIGURE 11.7
Frequency of breast cancer in dead relatives (mothers or sisters) of various groups of women: *Control,* no history of cancer; *Other Cancer,* a patient with cancer other than breast cancer; *Breast Cancer,* a patient with breast cancer. The frequency of breast cancer is several fold greater among mothers and sisters of breast cancer patients. (Data from Madge T. Macklin, in *Genetics and Cancer,* Univ. of Texas Press, Austin, 1959. pp. 408–425.)

cancer.) Second, even if a virus or other environmental agent is implicated, susceptibility to the agent seems to be a major distinction among different persons, and the susceptibility could be largely under genetic control. Skin cancer is much more common among very light skinned persons exposed to the ultraviolet rays of the sun than among dark complexioned persons. Therefore, among persons exposed to the sun, heredity would be a major factor in the susceptibility to cancer through its effects on skin pigmentation.

Twins and Heredity. Approximately one in eighty births is a multiple birth, primarily of twins. Twins have always had a special fascination to nontwins. Twins have often appeared in fiction (and some fiction that purports to be nonfiction) as having special affinities for each other, usually of a psychological nature based on extrasensory perception. We shall ignore the possibilities of extrasensory perception and consider instead the ways in which studies of twins have contributed to our knowledge of the genetic influence on human traits.

Galton was the first to recognize that twins provide a means of attacking problems of heredity. In a paper published in 1876, Galton noted that there are two kinds of twins: *identical* and *fraternal*. The identical twins come from a single fertilized egg that divides to form two embryos. They are therefore *monozygotic* (MZ). Fraternal twins come from two separate fertilized ova and are therefore *dizygotic* (DZ). Since MZ twins originated from a single cell, they are clones and are genetically identical. DZ twins have the same relationship to each other as any pair of single-born sibs. Approximately ⅓ of twin births are of MZ twins.

Since MZ twins are virtually all genetically identical, any differences between a pair of MZ twins should be nongenetic in origin. For traits influenced by both heredity and nonhereditary factors, one would expect that MZ twins would be more often alike or more often similar than DZ twins. Comparison of MZ and DZ twins is then one way to determine whether or not genetic variation contributes to the total variation observed in a particular trait.

Twin studies assume that the average environmental difference between members of a twin pair is the same whether the pair is monozygotic or dizygotic. For many traits this probably is true. Fingerprints are not known to be influenced by environment, which is why they are useful in distinguishing MZ from DZ twins, although fingerprints of MZ twins are never exactly the same. For other traits, especially behavioral traits, it probably is not true. Having a twin who is confused with you must be a different experience from having a twin who is not. Psychological studies with twins must therefore be interpreted with care and should have corroborative evidence from other kinds of studies if possible. In spite of these shortcomings, studies of twins are often the most effective way to assess the genetic variation contributing to complex traits.

Diagnosis of Twin Zygosity. The first requirement in studying twins is to classify them as MZ or DZ. This is ordinarily done by extensive blood typing and comparison of fingerprints, facial features, hair color, etc. Mixed sex pairs obviously are DZ but usually are avoided because of the obvious differences in environments of boys and girls. Any difference in any clearly inherited blood group or other Mendelian trait is sufficient to classify a twin pair as DZ. Most DZ twin pairs will differ by one or more Mendelian traits, but by chance some will not. If the pair is identical for all Mendelian traits, distinct differences in the complex traits, such as fingerprints, would still be a basis for classifying a pair as DZ. One could not subsequently study the traits so used, however. It should be possible with the extensive battery of tests now available to identify approximately 99% of DZ twins on the basis of a difference in one or more systems. Any twin pair not showing a difference is classified as MZ. In practice, twins who look very much alike, who often are confused with each other, even by friends and sometimes parents, are virtually always MZ. DZ twins should be no more alike than any pair of brothers or sisters. Sometimes they will look very much alike but rarely to the point of being confused with each other.

There is one certain way to diagnose zygosity of twins. That is by skin grafting. The body will reject any graft from another person unless that person is genetically identical. One can exchange grafts between members of a pair of twins. If the grafts take, the twins are MZ; if they are rejected, the twins are DZ. Needless to say, such a procedure would rarely be done. It was used on at least one occasion to prove that one of a pair of twins had been exchanged at birth with an unrelated infant. Some years after the birth of the children, the family with the "fraternal" twins happened to encounter the family with the "single-born" child, who very much resembled one of the twins. After noting that the children had all been born on the same day in the same hospital, the parents became suspicious that all was not right.

Twin Studies: Continuous Traits. A comparison of twin pairs for continuous traits is much like a comparison of other related persons. The difference is the expected degree of similarity. Dizygotic twins have half their genes in common on the average, as do any other pair of single-born sibs. The expected correlation coefficient for a trait whose variation is due entirely to additive genes is $+0.50$. Monozygotic twins, being genetically identical, should be identical for any genetic trait and have a correlation coefficient of $+1.00$.

Figure 11.8 shows the plot obtained when the heights of MZ and DZ twins are compared. The correlation coefficients are $+0.97$ for the MZ twins and $+0.77$ for the DZ twins. These particular twins were teenagers, and the data are not corrected for the effects of age. Some will not have attained their final height. This increases the correlation coefficients because both MZ and DZ twins will tend to resemble each other—compared to the rest of the group—since they are the same age. Nevertheless, the correlation coefficient for MZ twins is distinctly higher than that of the DZ twins. Inspection of Figure 11.8 confirms that the points for the MZ twins lie closer to the theoretical line for perfect correlations than do the points for the DZ twins. This supports the hypothesis that variation in stature is strongly influenced by heredity.

Twin Studies: Discontinuous Traits. The theoretical basis for using twins to study discontinuous traits is very simple: If a trait is entirely hereditary, MZ twins will always be concordant; that is, both will express it or neither will express it. DZ twins may sometimes be concordant but should not always be. If a trait is influenced by heredity but also by nongenetic factors, MZ twins will be concordant more often than DZ twins, but some MZ twins will be discordant.

One instructive example of the application of twin studies is found in Down syndrome. Prior to the finding that it is caused by trisomy for chromosome 21, it was known that the risk is higher in older mothers. Various hypotheses were advanced to account for the origin of Down syndrome based on the maternal age

FIGURE 11.8
Graph showing the height of twin A (arbitrarily selected) plotted against the height of the cotwin, twin B, for a series of MZ and DZ teenage twins. All points would fall on the 45° line if members of a pair were always the same height. The MZ twins lie much closer to this theoretical line than do the DZ twins. (Data collected by Philip J. Clark. The data have not been corrected for age.)

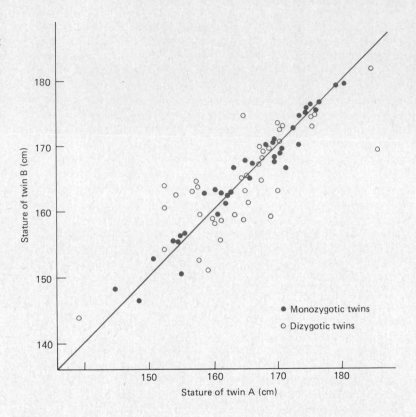

effect—the changing hormonal status of older women, the inability of an "ageing womb" to reject an abnormal fetus, and so forth. One observation destroyed the foundation on which these hypotheses were constructed: MZ twins are virtually always concordant for Down syndrome, whereas DZ twins rarely are. It is difficult to accommodate these facts in any hypothesis based on the environment of the embryos. The high concordance of MZ twins suggested an entirely genetic cause, although Mendelian traits should not change with maternal age. The low concordance of DZ twins, as well as the low risk of recurrence among subsequent single-born children, was inconsistent with a simple Mendelian trait also. Therefore some more complex genetic event was presumed to occur. That event turned out to be nondisjunction, as described in Chapter 9.

Table 11.2 lists several disorders that have been studied by twin concordance. These data are from a large twin registry in Denmark, consisting of all multiple births from 1870 to 1910. Only twin pairs with at least one affected person are included. Therefore, in concordant pairs, both are affected; in discordant pairs, only one is affected. For cancer, the concordance rates for MZ and DZ twins are not different. (The concordance is actually slightly higher in DZ twins, but not significantly so.) For cancer at all sites, the genetic components of risk are not large. Most other studies show a small genetic component. It may be much larger if

Only twins of the same sex are included. The twins were identified among all the twin births in Denmark between 1870 and 1910. At least one member of each pair listed in the table has been diagnosed with the indicated disorder.

TABLE 11.2
Frequencies of Twin Pairs Concordant or Discordant for Various Disorders, by Zygosity

Disorder	Monozygotic Twins		Dizygotic Twins	
	Concordant	Discordant	Concordant	Discordant
Cancer, all sites	17	126	43	249
Arterial hypertension	20	40	10	86
Epilepsy	10	7	6	31
Tuberculosis	50	35	42	183
Rheumatic fever	30	88	12	202
Bronchial asthma	30	4	24	53
Peptic ulcer	30	84	24	161
Diabetes mellitus	36	4	12	98

Data from B. Harvald and M. Hauge, in *Genetics and the Epidemiology of Chronic Diseases*, J. V. Neel, M. W. Shaw, and W. J. Schull, eds. U.S. Public Health Service Publ. No. 1163, 1965, pp. 61–76.

specific cancers only are studied. The present results suggest that most cancer may be due to nongenetic variation.

An interesting contrast is provided by tuberculosis and rheumatic fever. Both are due to infectious agents, but the twin results suggest that inherited susceptibility may be very important in determining which persons acquire the infection. It may be that in these cases, and in a number of other infectious diseases, most of the population is exposed sooner or later, but only the genetically susceptible get the disease.

There are lots of interesting studies one could do with rare classes of twins, such as MZ twins separated at birth. Such twins are too rare to be considered for most purposes. MZ twins in which only one is exposed to a particular environmental agent are somewhat more common. The unexposed twin is an ideal control for the exposed twin, since genetic variability would be removed from the study.

There are many anecdotal reports in the press of MZ twins separated in early childhood. Generally they grow up to be very much alike, sometimes even choosing spouses with the same first name, naming their children the same, and choosing the same profession. These reports, while interesting, cannot be taken seriously as proving anything. These particular twins are noted and written about precisely because they are unusual. The MZ twins whose patterns diverge are not interesting and not newsworthy. But they are just as important from the point of view of scientific research.

Adoptions

One additional comparison that often is possible is between adopted offspring and their biological parents and sibs. For any trait whose variation is primarily genetic, the adopted child should resemble the biological family rather than the adoptive family. On the contrary, a trait that is environmentally influenced should show strong similarity between the adopted child and the adoptive family.

Language is an obvious example of an environmental trait. We speak as our native tongue whatever language is spoken in the family in which we are reared. Even though there are hypotheses that the "deep" structure of language is inborn, there is no evidence that specific languages are.

We are not surprised, or should not be after the preceding sections, if adopted children resemble their biological parents in physical appearance. But there are many traits for which adoptive studies are especially valuable. For these traits the relative contributions of nature and nurture are poorly defined and family correlations and twin studies are subject to various criticisms. For the most part, these are behavioral traits: intelligence, special abilities, personality, mental health. These traits, of course, are the most difficult to study by other approaches because one can never be sure that early environment doesn't set into motion a path of development that persists in spite of later experiences. Examples of the methods of this chapter applied to behavioral traits are given in Chapter 18.

Review Questions

1. Height in humans is not usually considered a monogenic trait, for good reasons. What are the reasons?
2. Several kinds of dwarfism are known to be inherited in humans. Each kind has a simple monogenic pattern of inheritance, some dominant, others recessive. How does this fact fit with the explanation given in the first question?
3. In an isolated population in a small town in Norway, all persons have blond hair. A geneticist studying this population concluded that the heritability of hair color was zero, even though many other studies have demonstrated that blond hair is inherited. How do you explain her results?
4. What is the difference between fraternal and identical twins?
5. About one-third of twin births are monozygotic. A mother has twins, both of whom are female. What is the probability that they are dizygotic?
6. Of what value are twin studies in the determination of environmental influences on the phenotypic expression of genetic factors in humans?
7. Monozygotic twins are virtually always concordant for Down syndrome. However, cases are known in which one was afflicted and the other was not. How would you explain this? How would you test your hypothesis?
8. A great deal of time has been spent studying monozygotic twins who were separated at birth and lived apart. What is the special interest in such twins?

References and Further Reading

Allen, G. 1965. Twin research: Problems and prospects. *Progress in Medical Genetics* 4: 242–269. An evaluation of the problems involved in analyzing twin data.

Ehrman, L., and P. A. Parsons. 1982. *The Genetics of Behavior*, 2nd ed. Sinauer Associates, Sunderland, Mass. This general text has excellent chapters on the inheritance of continuous traits in humans and other animals.

Newman, H. H., F. N. Freeman, and J. K. Holzinger. 1937. *Twins, A Study of Heredity and Environment*. Univ. of Chicago Press, Chicago. 369 pp. A classic in the field that is still worth reading.

The kind of sex that we discuss in this chapter is covered neither in the literature of romance nor pornography. The definitions of sex vary, and there are many. Here we use only the biological definitions, and these have reasonably precise meanings that are in general quite simple.

To begin with, *sex is a mechanism whereby the genetic materials or substances from two different individuals are brought together, making it possible for genetic recombination to occur.* This recombination may occur by the mechanism of crossing over in the germ cells at the time of meiosis, as we have already discussed in Chapters 7 and 8, by independent assortment during meiosis when the homologous chromosomes separate, and by a variety of other mechanisms, such as conjugation, transduction, and transformation found in the prokaryotes as described in Chapter 14. The consequence of recombination by whatever mechanism is a great increase in the variety of genotypes among organisms that engage in sexual reproduction. It ensures that we humans along with other animals are each unique individuals genetically unless we happen to have an identical twin. And even identical twins may show differences that are genetic in origin. More importantly, such variety makes a population more likely to survive when the environment changes.

The animal and plant worlds have many varieties of sex. It will help us gain perspective on human sexuality if we view it in the context of sex as a general biological phenomenon.

Asexual Reproduction

Organisms that do not have sex as a necessary part of the reproductive cycle are said to be *asexual*. Bacteria such as *Escherichia coli* reproduce in great numbers by dividing in half in a process called *binary fission*. Each half or daughter cell gets a chromosome, otherwise it could not survive. This means that the increase in numbers is exponential, since the population doubles every division. Thus if a bacterium divides every 20 min (not uncommon for *E. coli* under proper conditions) a starting population of one cell will have had 72 divisions after 24 h. Provided there is adequate food, there will be 2^{72} or 4.72×10^{21} offspring. This is a

12

Sex,
Sex Determination,
and Reproduction

FIGURE 12.1
Asexual reproduction in bacteria and the consequences of mutation. When a mutation occurs in a cell, it passes only to the cells that descend from that mutant cell. As a consequence, different mutations stay in separate cell lines, and their actions together are not easily tested unless two different mutations occur in the same line.

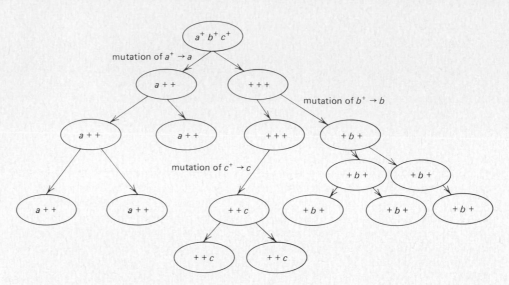

most effective way of reproducing but not of generating variability. Mutations will occur but they will remain within clones derived from the original cells in which the mutations arose (Fig. 12.1). This will be the extent of the variability produced unless sex intervenes, in which case cells with new combinations (or recombinations) of genes will be formed, as described in Chapter 14. Most organisims that reproduce primarily by asexual means also have a mechanism for sexual recombination.

Sexual Reproduction and Sexuality

We have so far used a rather rigid definition of sex as a mechanism for generating variability in DNA in a population of organisms. But as we stated in the introduction, sex has many facets. One of these other biological facets is a sexual difference in state or phenotype that we will call sexuality.

Sexuality is a state in which there are different sexes ordinarily thought of as male and female. Frequently, but not always, the different sexes may be morphologically different or *sexually heteromorphic.* If there are two different sexes with males and females morphologically distinguishable, it is called *sexual dimorphism.* This sexual difference has nothing to do with sexual preference in humans, a subject that we shall address further later in the chapter.

Sex and Life Cycles

The basic sex act is the fusion of two cells, ordinarily haploid cells in eukaryotes, to give a single diploid cell called a zygote. Among the lower eukaryote forms, chiefly the fungi and the protozoa, the two cells may be identical in appearance as shown in Figure 12.2a. The diploid cell formed by cell fusion or fertilization may undergo immediate meiosis and give rise to haploid cells that undergo mitosis and produce a haploid body or thallus that, among the fungi, is a mycelium made up of filaments formed by cells in a series (Fig. 12.2b). Or the haploid cells may remain separate as in some yeasts (Fig. 12.2c). Most of the fungi of which you are aware are of the type in which the dominant part of the life cycle is haploid. The diploid phase is represented only by the short-lived zygote.

Many yeasts, on the other hand, have diploid cells as the dominant phase of the cycle, as shown in Figure 12.2c. Here the diploids reproduce asexually by mitosis. On occasion, a diploid will undergo meiosis and produce haploid cells that may or may not fuse to form new diploids immediately.

Among the algae, some species have flagellated cells (Fig. 12.2d). These move about like sperm powered by their flagella. Two of these haploid cells can fuse to form a diploid zygote, which can either undergo mitosis to form a diploid line or undergo meiosis immediately as in the case of the fungi discussed above. The alga in Figure 12.2d is a unicellular variety that has a haploid asexual cycle, but many of the higher algae have a dominant diploid phase. The diploids may form large bodies, which we call "sea weed" (kelp is an example) and which form "sperm" and "eggs" not unlike animal sperm and eggs. These fuse to form a diploid called a sporophyte, as shown in Figure 12.3.

The examples discussed above and illustrated in Figure 12.2 obviously involve sex but not sexuality. Two yeast haploid cells fuse to form a zygote. But not just any two. They must be of different *mating types.* In the yeast ordinarily used in the laboratory and to brew beer and leaven bread, the two mating types are α and a. Only α haploid cells will fuse with a cells to form diploids. At meiosis, half of the four haploids are α and half a. But a and α cells are identical in appearance and can only be distinguished by testing to see if they will fuse. A step toward sexuality is found in those algae that form sperm and eggs (Fig. 12.3). However, the sperm-forming and egg-forming plants or elements are identical in appearance, and in fact the same individual will produce both kinds of gametes.

Sex Determination

Sexuality, as we defined it above, occurs in those organisms in which the sperm-bearing individual is different from the egg-bearing individual. This kind of sexuality is well developed in the higher plants and animals. But what determines who shall bear sperm and who eggs? The answer to this is rather complicated for most animals, including people. But let us try to answer this question of *sex determination* with some examples.

Perhaps the most unusual factor in sex determination from a human point of view is environmental. A well-studied example is the marine worm *Bonellia viridis.* Larvae may develop either as male or female. Mature females have a very large proboscis. If a larva is in the vicinity of a female, the larva attaches to the proboscis and develops into a male. If no female is present, the larvae develop into females. Male development can be induced in the laboratory by extracts of female proboscis or gut and by altering the mineral composition of the water. In this species, males

FIGURE 12.2
Reproduction in algae, fungi, and protozoa. These examples illustrate the diversity of processes in the biological world for solving the problem of genetic recombination. (a) Fusion of two haploid cells to form a diploid that undergoes meiosis almost immediately. Found in many simple algae. (b) A fungus that forms a haploid mycelium. (c) Life cycle of a yeast with two mating types. (d) An alga or protozoan with flagellated cells.

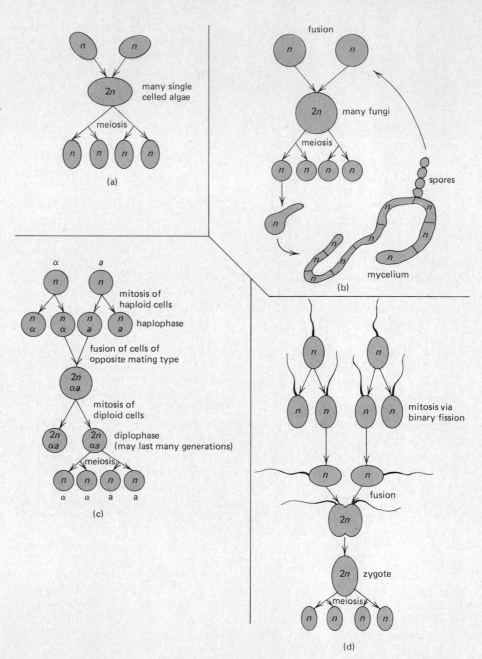

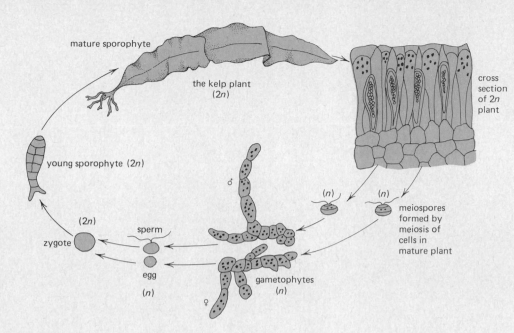

FIGURE 12.3
Reproduction and life cycle of kelp. (From H. J. Fuller, Z. B. Carothers, W. W. Payne, and M. K. Balbach, *The Plant World*, 5th Ed. 1972. Holt, Rinehart and Winston.)

are very small compared to females and eventually establish a permanent parasitic residence inside the female uterus.

Environmental influence on the determination of sex is found in some other invertebrates, especially among some species of snails. Among vertebrates it is also found in some turtles, an alligator, and some fish. For example, the common Atlantic silverside, a fish found off the eastern North American coast, definitely has two sexes, male and female, but the proportion of newborn of the two sexes may deviate from the 1:1 ratio generally expected among fish, depending on the temperature. Cold temperatures of 11–19°C tend to induce the newly hatched fry to develop into females, whereas warm water temperatures of 17–25°C induce a higher proportion of males. The deviation from the 1:1 ratio, however, is also dependent on the genotye of the parents. Some mothers will produce equal numbers of male and female progeny, regardless of the temperature. This example again illustrates the interplay of genotype and environment. Obviously sex is inherited in this fish, but environment can also play a role. For the most part though, sex in animals is determined by the genetic makeup of the individual, with varying degrees of influence by the external and internal environments.

Most animal species depend on chromosomal mechanisms for sex determination, including all mammals, birds, and insects, including the much-studied fruitfly *Drosophila*. Mammals and *Drosophila* have the XX/XY type of sex determination, in which males are heterogametic (produce X- and Y- bearing sperm). In birds and some insects, females are heterogametic, and the sex chromosomes are designated ZW in females and ZZ in males. The sex that produces gametes with only one kind of sex chromosome is called *homogametic* Thus human females produce only X-bearing eggs; male birds produce only Z-bearing sperm. Both are homogametic.

Sexual Development and Function in Humans

In view of the variety of mechanisms of sex determination, one must examine humans and their close mammalian relatives in order to understand the chain of events that lead to human males and females. In the following sections, we will consider some of the normal and abnormal conditions of sexual development that help answer some of these questions.

Hormones, Sex, and Reproduction in Mammals

As with many other aspects of normal physiological function, hormones play an important role in sexual development and function. Hormones are chemical messengers that are manufactured in the endocrine (ductless) glands, such as the gonads, thyroid, adrenals, and parts of the brain. They are transported throughout the body in the blood, inducing responses in tissues far removed from the gland in which they were synthesized. There are three types of sex hormones: *androgens*, *estrogens*, and *progestins*. All are complex chemical substances called steroids, as is cholesterol, from which they are manufactured in the body (Fig. 12.4).

Androgens.   The term androgen comes from the Greek *andros* (male) and *gennao* (produce), so named because androgens cause masculinization of an animal. In humans, this means increased muscle mass, beard growth, a deeper voice, and so forth. The primary androgen is *testosterone*, produced largely in the testes but to some extent also by the adrenal cortex. Testosterone is converted in various tissues to another androgen, *dihydrotesterone* (DHT). Both testosterone and DHT function by binding to an *androgen-receptor protein* in the cytoplasm of cells, being then transported to the nucleus. There they influence gene activity by mechanisms not fully understood. The two androgens are both necessary for normal male development.

Estrogens.   The term estrogen comes from the Greek *oistros* and Latin *oestrus*, meaning frenzy or insane desire. Estrogens are manufactured primarily in the ovary but also in the adrenals. Estrogen synthesis actually begins before implantation of the embryo in the uterine wall and is probably required both for male and female embryonic development. During the ovulation cycle, estrogens are released by the developing follicle, in response to the *follicle-stimulating hormone* (FSH) of the *adenohypophysis* (anterior pituitary gland) (Fig. 12.5). Estrogens stimulate growth of the uterus and other secondary sex characteristics of females, such as breasts and body hair.

Progestins.   The progestins (from Latin *gestatio*, carrying), principally progesterone, are synthesized by the *corpus luteum*, which is derived from the follicle after release of the ovum (Fig. 12.5). The *luteinizing hormone* (LH) of the adenohypophysis is responsible for ovulation and for stimulating the corpus luteum to secrete progesterone. The function of progesterone is to prepare the body for and to maintain pregnancy. It acts on the *uterine endometrium* to make it more suitable for implantation of the developing embryo. It also causes breast development and suppresses development of additional follicles that would lead to ovulation. The latter effect is one basis for birth control pills, which are synthetic derivatives of

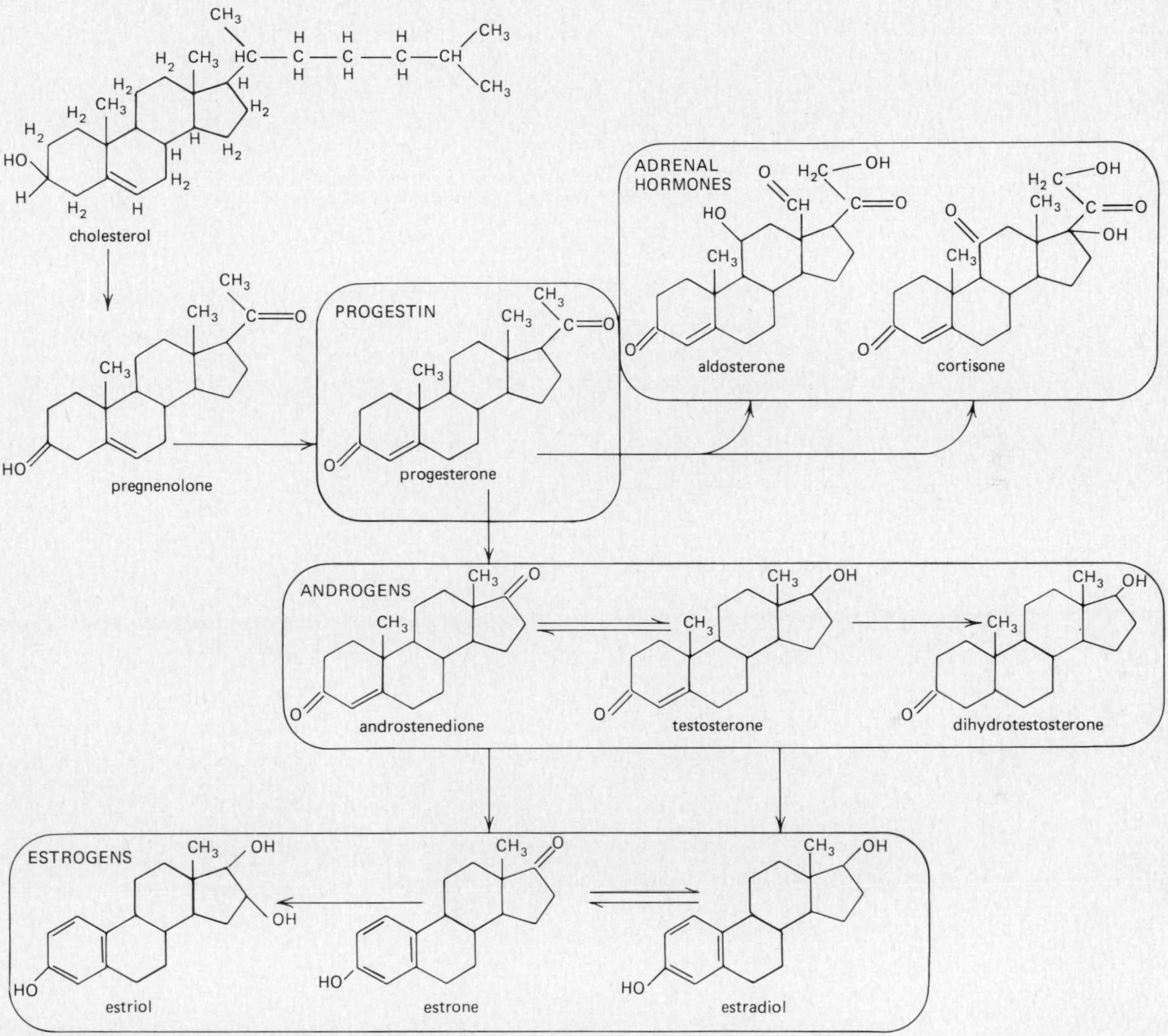

FIGURE 12.4
Structure and function of major steroid hormones. In the abbreviated structures used, each angle is a carbon atom and, except for cholesterol, most hydrogen atoms are omitted. The basic steroid structure consists of three six-carbon rings and one five-carbon ring. A number of intermediate steps have been omitted. Note that progesterone is the precursor to all other steroid hormones and that estrogens are derived entirely from androgens.

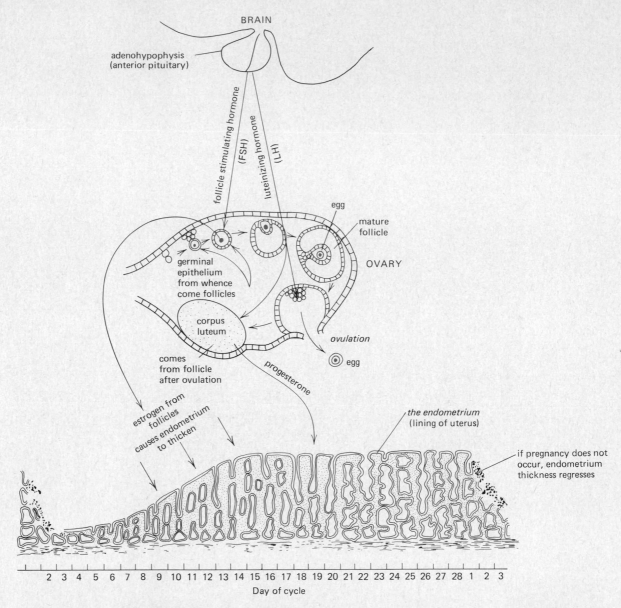

FIGURE 12.5

The human female menstrual cycle. A cross section of the endomentrium of the uterus is illustrated.

progesterone and estrogen. If pregnancy occurs, synthesis of progesterone is eventually taken over by the placenta. If pregnancy does not occur, the corpus luteum degenerates and menstruation occurs. A new cycle of estrogen secretion is induced by FSH.

Embryogenesis of Sex Organs in Humans

Before considering normal and abnormal sexual development in humans, it will be helpful to review the early development of gonads, genitalia, and related organs in mammalian embryos. Early in the development of the embryo, cells are set aside to form the *germ line*, that is, the cells whose destiny it is to give rise to sperm or ova. These primordial germ cells migrate from the yolk sac endoderm to the *genital ridge*, where they join with somatic cells to form the gonads. In the young embryo, the gonad is not readily identifiable as either testis or ovary and is often referred to as an "indifferent" gonad. However, the genotype of the gonad has already determined which it will become.

If a Y chromosome is present (more specifically, the short arm of the Y), the indifferent gonad is *induced* to become a testis. The nature of the inducer is unknown, although there is evidence that it may be a substance known as the H-Y antigen. Once induced, the testis consists of three main cell types: *germ cells*, *Sertoli cells*, and *interstitial* (Leydig) *cells* (Fig. 12.6). The synthesis of testosterone occurs in the interstitial cells in response to FSH from the adenohypophysis.

If a Y chromosome is not present in the early embryo, the gonad is not induced to become a testis and instead becomes an ovary. No ovary inducer has been demonstrated. Rather, ovarian development and hence female phenotype seems

FIGURE 12.6
Cross section of a testis. Mature sperm collect in the seminiferous tubules from whence they pass into the epididymis and are stored until they are ejaculated or until they degenerate.

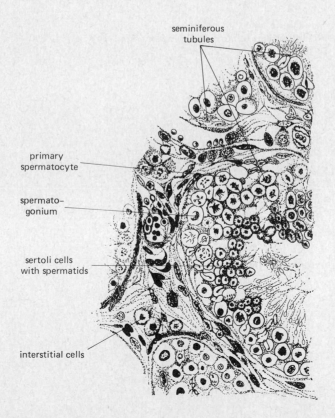

seminiferous
tubules

primary
spermatocyte

spermato-
gonium

sertoli cells
with spermatids

interstitial cells

to be the normal pathway in the absence of male induction. This has been studied in detail in rabbits, and various abnormal sexual states in humans indicate the same pattern.

The embryonic development of external genitalia and internal reproductive organs demonstrates the activities of the various factors that interact during this crucial period. At the stage of the indifferent gonad, there are two main accessory structures of interest: the *Müllerian duct* and the *Wolffian duct* (Fig. 12.7). In normal males, the Müllerian duct regresses as a result of the influence of a substance, the Müllerian inhibitor, secreted by the Sertoli cells in the testes. At the same time, testosterone induces the Wolffian duct to differentiate into the *epididymis, vas deferens,* and *seminal vesicles.* Testosterone also causes the external genitalia to develop into a penis and scrotum (Fig. 12.8). In females, the Müllerian duct differentiates into Fallopian tubes, uterus, and the upper part of the vagina. The Wolffian duct regresses, and the external genitalia acquire the female form (Fig. 12.8). The influence of hormones such as estrogens on these steps is complex and

FIGURE 12.7
Formation of internal sex organs from the primitive Wolffian and Mullerian ducts. Under the influence of secretions from the testes, the Wolffian ducts develop into male structures and the Mullerian ducts regress. In the absence of testicular secretions, the Mullerian structures develop into female organs.

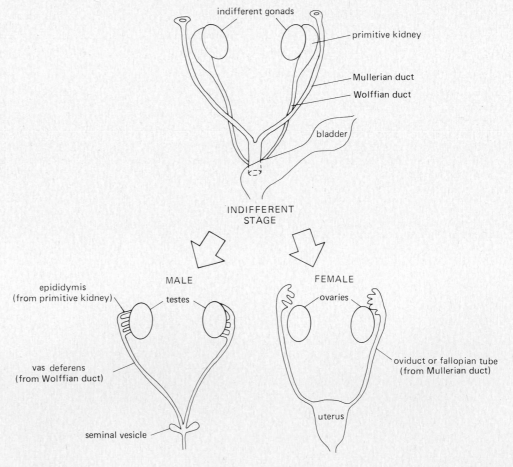

FIGURE 12.8
Differentiation of external sex
organs.

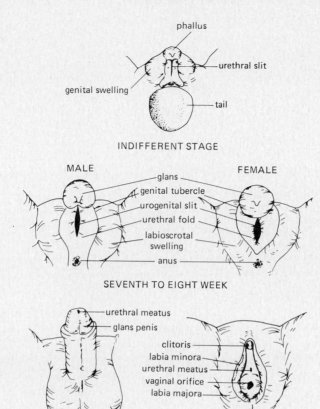

not fully understood. All persons, including males, produce estrogen in the
adrenal cortex so that it is virtually impossible to study development in the
complete absence of estrogen. A summary of the interactions of inducers and
hormones is given in Figure 12.9.

Sex Determination in Humans

The discussion of the normal events in embryogenesis of sexual organs and
secondary sex characteristics is helpful in understanding the various abnormal
conditions in which development has gone awry. Each step in normal development occurs because genes produce enzymes that catalyze specific metabolic

FIGURE 12.9
Summary of sex determi-
nation and major sex hor-
mone effects in humans.

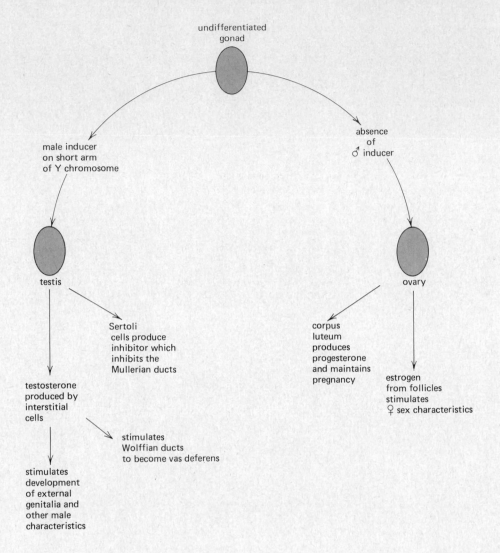

steps in the production of hormones. Or the genes may produce proteins that act
as receptors or in other ways necessary for normal function. Mutation of these
genes causes the corresponding step not to occur normally and in turn causes
abnormal sexual development. We shall consider some of the inherited defects in
sexual development, many of which have contributed to our understanding of
normal development.

It may be useful at this point to review some terminology often used in
describing persons with abnormal sexual development. The term *hermaphrodite*
was invented centuries ago to describe persons who are half male, half female
(from the Greek myth of Hermaphroditus, son of Hermes and Aphrodite). Typically
the hermaphrodite is pictured as being male on one side and female on the other.

Such persons seem to have been reported more often in ancient times than in modern, however. The true hermaphrodite today is defined as a person who has both a testis and an ovary. There may be various degrees of intersexual development. Since the gonads influence development of sexual characteristics by release of hormones into the blood, the genitalia and other secondary sex characteristics tend to be intermediate between male and female and are not separated into clearly masculine on one side and feminine on the other.

True hermaphrodites, as distinguished from pseudohermaphrodites, are known but are very rare. They are thought to arise very early in embryogenesis by nondisjunction of a sex chromosome in a primordial germ cell that gives rise to one of the two gonads. For example, an XY zygote might product a clone of XO cells, from which one of the gonads develops. One would expect a streak ovary, that is, a mass of fibrous tissue such as is found in patients with XO *Turner syndrome* (see page 229), and many true hermaphrodites do have a normal testis on one side plus a streak gonad on the other.

Pseudohermaphrodites are more frequently encountered than true hermaphrodites. Male pseudohermaphrodites are defined as persons who are genetically male with testes but who have a greater or lesser degree of female development. Female pseudohermaphrodites are genetic females, with ovaries, who develop some male characteristics. Many of the conditions to be discussed fall into these categories.

The existence of a variety of sex chromosome constitutions in humans has clarified many of the questions about sex determination. While it is easy to demonstrate that XY persons are male and XX are female, one would like to know how these different karyotypes produce their effect. In *Drosophila*, for example, C. B. Bridges showed in 1925 that the critical factor is the ratio of X chromosomes to sets ot autosomes (Table 12.1). Since humans also have an XX/XY sex determination, the same relation might be expected in them, but this turns out not to be so.

Sex in Drosophila *is not dependent on the Y chromosome, although the Y influences fertility. Rather, the ratio of X chromosomes to autosomes determine whether a fly is male or female. An XXY fly is a female, an XO is male. This is different from mammals, where the presence or absence of the Y is the determining factor.*

**TABLE 12.1
Determination of Sex in Drosophila**

Phenotype	Number of X Chromosomes	Number of Haploid Sets of Autosomes
Female (sterile)	3	2
Female (normal)	4	4
Female (normal)	3	3
Female (normal)	2	2
Intersex	2	3
Male (normal)	1	2
Male (sterile)	1	3

Drosophila data from C. B. Bridges, Amer. Natural. 59:127, 1925.

Klinefelter Syndrome

The first indications that the mechanism of human sex determination differs from that of *Drosophila* were reported in 1959 in studies of two disorders, Klinefelter syndrome and Turner syndrome. Persons with Klinefelter syndrome are males who are taller than usual and somewhat eunochoid in body form, that is, they may have some breast development (gynecomastia), small genitalia, and sparse body hair (Fig. 12.10). They need not appear very different from normal males, but they are sterile. They are positive for sex chromatin (Box 12.1), believed at the time (and subsequently proved) to indicate the presence of two X chromosomes. It was thought that they might be chromosomally XX females who, through some error, had developed as males. The ability to do human karyotype analysis was developed in the later 1950s, and Klinefelter patients were shown in fact to have 47 chromosomes consisting of the usual 22 pairs of autosomes plus two X chromosomes and a Y. Their karyotype formula is therefore 47,XXY. These results were clearly different from *Drosophila*, in which XXY flies are female.

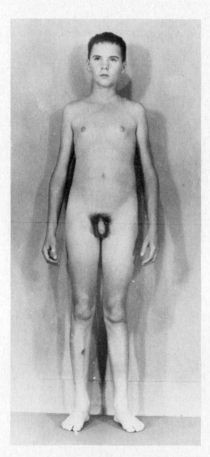

FIGURE 12.10
A person with Klinefelter syndrome. This XXY male has the tall stature and somewhat eunochoid build typical of Klinefelter syndrome. In addition, he has some breast enlargement (gynecomastia) often found in this syndome. (Courtesy of the March of Dimes Birth Defects Foundation.)

Box **12.1**

Sex Chromatin

In 1949, M. L. Barr and E. G. Bertram discovered a difference between nuclei of nerve cells from male and female cats. Female interphase nuclei often had a darkly staining body at the edge of the nucleus. Male nuclei rarely had such a body. This observation was extended to other species and other cell types and found to be a general rule for mammals. These darkly staining bodies, called *Barr bodies* or *sex chromatin*, are found in 50–80% of cells from females but only in 5% of cells from males. In fact, the male nuclei scored as positive may be mostly artifacts, due to prominent nucleoli that happen to lie at the periphery of the nucleus in a particular cell. A typical sex-chromatin-positive nucleus is shown in the illustration.

A variety of cell types can be used to assess sex chromatin. The most common method is to prepare a buccal smear by scraping the inside of the cheek vigorously with a tongue depressor and spreading out the cells scraped off onto a microscope slide. After staining with any of several nuclear stains, the cells can be scored by microscopic examination. A person with 30% or more positive cells is classified as sex chromatin positive. One with 5% or fewer positive cells is classified as sex chromatin negative. Intermediate values are found in persons who are mosaic XX/XY.

The number of sex chromatin bodies is always one less than the number of X chromosomes. Thus normal males have none and females one. But XXY males have one, XXXY males and XXX females have two, XXXXY males and XXXX females have three, and so forth.

The sex chromatin body is now known to be the inactive X chromosome (see page 117). According to the Lyon hypothesis, a cell has only one fully active X chromosome. Any additional are inactive. The inactive X chromosomes appear to condense near the nuclear membrane and are more readily stained in interphase than the active dispersed X chromosomes.

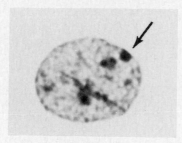

(Photo supplied by Angelyn Smith, The Austin State Hospital.)

Turner Syndrome

Persons with this disorder are females, of short stature, with streak gonads (presumably ovaries that have degenerated), webbed necks, and problems in certain types of spatial memory (Fig. 12.11). They are sex chromatin negative, and the frequency of X-linked colorblindness is similar to males rather than females. It

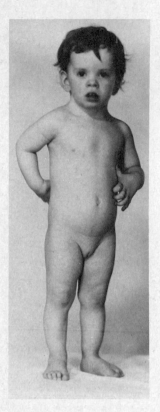
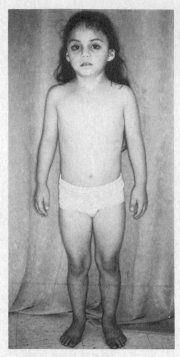

FIGURE 12.11
Patients with Turner syndrome. Note the short stature and the webbing of the neck.
(Left: courtesy of the March of Dimes Birth Defects Foundation. Center and right:
courtesy of Dr. Jan Friedman, The University of Texas Health Science Center at Dallas.)

was thought therefore that they might be derived from XY zygotes that had
developed as females. Again the karyotype analysis proved otherwise. They have
only 45 chromosomes, with one X and no Y. Their karyotype formula is 45,XO,
where the O indicates absence of an expected sex chromosome. An XO *Drosophila*
is definitely a morphological male, although sterile.

The Y Chromosome and Sex Determination

The combined results from Klinefelter and Turner syndromes indicate that the
critical factor in sex determination is presence or absence of a Y chromosome. A
zygote with a Y develops as a male, regardless of the number of X chromosomes.
Variants of Klinefelter syndrome have karyotypes of 48,XXXY and 49,XXXXY. Although the expression of the abnormalities is greater in these variants, the persons
unquestionably are male. On the other hand, a person without a Y always develops
as a female.

The existence of persons who have lost various portions of the Y chromosome enables one to locate the male-determining region. A number of males have been observed who have deletions of virtually all of the long arm of the Y. They are normal males with normal fertility. Apparently the long arm is largely if not entirely dispensable. This is in keeping with the fact that it is largely composed of heterochromatin and therefore is not expected to have many genes. In contrast, persons whose Y chromosome has lost the short arm are female. Some genetic element on the Y short arm is the key inducer of maleness.

The nature of the male inducer is not known. Evidence suggests that it may be the H-Y antigen (see Box 12.2). The Y-linked inducer may also be a regulatory gene whose product "turns on" the H-Y structural gene, which need not be on the Y chromosome. Whatever the correct situation may be, presence of the short arm of the Y induces male development; its absence permits female development.

—— Box 12.2

The H-Y Antigen

Vertebrates respond to foreign materials by making *antibodies*. The foreign materials (*antigens*) may be invasions of bacteria or viruses, injected chemicals, especially proteins, or transplanted tissues and organs. The antibodies are large protein molecules that combine very specifically with the particular foreign antigen that induced them (Chapter 15). If the antigen is a live cell, the combination with the antibody helps destroy the cell. This is the principal way in which we fight off bacterial infections.

Antibodies can discriminate between very closely related cells and proteins. Thus the relatively small genetic differences among members of a single species, such as *Homo sapiens*, are sufficient to invoke antibodies if blood or other tissues are injected or transplanted from one person to another without prior matching of antigenic types. The only completely safe transplant is between identical twins, since such twins have the same genotype.

Very highly inbred strains of mice, so-called *isogenic* strains, can, in general, exchange skin grafts without difficulty because the animals are genetically identical. There is one exception. Males have a Y chromosome and females do not. Therefore, if any genes on the Y chromosome produce antigens, they would be "foreign" to the females. This is tested by skin grafts. Small bits of skin from males are grafted onto females and vice versa. The female-to-male grafts are not rejected and are accepted permanently. The male-to-female grafts are eventually rejected—an indication that male cells have antigens that the females do not have.

Antiserum from females that have been *immunized* against male tissue has antibodies against the male antigen. This antigen is called H-Y antigen (H because a histocompability test was used to detect it, Y because it is dependent on the Y chromosome). The H-Y antiserum can now be used to detect the presence of H-Y antigens on other cells, since the antibodies will bind to H-Y positive cells and

Box **12.2**

(*continued*)

to no others. Several tests are used, based on the ability of H-Y antiserum to bind to cells and kill them. The H-Y antigen is found on a great variety of cells in the body. Furthermore, antisera against mouse H-Y reacts with H-Y positive cells from all vertebrates so far tested, including humans. Therefore one can test for human H-Y using mouse antisera rather than having to do skin transplants from human males to nonisogenic females, transplants that would be rejected promptly because of many other antigenic differences (Chapter 15).

The H-Y antigen has been helpful in understanding the sex chromsome status in various human sexual disorders. Persons with testicular feminiza-

tion are H-Y positive, as might be expected since they have a Y chromosome. Many XX males are H-Y positive, suggesting that a small piece of Y chromosome, containing both the male inducer and the H-Y gene, has been translocated to another chromosome not recognizable microscopically.

The H-Y gene, which may be either structural or regulatory, is located on the short arm of the Y chromosome. It is very close to and may be identical to the gene that induces the undifferentiated gonad to become a testis. Should this prove to be true, it will be an enormous gain in understanding the very early steps in sex determination.

Abnormal Sex Development in Humans

The many steps in sexual development are each to some extent under genetic control. These genes may mutate and produce modified or total loss of a particular step. The result often is interference with normal development. The following are a few examples, fortunately rare, of errors in sex development. Some are clearly genetic. Others appear to be exceptions to the rules we have derived. They require further study for complete understanding.

Testicular Feminization

This condition, first described about 50 years ago in humans and more recently in mice, is an example of male pseudohermaphroditism. It is due to a mutant gene on the X chromosome. The product of the normal gene is the androgen receptor in the cytoplasm of cells. Testosterone released by the testes is transported to tissues all over the body, both during embryogenesis and throughout the life of males. It reacts with certain "target organs," such as the Wolffian ducts (Fig. 12.7), causing them to develop or to exhibit male behavior. To do this, the testosterone binds to a specific androgen receptor in the cytoplasm of target cells and is transported to the nucleus, where it regulates the activity of various genes. Females also have androgen receptors, as shown by the masculinizing effect of injected testosterone or excess testosterone synthesized by some adrenal tumors.

The androgen receptor is a protein, the product of a structural gene on the X chromosome. If the gene mutates, the androgen receptor may no longer be present, or it may no longer bind testosterone. Or it may bind and transport testosterone less effectively. If the receptor is completely ineffective, the target organs cannot respond to testosterone. This means that during embryogenesis, the Wolffian ducts do not differentiate into male structures, and other organ changes characteristic of males do not occur. Affected persons are genetically XY with testes (usually located abdominally) that produce normal levels of testosterone. Externally they develop as females, with normal breast development and normal female genitalia. The vagina is short and ends blindly, and axillary and pubic hair is scant. Internally there are no Fallopian tubes or uterus. Affected persons typically come to medical attention because menstruation does not start.

If one refers to Figure 12.8, the complex of changes can be understood. Those events that depend on testosterone function cannot occur. On the other hand, the testes also secrete Müllerian inhibitor, which prevents the development of Fallopian tubes, uterus, and upper vagina. Some mutants produce androgen receptor that is only partially defective. Males with these mutations show various degrees of abnormal sexual development; some appear as masculinized females, others with more ambiguous external genitalia and other secondary sex characteristics.

Since the androgen-receptor gene is on the X chromosome, it should be transmitted like any other X-linked trait with an important exception: Persons who are XY and who inherit the mutant allele are sterile and cannot transmit it. XX females who have one X chromosome with a mutant allele are heterozygous and are completely normal. For many years, it could not be decided whether testicular feminization is an X-linked recessive producing sterility in affected males or whether it is an autosomal dominant expressed only in males, who would be sterile. Either explanation is consistent with a pedigree (Fig. 12.12). The issue was finally settled in mice, where the same gene is known to be very close to an X-linked coat color gene.

Persistent Müllerian Duct Syndrome

Müllerian inhibitor is a protein and, therefore, the product of a structural gene. This gene can mutate to a form producing no inhibitor or a defective inactive inhibitor. The tissues that normally respond to inhibitor may also lose their ability to respond because of mutation. It is therefore not surprising to find persons who are deficient in Müllerian inhibition. It would be expressed only in males, since only testes produce Müllerian inhibitor. Affected persons have the expected phenotype. They are essentially normal males with Wolffian duct derivatives but with undescended testes. Most significantly, they also have Fallopian tubes and a uterus in the abdomen. Inheritance seems to be as an autosomal recessive, but information is scanty.

FIGURE 12.12
Pedigree of testicular
feminization showing typ-
ical X-linked recessive inheritance. (From G.
Pettersson and G. Bon-
nier, *Hereditas* 23: 49,
1937.)

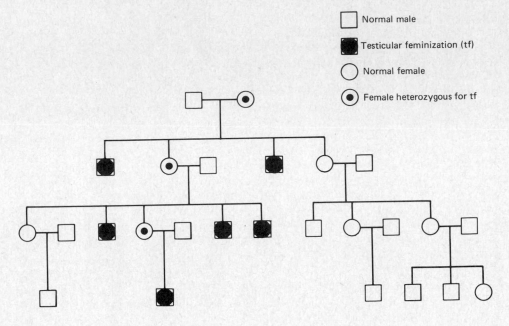

XX Males

Approximately 1 in 10,000 males has a karyotype of a normal 46,XX female. The affected persons somewhat resemble Klinefelter males, with small testes, increased height, and sterility. Most cases are isolated (occur in families with no other affected related persons known), but a few have been observed to cluster in families, suggesting that some of the cases at least have a genetic basis. If most cases were due to a dominant gene that produces sterility, only occasionally would one see familial clustering. This condition may be related to one in mice known as *sex reversal*. A dominant mutation at the autosomal *Sxr* locus causes XX female mice to develop testes, but spermatogenesis does not occur. There may be other causes also for the development of XX males, including the possibility that some are mosaic XX/XXY, with the Y chromosome-bearing cells involved in gonad formation but not in formation of blood cells and other easily karyotyped tissues.

XY Females

These rare persons have streak gonads rather than testes and they have Müllerian development, including Fallopian tubes and complete vagina. Breast development is absent or minimal. On chromosome analysis, they have a normal male 46,XY karyotype. Some families show transmission of the trait consistent with an X-linked gene expressed in hemizygous males (the XY females) but not in heterozygous XX females. There is no indication yet where the primary genetic defect lies.

XYY Males

Approximately 1 per 1000 males born has two Y chromosomes with a karyotype of 47,XYY. They arise because of nondisjunction of the Y chromosome in the second meiotic division, which produces a sperm with two Y chromosomes. Much attention has been given to these XYY males, since the first reports strongly suggested that this karyotype is associated with criminal behavior. It is now known that many XYY males do not exhibit antisocial behavior, although there does seem to be an excess of XYY males who do. The behavioral aspects of this condition will be discussed further in a later chapter. XYY males are taller than their XY sibs but othewise are not physically distinctive. They are fertile. With three sex chromosomes, one would expect a high frequency of XY and YY sperm, producing trisomic offspring. However, the offspring of XYY males do not have higher than usual frequencies of XXY and XYY chromosome complements.

XXX Females

Triplo-X females arise through nondisjunction of the X chromosomes in a female to give an XX ovum, which is fertilized by an X-bearing sperm. Or they could arise by nondisjunction in the second meiotic division of a male to give rise to XX sperm. XXX females were discovered in *Drosophila* many years ago and were given the name *superfemales*, although they are sterile rather than being super. Human XXX females, however, are normal, fertile females without other distinguishing characteristics. One might expect half the ova from a triplo-X female to have two X chromosomes rather than one. However, the offspring of XXX females do not have an elevated frequency of XXX and XXY offspring.

Females with four and five X chromosomes have also been observed. They are mentally retarded but sexual development is not grossly abnormal. The normality of females with extra X chromosomes is consistent with the observation that only one X chromosome is fully active in a cell. The number of sex chromatin or Barr bodies also is one fewer than the number of X chromosomes in all cases. Triplo-X females have two, tetra X females have three, and so forth.

Aberrant Sex Phenotypes Due to Metabolic Blocks

Reference to Figure 12.4 will indicate some of the metabolic steps involved in the biosynthesis of steroid hormones. In fact, many of the intermediate steps have been omitted. Each step is catalyzed by an enzyme, and each is therefore subject to failure because of mutation. As a consequence, a number of metabolic blocks occur in the biosynthesis of steroids. Most of these also influence the production of the steroid hormones of the adrenal cortex, hormones that are crucial in maintaining salt balance and in regulating the metabolism of many tissues. The abnormal sexual development is therefore only a part of the problem. The defects in development depend on the particular block involved. The results typically are some degree of intersexuality. Many of these conditions are inherited as simple recessive traits.

Sex and Behavior

Males and females do, on the average, behave differently. How much of the difference is inborn and how much is learned is difficult to prove. In other primate species, males are more aggressive than females. By analogy, the greater aggressiveness of human males very likely has a biological basis. This does not have any implications for differences among males or among females.

The initiation of sexual activity is clearly dependent on secretion of hormones at the time of puberty. The amount of sexual activity in male experimental animals and presumably in human males also is in part a function of the level of testosterone. But in humans, one would expect many other factors to be just as important, provided the hormonal balance is adequate for sexual function.

One question that often arises is the extent to which homosexuality is inborn. More precisely stated, do some persons have genes that predispose them to be homosexual? Despite many efforts to demonstrate differences, no differences in sex hormones or in any other aspect of metabolism have ever been found to distinguish homosexuals and heterosexuals. There is also a compelling genetic argument against expecting such differences. If there were genes that favor homosexuality, they should be associated with lower than normal numbers of offspring. Therefore, natural selection should keep such genes at very low frequencies. Arguments can be made to circumvent this, but they are somewhat contrived. It seems better to assume that homosexuality is entirely environmental until there is solid evidence to the contrary.

Review Questions

1. What is the biological definition of sex?
2. How does sexuality differ from sex?
3. What is the difference between sexual heteromorphism and mating type?
4. What is meant by sex determination?
5. How do the sex-determining mechanisms differ in humans and Drosophila?
6. The human male is heterogametic. What does this mean?
7. What seems to be the principal factor that determines sex in humans?
8. Persons with Klinefelter syndrome are sex-chromatin positive like normal females. Nevertheless, they are males. Why?
9. What is the basis for the disparity between sex-chromatin status and sex phenotype of persons with Turner syndrome?
10. By what mechanism(s) do Turner and Klinefelter syndromes arise?
11. What is a true hermaphrodite? What is the most likely origin?
12. Describe and contrast the conditions "testicular feminization" and "Mullerian duct syndrome."
13. Identify the sex of the individuals with the following sex chromosome constitutions: X, XXX, XYY, XXY, XXXX.
14. The normal human female is a functional mosaic. What does this mean and why does it occur?

References and Further Reading

Bridges, C. B. 1925. Sex in relation to chromosomes and genes. *Amer. Naturalist* 59:127–137. Historically an important paper showing the relationship of genes and sex in *Drosophila*.

O'Flaherty, W. D. 1980. *Women, Androgynes and Other Mythical Beasts.* Univ. of Chicago Press, Chicago. An interesting exposition of the role of sexual metaphors and animal symbols in the development of religious concepts.

Maynard Smith, J. 1978. *The Evolution of Sex.* Cambridge Univ. Press, Cambridge. 222 pp. A theoretical treatment of why we have sex.

Naftolin, F. (ed.). 1981. Understanding the bases of sex differences. *Science* 211:1263–1324. An excellent series of technical reviews on the current knowledge of sex determination and sex influence on development in people.

Simpson, J. L. 1977. *Disorders of Sexual Differentiation.* Academic Press, New York. 466 pp. A comprehensive review of the biological errors that interfere with normal sexual development.

Vallet, H. L., and I. H. Porter (eds.). 1979. *Genetic Mechanisms of Sexual Development.* Academic Press, New York. 497 pp. A series of articles by outstanding authorities on normal and abnormal development.

Williams, G. C. 1975. *Sex and Evolution.* Princeton Univ. Press, Princeton. 200 pp. A discussion of the importance of sex, or lack of it, in the evolution of various types of organisms.

In the nine month period after fertilization of a human egg, the zygote grows into the billions of cells that are a human being. We know relatively little about this fascinating process which we call *embryogeny*, the development of the embryo. We can ask questions about it though, and that is the first step in finding out what is going on. Obvious questions are ones such as these: Why does a human ovum develop into a human being? What guides it? To this central question we may add subsidiary ones such as: What factors operate to cause cells to differentiate into liver cells, nerve cells, retinal cells, skin cells, and so forth? What guides these differentiated cells to organize into organs? What determines that an organ or a part or even the whole should grow to a certain size and then cease growth?

These are just a few of the questions that come to mind. None of them has a satisfactory definitive answer. But we can make a general statement with assurance of its essential correctness. The genes in the fertilized egg donated by the mother and the father are the chief directors and administrators of the development processes. We know relatively little about the process except that a central role must be played by what we call *regulatory genes*. These genetic elements control the activities of *structural genes* that specify the structure of proteins with specific functions, such as enzymes, hemoglobins, collagens, and histones.

The study of development from a fertilized egg is a study of gene action. Therefore this chapter will be devoted primarily to that topic. First, however, we need to take a broad view of some of the events that go on during the nine month gestation period when a single cell becomes a human being.

Embryogeny of a Human

The first event is fertilization of an egg by a sperm in the oviduct or Fallopian tube of the mother. This is called *conception*. As shown in Figure 13.1, the fertilized ovum proceeds down the tube into the uterus. During this procession, it begins to divide and by the fourth day, having reached the uterus, it is now at the *blastocyst* stage, consisting of about 100 cells. Now follow the further processes by referring to Box 13.1.

Shortly after arriving in the uterus, the blastocyst begins to *implant* in the *endometrium*, or the lining of the uterus. This implantation consists of the blastocyst sinking into the endometrium and beginning to form the *extraembryonic membranes*. These membranes form the *amniotic cavity*, within which the embryo develops in a watery environment, and the *yolk sac* and *placental membranes*. Together these constitute the extraembryonic membranes.

The placental membranes interdigitate into the uterine wall and the blood vessels of the mother that are within the uterine wall. This forms the *placenta*, a combination of embryonic and maternal tissues. As a result, soluble food, such as amino acids and sugars, and oxygen, vitamins, and hormones may diffuse into the extraembryonic vessels. The blood from these now enters the embryo via the *umbilical cord* and feeds the developing embryo, which has no other significant source of nutrient. This process is in contrast to that of the chick embryo, which

13

Human Development and Gene Action

develops in a closed egg using the highly nutritious yolk with which the egg is endowed. The human embryo develops a yolk sac too, but its functions are not primarily to provide nutrition, since the human egg contains almost no yolk.

As the embryo develops, it produces waste products of the same sort that we adults do. It uses oxygen in respiration and therefore forms CO_2 and H_2O. It breaks down amino acids, purines, and pyrimidines and forms nitrogenous waste products. These are eliminated through the placenta and disposed of by the mother via her lungs and kidneys.

Box 13.1

A Brief Log of Events in Human Development

General Stage	Approximate Time in Days from Fertilization	Events
Ovum	0	Fertilization in Fallopian tube.
	3	Cleavage and blastocyst formation in Fallopian tube.
	4–7	Implantation in uterine wall and formation of amniotic cavity.
Early embryo	7–21	Placental membranes develop; cell differentiation is underway, leading to the formation of the primary embryonic tissues—the mesoderm, endoderm, and ectoderm.

Box **13.1**

(continued)

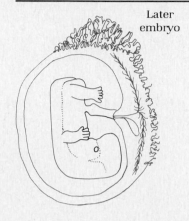

| Later embryo | 22–56 | Embryo grows very rapidly from about 2 mm in length to about 40 mm at 56 days.
During this period all major organ systems are laid down. By the end of it the embryo has started to assume recognizable human form: The heart is functioning and pumping blood into and out of the placenta, and limbs and fingers are evident. |

| Fetus | 57–266 | During the fetal period, the developing baby grows in crown-rump length from about 40 to 360 mm and its weight increases from about 5 to 3400 g.
The organ systems reach their functional state ready to assume their full responsibilities at the time of birth when the connection between the mother and infant is severed. The placenta and other extraembryonic membranes are passed out of the uterus as the afterbirth. |

The two week period following the beginnings of implantation is one in which the three major primordial embryonic tissues—the *mesoderm*, *endoderm*, and *ectoderm*—are differentiated within the mass of cells that form the embryo (Fig. 13.2).

The mesoderm gives rise to muscle, connective tissue, cartilage, bone, the entire circulatory system, the kidney tissue, and various other things (Fig. 13.3). The endoderm goes into the formation of the digestive system, liver, certain glands, and the lining of the lungs. The ectoderm provides the entire nervous system, includ-

FIGURE 13.1
Fertilization of the human egg and formation of the blastocyst. At the time of release from the ovary, the egg has not completed oogenesis. It has completed only the first meiotic division and has one polar body attached to it (see Fig. 5.8). Upon fertilization in the oviduct, the second meiotic division is completed, and the haploid egg nucleus and haploid sperm nucleus fuse. Technically this is the moment of conception, because the nucleus of the offspring is formed at this time, not at the time of fertilization.

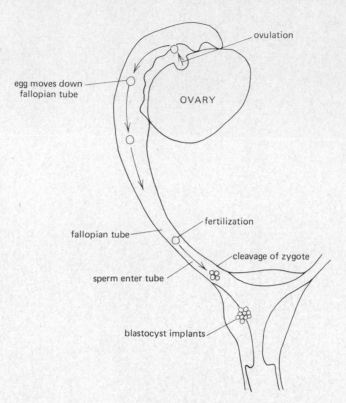

FIGURE 13.2
Diagram of an early human embryo and its extraembryonic membranes.

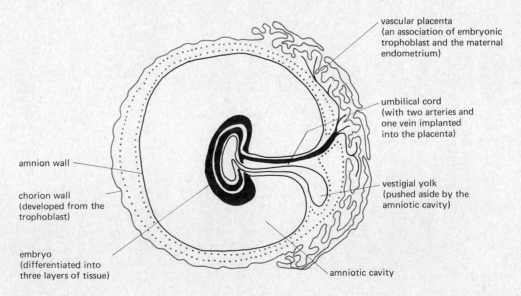

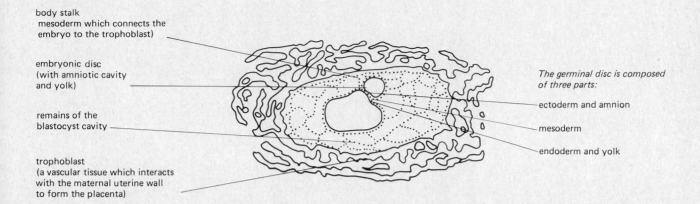

body stalk
 mesoderm which connects the
 embryo to the trophoblast)

embryonic disc
(with amniotic cavity
and yolk)

remains of the
blastocyst cavity

trophoblast
(a vascular tissue which interacts
with the maternal uterine wall
to form the placenta)

The germinal disc is composed
of three parts:

ectoderm and amnion

mesoderm

endoderm and yolk

TROPHODERM
forms the placental tissue (trophoblast together
with the maternal endometrium).

ECTODERM
forms the amnion and umbilical cord, skin,
perspiration glands, hair, mouth, nasal cavity,
sense organs, mammary glands, neuroglial cells,
eye neurons, brain cells, and other neurons.

MESODERM
forms the vascular core of the umbilical cord, red
and white blood cells, cardiac, striated, and smooth
muscle cells, bone, cartilage, kidney, fat, genital
ducts, blood vessels, body cavities, mesenchyme cells.

ENDODERM
forms the vestigial yolk, endocrine glands, lining of
the digestive tract, liver, vagina, lung, and throat.

(Not shown: germinal cells which give rise to
gametes in the gonads.)

FIGURE 13.3
The primordial embryonic tissues and their derivatives.

ing the brain, and from it come our skin, nails, hair, outer layer of teeth, the mucous membranes of the mouth, and parts of our primary sense organs. During this early embryonic period, we don't look like a human, but the stage is set for the later embryonic period.

The fourth to the eighth weeks of development (ca. 22–56 days after fertilization) is a period of organogenesis. From the cells that have begun to differentiate in the first two weeks, tissues are formed that begin to organize into functional organs. The heart and its connecting blood vessels fill with blood and begin to operate as a circulatory system that transports nutrients to the embryo and eliminates waste products. By the end of this late embryonic period, approximately two months after conception, the embryo has begun to assume the appearance of a human.

The beginnings of the fetal period is arbitrarily set at about 57 days. The embryo is then called a *fetus* because of its advanced stage of development. Its organs, muscles, bones, and other parts are mostly well developed by this time, but many other subtle changes are taking place in the developing nervous and circulatory systems, for example. By the fourth month the fetus is beginning to move and the mother may feel its movements. It is growing at a fast rate and increases in length from about 5 or 6 in. to the 18–19 in. it will generally achieve by the time of birth, roughly 266 days later.

With this rather superficial and rapid survey of events occurring between fertilization and birth in mind, let us consider certain aspects in more detail in order to illustrate gene action in the developmental process.

Gene Action in Human Development

As we have mentioned on several occasions in the previous chapters, development in a eukaryotic organism such as a human is characterized by the action and nonaction of sets of genes over the period of time starting with the zygote and continuing right up to the time of birth. At any one time, out of the many thousands of genes in each set of chromosomes, only a few percent may be active in transcribing RNA. Sets of genes active at one time may be different from sets active at another. Also, as differentiation proceeds, it becomes obvious that one kind of differentiated cell has a different complement of active genes than another kind. Of course, some genes may be active at all times. These would be those responsible for transcribing ribosomal RNA and the ribosomal proteins, tRNA, and enzymes essential to all cells, such as enzymes necessary for glycolysis and respiration. But actually, the number of these genes is definitely in the minority. An analogous situation might be a symphony orchestra. Seldom, for most symphonic works, do all instruments play at once. The individual instruments play or do not play, depending on the score, to give the symphonic sound of the participating musicians. So it is in a developing animal in which the genes are the musicians and the creature is the composition. The score is in the total genome.

Hemoglobins in Development

The blood of humans and other vertebrates is probably their most studied part simply because it is the most easily accessible. Giving a pint of blood is a simple procedure, and a chemist can make thousands of analyses with a sample as large as that. As a result, we know a great deal about the blood of humans and other vertebrates. We especially know a great deal about the hemoglobins that constitute the bulk of the content of our red blood corpuscles.

We have already alluded to the fact that in the human adult the primary hemoglobin is a tetramer composed of two α globin chains and two β globin chains. Therefore we designate it as $\alpha_2\beta_2$ or Hb A. A small fraction (about 2.5%) of our adult hemoglobin is Hb A_2, or $\alpha_2\delta_2$. It has a δ-globin in place of the β-globin of Hb A. However, if we examine the blood of fetuses and embryos, we find a significant difference from the blood of adults. The components of the red blood corpuscles for the different periods are outlined in Figure 13.4, which also gives the time of appearance and disappearance of the different kinds of hemoglobin.

First, it should be noted that at least seven genes, ζ, ϵ, $^G\gamma$, $^A\gamma$, δ, β, and α are

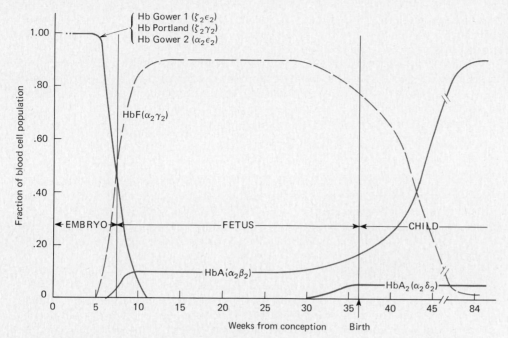

FIGURE 13.4
The hemoglobins of the blood at different stages of human development.

required to code for the known seven kinds of polypeptide chains. The linkage relations of these are now known. The genes ϵ, $^G\gamma$, $^A\gamma$, δ, and β are all on chromosome 11, closely linked together and apparently in that order. The ζ and α genes are not linked to the other group. The α gene is on chromosome 16 and is generally found to be present in duplicate. The ζ gene is linked to it and has a closely related structure.

The essential fact that we wish to emphasize here is that all seven kinds of Hb polypeptides do not appear simultaneously. The ζ and ϵ chains appear first in the very earliest embryos analyzed. These chains may be formed in the first embryonic blood-forming or hematopoietic center, the yolk sac (Box 13.1). By about the fifth week, the α and γ chains begin to form, probably in the developing liver. The result is that the late embryo has a mixture of four kinds of hemoglobin, $\zeta_2\epsilon_2$, $\zeta_2\gamma_2$, $\alpha_2\epsilon_2$, and $\alpha_2\gamma_2$. The last one to appear is called fetal hemoglobin, Hb F, and it becomes the dominant hemoglobin throughout the life of the fetus, while the other three forms disappear early in fetal life as shown in Figure 13.4. About the time of the beginning of fetal existence in the eighth week, another chain, designated β, appears. It maintains a low level in the form of Hb A $(\alpha_2\beta_2)$ throughout fetal life, during which time the Hb F form containing the γ chain constitutes about 90% of the total hemoglobin. Shortly before birth, Hb F begins to decline and Hb A increases. Concommitantly, a new chain, δ, begins to appear. After birth, the fetal form declines rapidly and is replaced almost entirely by Hb A and Hb A_2 during the first year of life after birth.

During the early fetal period, the liver and spleen are the main hematopoietic centers, to be followed by the bone marrow as it forms in the bones of the fetus. By the time of birth, the liver and spleen lose their ability to make red blood cells in humans, and this activity is carried on only in the red bone marrow of the long bones of arms and legs, the sternum, and the cranium.

Here in this panorama of hemoglobin, we have a most instructive example of gene action in the human animal. It is obvious that the ζ and ϵ genes become active in the cells of the yolk sac to transcribe mRNA about the second week. Shortly thereafter, the α and γ genes are turned on, probably in the primordial liver. Notice that there are two factors here.

1. Certain genes and not others are turned on.
2. The genes are not turned on in all cells but only in certain cells.

At about the same time in the late embryo, the β gene becomes active. Meanwhile the ζ and ϵ genes are turned off. Late in fetal life, the δ gene is turned on to remain on all through life along with the α and β genes. By the end of the first year after birth, the γ genes are only barely active. Hb F $(\alpha_2\gamma_2)$ constitutes only about 1% or less of adult hemoglobin except in certain pathological conditions we will discuss later.

These events are part of the process of differentiation in the developing human. The turning on and turning off of these Hb genes is but one example of similar events among the genes controlling all the other activities that go on in the embryo and fetus. Obviously the whole process of development is a series of coordinated processes under regulatory control. Some factors operate to regulate the activities of genes that transcribe RNA to form active proteins. What these regulatory factors are we do not know. But they must operate in a rather rigid time frame. If genes are turned on or off at the wrong time relative to other genes, the whole developmental process can be thrown into chaos, resulting in the death of the embryo or fetus or the birth of a child with defects. These defects may be mild or they may be very serious.

Upsets in the timing of regulation may be caused by genetic factors or by environmental factors, such as chemicals ingested by the mother and passed on to the embryo via the placenta. We will discuss these factors presently, but first let us consider some other examples of gene action that influence formation of the body parts rather than specific molecules.

Cleft Palate

About 0.04% of the general population has cleft palate. This high incidence makes it one of the most common *congenital defects* in humans. By congenital defect, we mean a defect that appears during embryogeny and, if the embryo survives to birth, is present at birth. It may result from genes inherited from one's parents, certain environmental factors, or both interacting to produce abnormal development at some stage during development. Hundreds of different kinds of congenital defects are recognized. We have chosen cleft palate as an example because it is so common and has been studied extensively.

During the development of the human head, two "shelves" of tissue, called the palatal shelves, form in the area of the mouth cavity. One forms on each side of the head between the primordial nasal and mouth cavities as shown in Figure 13.5.

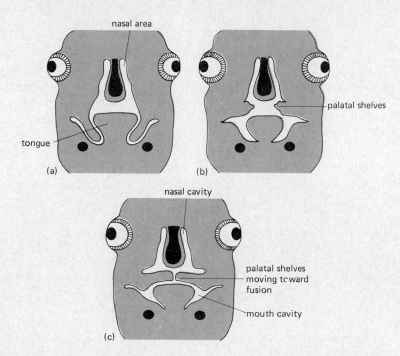

FIGURE 13.5
The formation of the palate by fusion of the palatal shelves. (Redrawn from L. Saxén and J. Rapola, *Congenital Defects*, Holt, Rinehart and Winston, 1969.)

(See also Fig. 11.4.) They grow together and finally form the palate or roof of the mouth. Cleft palate is a condition in which complete fusion of the palatal shelves does not occur. Fortunately, the condition can be corrected by surgery, and the affected persons then can lead normal lives.

Although the frequency in the population as a whole is 0.04%, it is even higher in certain families. For example, if either parent has cleft palate and one child has already been born with cleft palate, the probability that the next child will have cleft palate is very high, 15%. Genetic factors are clearly involved, but so are environmental factors. It is known that hydrocortisone, large doses of vitamins A and K, and X rays predispose embryos of certain strains of mice but not others to develop cleft palate. Whether this is true also for the human embryo is not known, however. Another factor that may be important is the sex of the individual. Sixty percent of the affected newborn humans are males, a fact that indicates a hormonal influence.

Modification of Development of an Embryo by Gene Action and External Agents

It is estimated that 50–80% of human fertilized eggs are aborted at some stage during pregnancy. Many of them are aborted before the potential mother knows she is pregnant, and these conceptions, of course, escape detection. Therefore, these estimates may be too low. Of the recognized pregnancies, as many as 15% may abort during the second or third trimester or be stillborn. Obviously there is a rocky road to be traveled between conception and birth. Those of us who have made it are the lucky ones, except for those born with severe congenital defects that cannot be repaired and that render the individual incapable of living a reasonably satisfactory life as a member of society.

Factors Involved in Spontaneous Abortions, Stillbirths, and Live Births with Defects

Both genetic and environmental factors are involved in the induction of defects in a developing embryo and fetus. Depending on the character of the defect, one will generally be more significant in its influence than the other. Some defects will result in the death of the developing child before birth, others are manifested in a live child at birth or may become expressed several months or years after birth. The study of congenital defects is technically called *teratology*. It has become a highly important field of study engaging the efforts of geneticists, toxicologists, pharmacologists, and research physicians. We shall first here consider the genetic factors involved in the origin of defects while bearing in mind that environmental factors always also play some role to a greater or lesser extent in the expression of the defective phenotype. (Actually we are all born with defects, but the so-called "normal" infant's defects do not show up ordinarily until some time between the ages of 50 and 100 years, when they result in a terminal episode called death.)

Genetic Teratology.   We can best approach a discussion of this complex topic by reminding the reader that two general types of inherited alterations exist in genotypes: gene mutations and chromosomal mutations or alterations in number. As we have pointed out in Chapter 9, the two kinds have no clear line dividing them, but it is clearly convenient to refer to these categories.

Gene mutations may be of at least two general subtypes: (1) mutations that alter the code in structural genes, causing them to code for altered proteins that may or may not be able to carry out their assigned functions, and (2) mutations that alter the activities of regulatory genes or those parts of structural genes that are regulatory. These regulatory elements can modulate or tune the activity of structural genes and quantify and time their productive output of specific proteins. The basics of the action of regulatory elements in these all important roles are not understood, but obviously they must have something to do with the regulation of transcription or translation. Their role is regulatory and temporal while that of the structural genes is constitutive.

Structural Gene Mutations.   Mutations in structural genes are obviously easily understood causes of congenital defects. Rarely, a zygote will start out homozygous for a gene coding for a defective protein. Depending on the role of this protein as an enzyme, or carrier of oxygen, or in some structural capacity, death can occur as early as the one cell stage or at some stage after birth, as in the case of Lesch–Nyhan syndrome (page 000). Some inherited defects such as phenylketonuria and galactosemia can be overcome by proper diet because we understand the metabolic basis, but others that cause death *in utero* cannot easily be analyzed and are always spontaneously aborted.

Regulatory Gene Mutations. Mutations that alter the regulatory role of the genotype are the most difficult to study. These genes control the synthesis of specific proteins by controlling the activity of structural genes or their RNA messengers. Even if the structural genes are capable of producing active proteins their products will be ineffective if synthesized at the wrong time or not at all. A number of these kinds of genes have been studied in mice, but they are difficult to study in humans. No doubt exists that mutant forms of these genes are the cause of a portion of embryonic and fetal deaths by altering the program of developmental events.

The sequence of formation of hemoglobin types in the human was discussed earlier in this chapter to illustrate regulation in development. Although most adults have about 98% $\alpha_2\beta_2$ and 2% $\alpha_2\delta_2$, a small minority do not. These persons may have small to relatively large percentages of fetal hemoglobin, Hb F ($\alpha_2\gamma_2$), ranging from 5 to 100%. Persons with 100% Hb F are homozygous for a gene that produces lesser amounts in heterozygotes. The condition, known as hereditary persistence of fetal hemoglobin (HPFH), is found mainly in persons of black African and Greek ancestry.

The nature of the gene defect in HPFH has been partly established through analysis of the DNA structure in persons with various mutations that affect hemoglobin synthesis. By the use of recombinant DNA techniques, discussed more extensively in Chapter 14, it is possible to "map" point mutations and small deletions, the latter much too small to be visible under a microscope. Figure 13.6 summarizes some of the results obtained with mutants of this complex. Some of the HPFH mutations are deletions of a long segment of DNA that includes both the δ and β structural loci and that extends well into the region between the γ genes and the δ gene. This deletion allows the γ genes to continue to function, producing large amounts of γ mRNA.

Another group of hemoglobin mutants are known by the generic name *thalassemia*. There are two general classes: α-thalassemia, in which there is deficiency of α chain production, and β-thalassemia, in which there is deficiency of β chain production and often δ chain production also. The α- and β-*thal* mutations are located in the α- and β-globin gene complexes, respectively. β-Thalassemia is especially prevalent in Italy, Greece, and Africa and in persons whose ancestors came from these areas. α-Thalassemia is more common in Asia. Persons heterozygous for a β-*thal* mutation are anemic, but the condition is not life-threatening. Persons who are homozygous cannot manufacture Hb A and are therefore severely anemic. Fetal hemoglobin levels may increase somewhat but not nearly enough to compensate for the lack of Hb A. Homozygotes do not ordinarily survive to adulthood. The genetics of the α-thalassemias is somewhat more complicated because there are two loci in tandem that code for α chains.

The β-thalassemia mutations can be divided into those that, when homozygous, permit a very small amount of β-globin synthesis (β^+-*thal*) and those that permit no synthesis of β-globin (β^0-*thal*). The β^+-*thal* mutations that have been analyzed at the gene level all appear to be point mutations; that is, none of the DNA seems to be missing. On the other hand, β^0-*thal* mutations seem most often due to deletions of the β-globin gene, sometimes including the δ gene as well.

A deletion characteristic of β^0-thalassemia is illustrated in Figure 13.6. Of interest is the fact that the β^0-*thal* deletion extends into the 5' direction only a short

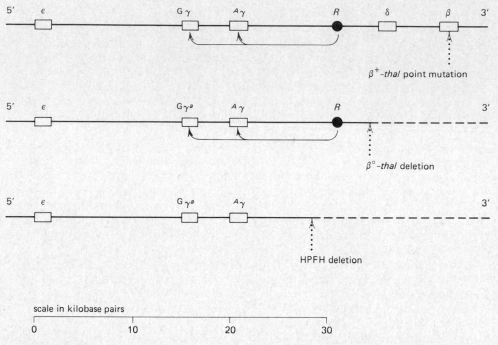

FIGURE 13.6

The human β-globin gene complex. During fetal development, the two γ-globin genes are active. Normally the γ genes cease being active about the time of birth, and the δ and β genes become active. If the δ and β genes are deleted, as in many cases of β⁰-thalassemia, the γ genes still turn off, but there are no δ and β genes to become active. However, if the deletion extends farther into the region between $^A\gamma$ and δ, as in hereditary persistence of fetal hemoglobin (HPFH), the γ genes no longer turn off. Therefore, a regulatory gene (*R*) that controls γ-globin production must lie 5′ from the δ gene.

A number of β⁰-*thal* and HPFH deletions have been mapped. The results follow the pattern indicated here, although somewhat different segments may be deleted in independently arising mutations. The 3′ boundaries of the deletions have not been established.

distance past the δ gene. In persons who are homozygous for such deletions, the γ genes are switched off at birth, leaving the infant unable to manufacture either β globin, because the β genes are missing, or γ globin, because the γ genes are turned off. By contrast, the HPFH deletions extend farther in the 5′ direction and permit continued synthesis of γ globin. Therefore, the additional length of DNA deleted in HPFH must include a regulatory gene *R* that controls the activity of the γ genes by inhibiting them.

No information is available on how the *R* gene knows when to switch the γ genes off. In persons who are homozygous for β⁰-*thal*, the *R* gene switches normally, leaving the persons acutely anemic. Therefore it does not respond to hemoglobin need. Rather, it is responding to a signal that is part of the program of fetal development.

The β^+-thalassemias have also proved interesting from the point of view of quantitative gene action. Some have been demonstrated to have nucleotide substitutions in the intervening sequences of the β-globin gene. The transcription of the gene to form hnRNA appears to occur normally, but the hnRNA cannot be processed normally into messenger RNA (Chapter 6). Other β^+-*thal* mutations are nucleotide substitutions that lead to amino acid substitutions in the β-globin polypeptide, substitutions that greatly impede the rate of translation of the mRNA.

Very little is known about gene regulation in mammals, either the regulation that is part of differentiation and development or the regulation that permits an organism to make appropriate responses to environmental change. This is a very important research area of biology and medicine, an area that is surely to be significant in the prevention and management of inherited defects.

Chromosomal Abnormalities.   In Chapters 9 and 12, we discussed various kinds of chromosomal abnormalities that are known to occur in humans who deviate from the normal karyotypes of 46,XX and 46,XY. Children born with trisomies of the smaller chromosomes such as 21, 22, 13, 14, 8, 9, and 18 may survive to birth and beyond, but they are always defective physically and mentally (Fig. 9.12). Trisomies are known for other chromosomes too, but except for the sex chromosomes, fetuses with them invariably die *in utero*. Trisomy 16 is the aberration found most frequently, in fact, in spontaneously aborted fetuses. Many other aberrations occur in addition to trisomy 16 that do not allow the embryos and fetuses to reach full term. About 40–50% of spontaneously aborted fetuses have chromosomal anomalies such as trisomy, monosomy, triploidy, and tetraploidy (Table 9.1). Obviously, it is important to have a normal karyotype with no aneuploidy of either autosomes or sex chromosomes. The reasons for this are not understood, but again it probably involves regulation of gene function. The regulatory regime of the cells of the body is dependent not only on the proper state of the regulatory genes themselves but also on their dosage. To function in the direction of normal development, they must be present in the diploid condition. The only exceptions to this rule are the sex chromosomes. The X chromosome is present twice in females and once in males. But even then, to get a male the Y must be present also.

Environmental Teratology.   This topic has come into prominence in the last few years principally because it has become clear that a pregnant woman can pass into her unborn child many substances that she ingests. There is no direct connection between the embryonic and maternal blood systems. All exchanges between them go through membranes. However, toxic substances and even some proteins can pass from the mother's blood into the embryonic or fetal blood through the placenta. The mother may detoxify some substances before they reach the placenta, but many are not detoxified and these may have effects ranging from very mild to severe, including abortion or defects in the newborn infant that persist into the adult.

The most dramatic example of this is the effects of the drug thalidomide, a mild sedative at one time widely used in some countries of the world. It was found that those mothers who took thalidomide during pregnancy had a marked increase in malformed babies. These children had (and still have) moderate to severe limb defects ranging from minor abnormalities in fingers and toes to flipper-like hands or feet to total absence of limbs. Also they had in some cases internal malforma-

FIGURE 13.7
The correlation between the sale of thalidomide in Hamburg, Germany, and the incidence of thalidomide-type limb abnormalities among newborn. (From W. Lenz, *Embryopathic Activity of Drugs*, J. M. Robson, F. M. Sullivan, and R. L. Smith, eds. Little, Brown and Company, 1965, p. 182.)

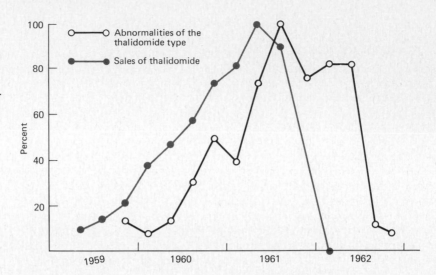

tions of some organs. Figure 13.7 shows the correlation between sales of thalidomide and the incidence of abnormalities of these types in Germany over the period 1959 to 1962. When the drug was taken off the market in 1962, the incidence of the syndrome dropped precipitously shortly thereafter. Fortunately, thalidomide was not available in all countries. In the United States, an official of the Food and Drug Administration refused to approve it for use because she did not believe that it had been tested adequately. In these countries, the thalidomide syndrome was virtually unknown. However, a considerable number of children born in countries in which the drug was used are now in their twenties and suffering from the effects of the drug.

It is becoming increasingly apparent that pregnant women should take drugs only when absolutely necessary and refrain from excessive drinking of alcohol and smoking. Excessive consumption of the drug alcohol has been linked to certain defects in offspring, as has excessive smoking. The effects of drugs and smoking will of course vary with the genotype of both the mother and the developing embryo, but we have no way of knowing at the present who is sensitive to the environmental conditions altered by these sorts of substances and who is not. The best course is to be prudent for the sake of the embryo.

Review Questions

1. Distinguish between the fundamental roles of structural and regulatory genes.
2. It is said that the development of a complex eukaryote such as a human is mostly a matter of gene regulation. What is the basis for such an assertion?
3. Where in the female reproductive system does fertilization occur?
4. What is the crucial step in development after fertilization?
5. Why are the extraembryonic membranes of the embryo and fetus of such great importance?
6. Why is the diet of the mother of great importance to the well being of embryo and fetus?

7. What general principle of development is illustrated by the hemoglobins?
8. What is a congenital defect? What is the study of these defects called?
9. Why is an understanding of human embryogeny important to an understanding of congenital defects?
10. What is the basic difference between the thalassemias and sickle cell anemia?
11. What role do chromosomal abnormalities play in spontaneous abortions?
12. What are some of the relationships that exist between environmental and genetic teratology?

References and Further Reading

Arey, L. B. 1974. *Developmental Anatomy*, 7th ed. Saunders, Philadelphia. A standard classic in the field of human development.

Bunn, H. F., B. G. Forget, and H. M. Ranney. 1977. *Human Hemoglobins*. Saunders, Philadelphia. Treats, among other things, the changes in hemoglobins during human development.

Fowler, R. E., and R. G. Edwards. 1973. The genetics of early human development. *Progress Med. Genet.* 9:49–112. Emphasizes the role of genes in the regulation of development.

Moore, J. A. 1972. *Heredity and Development*, 2nd ed. Oxford Univ. Press, London and New York. 292 pp. An elementary text that ties together genetic and developmental processes.

Saxén, L., and J. Rapola. 1969. *Congenital Defects*. Holt, Rinehart & Winston, New York. An easy-to-read treatise on this most important subject for those who wish an introduction.

Weatherall, D. J., and J. B. Clegg. 1979. Recent developments in the molecular genetics of human hemoglobins. *Cell* 16: 467–479. More about hemoglobin changes during human development.

Bacteria and viruses are so bound up in our everyday lives that their genetics is almost as important to us as our own. In fact, certain viruses may even be part of us in a sense and quite inseparable from our so-called human genotype. Bacteria and their viruses, generally call *phage*, together with *plasmids*, form a sort of triumvirate in which they function together for the benefit of each, but frequently to the distress of humans and other animals.

Bacteria

Bacteria are not only around us in the air we breathe, the food we eat, the water we drink, the things we touch, but also on us and in us. Our skin is a haven for bacteria, especially those of the genus *Staphylococcus*. Inside, we are sterile under normal circumstances except for our large intestine and mouth. The large intestine in particular teems with bacteria, including *Escherichia coli*. It belongs to that group of bacteria called colon bacteria because of their association with the lower part of the large intestine, or colon. We shall use *E. coli* as the example bacterium, since it is practically part of us and it is almost as much studied as humans themselves.

Structure and Function

Figure 4.1 depicts the essentials of a bacterial cell. It is significantly different from eukaryotic cells. It has no nucleus, no mitochondria, Golgi apparatus, or endoplasmic reticulum. Nonetheless, it carries on most of the same functions as the eukaryotic cell, but with a somewhat different organization internally. It carries out the same essential functions as mitochondria but has no mitochondria. It is in fact about the size of a mitochondrion.

Reproduction and Genetics

Bacteria reproduce by binary fission; that is, they just divide as shown in Figure 14.1 to produce two daughter cells, each of which receives at least one of the circular chromosomes shown in the figure. These circles of DNA are about 1.1 mm long in *E. coli.* Each is similar or identical to the other, so *E. coli* effectively has only one kind of chromosome. Its information content is much less than our haploid set of 23, obviously, since we have about 900 times more DNA. Still, it is adequate to enable an *E. coli* to make more *E. coli* and to carry on an existence that is about as independent as ours.

Bacterial reproduction by fission results in clones that are identical in genotype except for mutation (Chapter 9). This asexual mode of reproduction accounts for the increase in numbers of bacteria. Under ideal circumstances, the increase may be very rapid, since fission may occur every 20 min. This means that one bacterium can produce 4.7×10^{21} offspring in 24 h! In order to reproduce in this way, *E. coli* must be able to reproduce its DNA too. This it does in essentially the same way that we do except that the *E. coli* chromosome is circular, and it has to accommodate this peculiarity, which seems general among the prokaryotes. There is no mitosis as such in bacteria because there are no centromeres or spindle fibers.

14

Bacteria, Viruses, Plasmids, and Genetic Engineering

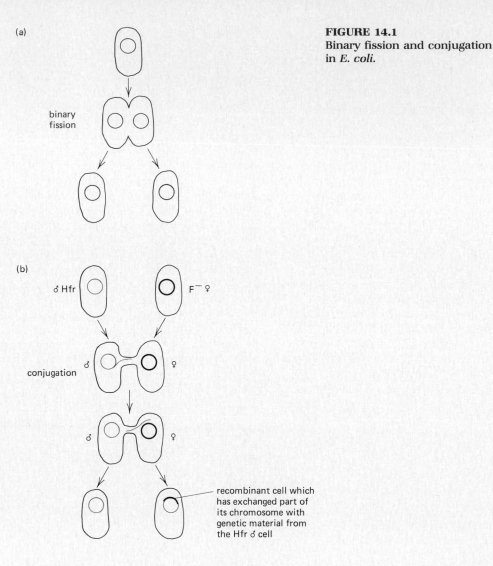

FIGURE 14.1
Binary fission and conjugation in *E. coli.*

  E. coli, and probably most other prokaryotes, have sex, and as a consequence they have recombination of their genetic material. *E. coli* has at least two types of mechanisms for accomplishing this: conjugation and transduction. Conjugation involves the breaking of the chromosome to form clones of cells with linear chromosomes, as shown in Figure 14.1B. These cells are called *Hfr* and are considered male, whereas those that maintain the circular chromosome configuration are called *F⁻* or female. Conjugation consists in a male cell fastening to a female as shown in Figure 14.1 and passing part or all of its chromosome into the female. At this phase, the receptor female is a "diploid" or partial diploid. This condition does not remain long, however, because the donated DNA strand integrates into the recipient's DNA, replacing the strand homologous to it. The recombinant cell now has a different genotype than either parent. The process may be considered akin to the process of crossing over in the meiotic phase of a eukaryotic cell, but, of course, there is no meiosis in prokaryotes.

Many bacteria have the ability to take up into their cells pieces of DNA from other members of their species. DNA taken up in this way may replace DNA homologous to it and result in a genotypic change (Fig. 14.2). This is another form of genetic recombination and is given the name *transformation*. So far as we know, transformation only occurs in the laboratory. Bacteria may or may not use it for recombination under natural conditions. One reason it is important is that in 1945 it was shown that DNA alone from a bacterium of one genotype could *transform* a bacterium of another genotype to the donor's genotype. This was the first conclusive demonstration that DNA is the genetic material. Another reason is that we now are able to transform cultured mammalian cells using the same general approach.

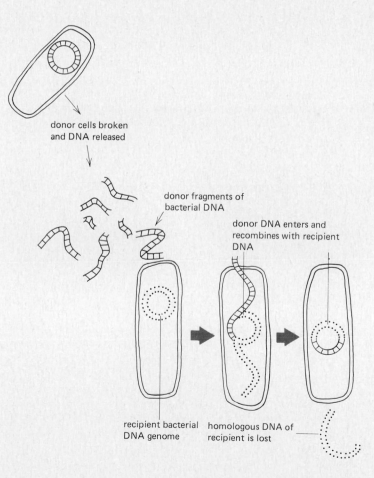

FIGURE 14.2
Transformation in bacteria. Essentially the same process can occur with eukaryotic cells in culture. Bits of DNA can be taken up just as in the case of bacteria and incorporated into the chromosomes, changing the genotype of the cell.

donor cells broken
and DNA released

donor fragments of
bacterial DNA

donor DNA enters and
recombines with recipient
DNA

recipient bacterial
DNA genome

homologous DNA of
recipient is lost

Plasmids and Phage

Plasmids

E. coli may contain rings of naked DNA that replicate independently along with the host cell and hence maintain their number per host cell during successive cell divisions. These rings are called *plasmids*, and their circumferences vary from less than a micrometer to about 100 μ. Some plasmids remain independent of the host DNA; others may integrate and become part of the host DNA (Fig. 14.3). These latter plasmids are called *episomes*, and they may play an important role in the life of the bacteria. But plasmids are important to us too because

1. They can have an important influence on the phenotype of the bacterium that may be quite injurious to us.
2. They can be manipulated in a number of ways in the laboratory to do useful work for us as a form of genetic engineering.

As we stated in the introduction, *E. coli* is ordinarily looked upon as a relatively harmless microbe that essentially is symbiotic with us. By this we mean it can live in our colon where we provide it with food and warmth, and in return it may provide us with vitamin supplements, for it can synthesize vitamins we can't. However, *E. coli* can become dangerous if it is infected with certain plasmids. A class of these, called R factors, are able to confer on *E. coli* (and related bacteria associated with humans) resistance to one or more of the antibiotics in common use. This sometimes occurs in hospitals, with unfortunate results, because a

FIGURE 14.3
The integration of episomes into the bacterial chromosome. Some episomes, such as the phage λ, are true viruses, since they have protein coats around the DNA. Some animal viruses are also known that act like episomes.

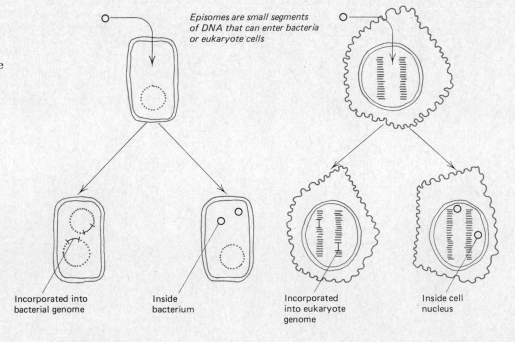

Episomes are small segments of DNA that can enter bacteria or eukaryote cells

Incorporated into bacterial genome

Inside bacterium

Incorporated into eukaryote genome

Inside cell nucleus

harmless bacterium may become pathogenic when infected by plasmids that produce proteins called colicins. If it is also infected by R factors, the bacterium can be very dangerous because antibiotics will not suppress its growth. The colicin producing plasmids are called *Col* factors. The colicins produced act as antibiotics toward other bacteria and in addition may transform a harmless bacterium into a killer. Infant diarrhea causes the death of millions of infants every year. The main causative agents are Col-infected *E. coli* and its coliform relatives. Also, urinary tract infections that can cause much misery, especially among older persons, are frequently the result of *Col*-infected bacteria, some also with R plasmids granting them immunity to commonly used antibiotics. Hence the infection can frequently become difficult to treat.

Besides disease-related plasmids that seem to play no useful role for either human or bacterium, bacteria may harbor plasmids that function in the sex life of the bacteria and presumably therefore are beneficial. One of these plasmids infecting *E. coli* is the F´ factor. Its presence in an *E. coli* cell confers upon the cell ability to function as a male. It is one of those plasmids capable of integrating into the DNA of the host cell. When it does, the host chromosome breaks at the site of integration and becomes an *Hfr* chromosome now capable of conjugating with a cell not carrying the F´ plasmid, the "female" referred to above. F´ is an episome, since it can become part of a host chromosome.

Phage

In addition to plasmids, bacteria may be infected with *bacteriophage* or, as they are called for short, *phage*. They may also be called bacterial viruses. Phage may be compared to plasmids, since they have short stretches of DNA comparable to plasmid DNA. But here the similarity ends, for phage always have a protein coat and sometimes proteinaceous appendages (Fig. 4.2), and they generally have rather complicated life cycles that frequently are a significant part of the host bacterial cell. Like viruses in general, they are incapable of reproducing except inside a host cell. Plasmids incidentally are also incapable of reproducing outside a cell.

Phage lamba (λ) is one of the thoroughly studied phage infecting *E. coli* and will serve as an example. It has a DNA molecule about 15.5 μm in length, which functions as its total genetic material. Many plasmids have about the same amount of DNA, but phage λ is different from them because it has a protein coat. This confers on it the status of phage. It also has a tail that enables it to infect an uninfected *E. coli* cell. Infection consists in the phage DNA entering the bacterial host cell.

Once in, the DNA may follow either one of two paths.

1. It may reproduce more DNA molecules like itself and then cover these with protein and generate more λ phage particles that finally break out of the host cell, which is now mostly a shell since its insides have largely been converted to phage. This is called the *lytic* cycle (Fig. 14.4).
2. An alternative to the lytic phase is *lysogeny*. The phage DNA, instead of replicating independently and forming more phage, integrates itself into the

FIGURE 14.4
The lytic and lysogenic
cycles of phage.

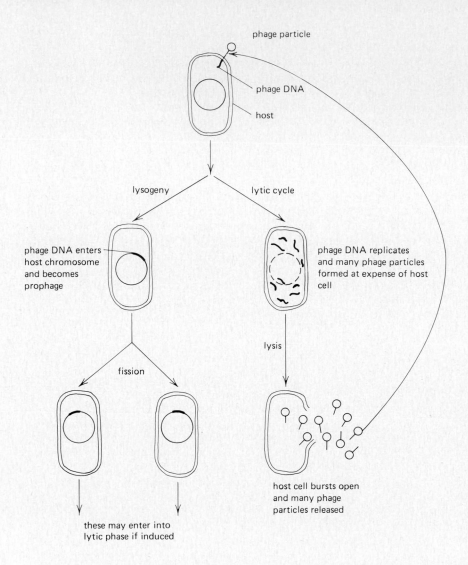

host chromosome and essentially becomes part of the host genome, replicating
with it. In this state, the phage DNA is referred to as *prophage* and it is not
necessarily harmful to the host cell. However, under certain conditions,
prophage may leave the host chromosome and enter into the lytic cycle.

What is most important for us to understand is that when a prophage leaves the
host chromosome it may carry with it a piece of the host chromosome DNA. This
would be of little consequence except that the bearer can now infect *another* host
cell and incorporate into *its* DNA that which it picked up from the previous host.
Since the phage-introduced DNA replaces the segment of DNA homologous to it,
what is effectively accomplished is a *recombination* of host DNA. This mode of

transferring a segment of DNA from one bacterial cell via a phage to another is called *transduction*, as opposed to conjugation.

Transduction is apparently common among various kinds of bacteria, including those that infect humans, and this has important consequences, because phage can therefore cause mischief similar to that caused by plasmids by carrying DNA segments that convert nonvirulent or partially virulent bacteria into highly dangerous strains resistant to antibiotics in general use. A great many of the troubles that arise in modern hospitals caused by the sudden appearance of resistant bacteria, such as *E. coli*, *Staphylococcus aureus*, and *Streptococcus*, are undoubtedly in considerable part the result of the activities of plasmids and phage as we have described.

It should be noted that the transformation process we described earlier is just another aspect of the general process of transferring bits and pieces of DNA from one bacterial chromosome to another. Because of transformation, transduction and conjugation, DNA is highly promiscuous in the world of bacteria, and this is of great importance to the world of the eukaryotes upon which many of these bacteria prey.

The Viruses of Eukaryotes

The bacterial viruses or phage are important to us, although they do not infect us (we think). Their importance is in the powerful effects they have on the genetic material of their hosts, which may infect us. Of more direct importance to us are the animal viruses that are agents of many of the diseases that infect humans and other animals as well as plants of all kinds.

The phage of bacteria we have discussed above are DNA phage, but other phage exist in which the genetic material is RNA rather than DNA. The same is true of animal and plant viruses: some are DNA and some RNA. They are one or the other, not both. We consider the RNA animal viruses first.

The Retroviruses

This group of RNA viruses is one of the most studied and is known to cause a number of important viral diseases in the vertebrate animals. They carry a single strand of RNA and an enzyme, *reverse transcriptase*. The intact virus particle outside a host cell is called a *virion* and in this state it is quite incapable of reproducing.

When it enters a host animal cell, the virion can reproduce by a series of steps, the most important of which is catalyzed by the reverse transcriptase, also called *RNA-dependent DNA polymerase*. This polymerase, unlike the polymerase discussed earlier in Chapter 5, transcribes, not a DNA strand into a complementary RNA strand, but the opposite, an RNA strand into a DNA strand—hence the *reverse* part of the name. As shown in Figure 14.5, the DNA strand so formed becomes

FIGURE 14.5
The life of a retrovirus.
(a) The virion. (b) Infection of the host cell by viral RNA. (c) Reproduction of viral RNA into complementary viral DNA by reverse transcriptase. (d) Integration of viral DNA copy into host DNA and subsequent formation of more virus.

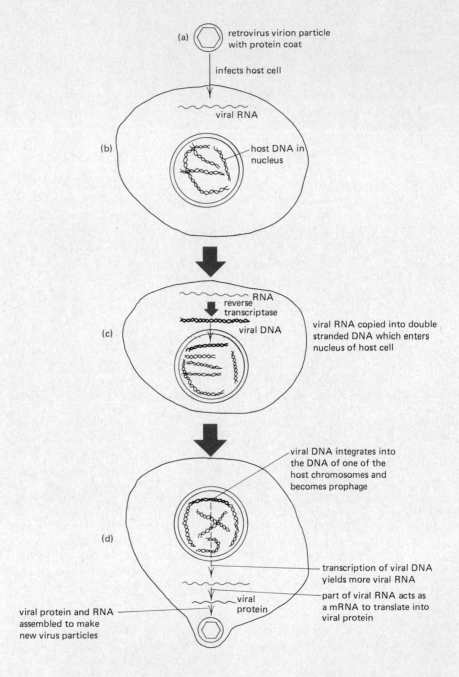

(a) retrovirus virion particle with protein coat

infects host cell

viral RNA

(b) host DNA in nucleus

RNA
reverse transcriptase
viral DNA

(c) viral RNA copied into double stranded DNA which enters nucleus of host cell

(d) viral DNA integrates into the DNA of one of the host chromosomes and becomes prophage

transcription of viral DNA yields more viral RNA

part of viral RNA acts as a mRNA to translate into viral protein

viral protein

viral protein and RNA assembled to make new virus particles

double stranded, moves into the nucleus of the host cell, and integrates into the DNA of one of the host chromosomes.

Once becoming part of the host genome, the viral DNA genome, now called a *provirus*, replicates with the host chromosomal DNA. If the provirus is integrated into a host germ cell, it can pass through successive generations of the host's lineage. Under certain conditions, the provirus in a cell line may leave the host chromosomes and transcribe more viral RNA, which becomes virions. On leaving the host cell, these virions are now ready to infect other hosts not previously infected. The process is not unlike lysogeny in the bacteria, described in the previous section.

Many mammals and birds have been found to carry identifiable retroviruses. Some have been directly linked to cancers, including leukemia. These are frequently referred to as *oncoviruses* (*onco* = tumor). But they may also generate diseases leading to degeneration of the nervous system and the blood-forming tissues. Many of them appear not to be pathogenic, and the general consensus among those who work with them is that they are endogenous companions of vertebrates in general, including ourselves. A link between retroviruses and human cancer has not been established, but the fact that they have been shown to be oncogenic in mice and chickens leads to the strong suspicion that they may in fact be responsible for at least some human cancers.

Viruses Found in Humans

RNA viruses infect the human population and can cause epidemics of major diseases. Chief among these are the influenza viruses that afflict us. The many strains have varying degrees of virulence. They replicate via an intermediate DNA formed by a reverse transcriptase, but the DNA never seems to be integrated as a provirus as in the case of the retroviruses.

DNA viruses also abound in humans. One of these, the Epstein–Barr virus (EB virus) commands attention because it is closely linked to a human malignant cancer called Burkitt's lymphoma. This tumor has a high incidence in parts of Africa and New Guinea. A survey of the children in these areas shows that most are infected with the virus, and those infected have a thirty times greater chance of contracting Burkitt's lymphoma than those who do not. A synergistic factor seems to be malaria, which is also endemic in the same areas. Children with both malaria and the virus have the greatest chance of developing the cancer.

The Epstein–Barr virus is practically ubiquitous in humans and in subhuman primates. Besides Burkitt's lymphoma, its presence has also been linked to the incidence of nasopharyngeal carcinoma (cancer of the area of the throat where the nasal passages empty). Whether the virus DNA integrates into the host genome in the same way that a provirus of a retrovirus does is still not proved, but the observation that it can "hide" and not be detectable as a virion and then suddenly appear in the absence of obvious exposure to reinfection lends credence to the possibility that we, like other animals, can carry within our genomes DNA sequences not our own.

The EB virus belongs to that group of viruses known as the *herpes* viruses. They cause herpes (fever blisters), genital herpes, shingles, and perhaps mononucleosis and measles. They are a notorious and nefarious lot who are best avoided except for study in the laboratory.

Viruses are for the most part highly variable. New strains of the human influenza virus are constantly arising and plaguing us all. We may become immune to one strain only to have a new one arise and give us the flu anew. Part of the reason for this variability is mutation, but part (and perhaps the most important part) is the result of recombination. Recombination is a fact of life for all life.

Plant Viruses

Plant viruses are important to us humans because they can infect important staple food plants and cause significant reduction in yields. At least eight different viruses are known to infect the maize or corn varieties grown in the United States. None of these is considered as serious a threat as the many fungi that cause most of the corn diseases. However, they can be a serious potential threat, and vigilance must be maintained toward them. One factor that has not been mentioned in these discussions is the resistance of the host to the infecting agent and its effects. Some humans are quite resistant to the bacteria that cause typhoid, pneumonia, and so forth, while others are not. The same is true for viruses. Plants are no exception. The great danger with cultivated plants, however, is that thousands to millions of acres may be planted with a single strain of corn. If the strain turns out to be vulnerable to a particularly virulent virus, the results could be a catastrophic economic loss. We shall return to this point again in later chapters and discuss it more fully.

Recombinant DNA and Cloning

Up to this point in our discussion we have discussed recombination as it occurs between members of a single species. In general, recombination occurs by crossing over among eukaryotes, described in Chapter 8, and by a similar process among bacteria and viruses, described in this chapter. Swapping of pieces of DNA among members of a species is a universal phenomenon. This swapping is sometimes aided by bacterial viruses, as in transduction. Whether a similar process also occurs in animals is yet to be determined, but it will not be surprising if it does. We may at this instant be swapping pieces of DNA via the viruses we carry.

The natural order of things seems to be that DNA is everywhere about us as well as within us, and many things are happening because of this, including our getting colds, the flu, and other viral diseases. Now enters the human element again with the invention of a technique to make recombinations between DNAs from widely *different* species. This has been made possible by (1) the discovery of plasmids in bacteria and (2) the discovery of a group of enzymes in bacteria that are now referred to as *restriction endonucleases* (RE).

A large number of these enzymes have been isolated from bacteria such as *E. coli*. They all have one thing in common: they break nucleotide phosphate bonds in DNA. Type I restriction endonucleases break DNA at random and are not useful in genetic studies. However, type II restriction endonucleases break the DNA bonds at very specific places, determined by the sequence of nucleotides in the DNA (Fig. 14.6). The nucleotide sequences often are *palindromes;* that is, the sequences on the two strands of DNA are identical but are aligned in opposite directions. Thus the sequence 5'-GGCC-3' on one strand is paired with 3'-CCGG-5'.

Some restriction endonucleases break the two strands at bonds that are at the same location. For example, the enzyme Hae III breaks the sequence 5'-GGCC-3' between the G and C, producing 5'-GG + CC-3'. The products would be completely double stranded (Fig. 14.6). The most useful restriction endonucleases in

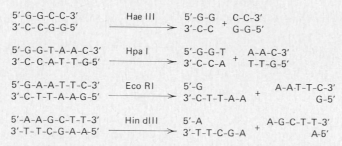

Examples of recognition sequences of restriction endonucleases:

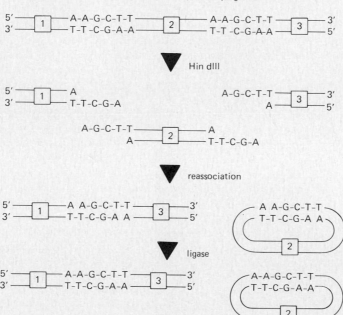

Fragments of DNA cleaved by a restriction endonuclease that leaves single stranded tails tend to reassociate, although not necessarily in the same order as they originally occurred. They can be rejoined by ligases.

FIGURE 14.6
The actions and proper-
ties of restriction en-
donucleases. Each
different kind of en-
donuclease recognizes a
different sequence of
bases in DNA, where it
can attach and break the
chain.

genetic engineering are those that produce single-stranded tails. For example, Hin dIII produces a four-nucleotide tail. These tails are complementary to each other and will tend to stick together. In the presence of enzymes called *ligases*, the nucleotides can be rejoined.

When mixed with DNA from a natural source, a restriction endonuclease will break the DNA at each point at which its specific recognition sequence of nucleotides occurs. The DNA from a mammal is very large, and millions of fragments might be produced. However, the DNA in a virus is very much smaller, and only a few fragments are produced. Particular combinations of plasmids and restriction enzymes can sometimes be found that produce only one break per plasmid, converting the circular DNA of the plasmid into an open chain. DNA from other sources can be mixed with the plasmids, and ligases are then used to re-form circular DNA. In this way, foreign DNA is introduced into the plasmids, which will still replicate in the bacterial host. If the foreign DNA is of human origin, human genes will be inserted into the plasmid, and many copies can be made in the bacteria.

Human DNA is readily available in any hospital with a maternity ward in the form of afterbirths. These extraembryonic membranes envelop and feed the embryo and fetus *in utero*, and they are, of course, cellular and contain large amounts of DNA. This DNA is extracted by appropriate means and cut up into short segments about 10,000 nucleotide base pairs long (Fig. 14.7).

The plasmid DNA pieces have their ends modified as shown in Figure 14.7, and the human DNA pieces are also modified with strings of bases complementary to those added to the plasmids. Usually poly-A DNA is added to the plasmid ends, and poly-T is added to the DNA to be inserted. The poly-A and -T groups are added by using special enzymes (terminal transferases).

Next the modified plasmids and human DNA segments are mixed together and formed into circles, again in the presence of appropriate enzymes, as shown in Figure 14.7. The resultant population of plasmids is heterogeneous in the sense that many different pieces of human DNA approximately 10,000 base pairs in length have contributed to making as many different plasmids with human DNA. These are all used to infect *E. coli* cells. The cells are grown and the newly created plasmids grow with them.

Now comes the hardest part. The reason for making plasmids with human DNA in them is to be able to replicate specific human DNA sequences within the bacterial cells. This process is called *cloning*. For example, the human growth hormone, somatotropin, is a protein that, of course, has a structural gene making its messenger RNA. If the gene for somatotropin is desired, it is necessary to select for it by appropriate means by selecting for the *E. coli* cells that carry the plasmids incorporating the somatotropin gene DNA. We shall not go into this complicated process here. Suffice it to say it has been done, and *E. coli* strains that make large amounts of somatotropin DNA have been selected.

The surprising payoff from this rather complicated scenario, however, is not the ability to make human DNA outside the human body, but the fact that the infected bacterial cells themselves make human somatotropin. The plasmid's human DNA is transcribed to make somatotropin messenger, which is in turn translated to make somatotropin that accumulates in the bacterial cell. This may then easily be extracted and purified. The process of synthesis is carried out by the bacterial

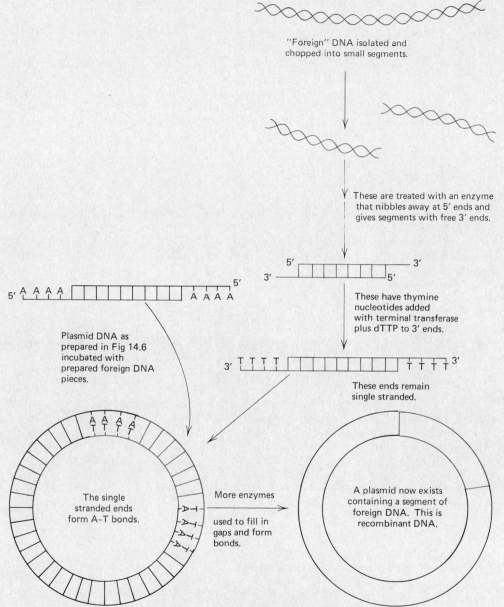

FIGURE 14.7
The incorporation of foreign DNA into plasmids.

"Foreign" DNA isolated and chopped into small segments.

These are treated with an enzyme that nibbles away at 5' ends and gives segments with free 3' ends.

These have thymine nucleotides added with terminal transferase plus dTTP to 3' ends.

These ends remain single stranded.

Plasmid DNA as prepared in Fig 14.6 incubated with prepared foreign DNA pieces.

The single stranded ends form A–T bonds.

More enzymes

used to fill in gaps and form bonds.

A plasmid now exists containing a segment of foreign DNA. This is recombinant DNA.

enzymes and can be very efficient; it is thus possible to produce literally tons of somatotropin in an industrial plant, if so desired.

Human somatotropin is now being made for human use where needed. Other important human proteins such as insulin are also starting to be made. Theoretically it is possible to make any kind of human protein in large amounts by these means. The future to this kind of technology is bright, and many of the companies that make drugs are actively engaged in recombinant DNA work. This is but one aspect of what we currently call *genetic engineering*. Other kinds of this meddling with DNA will be discussed in later chapters.

Review Questions

1. What advantages does asexual reproduction have over sexual reproduction? What disadvantages?
2. What kinds of sexual reproduction occur in bacteria such as *E. coli*?
3. Distinguish among phage, viruses, plasmids, and episomes.
4. What makes R factors so important?
5. What is the difference between DNA transformation and transduction? What are the similarities?
6. Some of the viruses that infest humans are RNA viruses. How do these reproduce?
7. Retroviruses can actually become part of us. How?
8. What is the significance of the Epstein–Barr virus so far as we humans are concerned?
9. What unique properties do restriction endonucleases possess? How are they useful in genetic engineering?
10. What role do plasmids play in modern genetic engineering?
11. A strain of bacterium causes a great deal of trouble in a hospital, not only because it is pathogenic but also because it is resistant to the antibiotics that killed off most members of its species. What explanations can you advance for the origin and "success" of this new strain?

References and Further Reading

Abelson, J., and E. Butz (eds.). 1980. Recombinant DNA. *Science* 209:1319–1438. A special issue that contains a number of review articles with current information on gene sequencing, transposable elements, and genetic engineering.

Anderson, W. F., and E. G. Diacumakos. 1981. Genetic engineering in mammalian cells. *Scientific American* 245(July):106–121. A good description of the use of mammalian cells in transformation by microinjection and viral transduction by DNA

Bishop, J. M. 1982. Oncogenes. *Scientific American* 246(March):80–93. Cancer and the viruses that are associated with the genes that cause cancer.

Diener, T. O. 1981. Viroids. *Scientific American* 244(January):66. These short strands of naked RNA are known to cause plant diseases and may be important factors in human disease.

Simons, K., H. Garoff, and A. Helenius. 1982. How animal virus gets into and out of its host cell. *Scientific American* 246(February):58–66. A description of animal virus life cycle with excellent diagrams.

The cells of all organisms—whether humans, plants, bacteria, or viruses—are covered with an enormous variety of proteins and other large molecules that are characteristic of the species of origin, the individuals within those species, and the tissues within those individuals. In vertebrates, a system of defense against invasion by bacteria, viruses, and other foreign cells has been developed, based on recognition of the differences between "foreign" molecules and those normal for the host. This *immune* system is responsible for our ability to resist infections. Rare persons who inherit a defective immune system are subject to overwhelming invasions of bacteria and viruses. The immune system also is responsible for blood transfusion reactions and the rejection of organ grafts. (Someone else's heart is, after all, foreign.) The immune system is the basis of allergies to food and pollen and hypersensitivity to insect stings.

The Nature of Immunity

An immune response is initiated by an *antigen*, which may be defined as any substance that induces an immune response. Ordinarily, antigens are large molecules, such as proteins, but they may be small. They may be naturally occurring or synthesized in a laboratory. Not all foreign molecules function equally well as antigens. Some almost always induce a response; others almost never do.

There are two general types of immune response, both due to lymphocytes (white blood cells). *Humoral* immunity signifies the presence of soluble *antibodies* circulating in the blood, antibodies that are made by B lymphocytes. Antibodies are proteins that combine specifically with the inducing antigen or one closely related but not with other antigens. *Cellular* immunity refers to the ability of certain white blood cells, the T lymphocytes, to bind to specific antigens, probably through antibodies attached to the surface of the white blood cell. Humoral immunity is a primary defense against invading bacteria and viruses. Cellular immunity is thought to be more important in the defense against fungi and possibly cancer. Cellular immunity is also involved in allergy and the rejection of transplanted tissues and organs.

An immune response occurs only after exposure to an antigen. It was thought for a long time that the structure of the antibody produced in response to a specific antigen must somehow be determined by the antigen. With the demonstration that the primary structure of proteins is determined by the genes and that the secondary and tertiary structures are a consequence of the primary structure, it became difficult to reconcile these facts with antigen control of antibody structure. The explanation is provided by the *clonal selection theory*. This theory states that each antibody-producing cell can secrete only one kind of antibody, but there are millions of cells to provide an enormous diversity of antibodies. If an antigen is introduced into the body, those cells that produce the corresponding antibodies are stimulated to grow and divide to form a clone of much more numerous cells that produce the same antibody. It may take several days for the number of cells to

15

Immunogenetics

build up, during which time an invading bacterium or virus may cause disease. After several days, however, the production of antibodies is usually sufficient to combat the invaders successfully.

Vaccination.   Vaccination is the deliberate introduction of disease antigens into the body in order to stimulate an immune reaction. These antigens may be killed viruses or bacteria or they may be attenuated strains, that is, strains that do not produce disease (or that produce a very mild form of the disease) but that share sufficient antigens with the disease-producing strains so that antibodies against them are formed. Successful vaccination has also been accomplished in domestic animals by injecting viral protein rather than intact virus.

Vaccination may also be against toxic products of an infectious organism. In the case of both diphtheria and tetanus, the detrimental effects of the infections are due to the secretion by these bacteria of very potent protein toxins. The antigen (toxoid) used in vaccination is toxin that has been deactivated by chemical treatment but that still has most of the structure of the active toxin. It therefore can still act as an antigen. Antibodies to the toxoid combine with toxin as soon as it is released during an infection. The antibodies thus prevent the expression of the disease and allow the body time to cope with the infection. Many immunizations diminish with time, and it is desirable to get "booster" shots if a person is injured (in the case of tetanus) or exposed to other infectious agents.

Vaccination against influenza and cold viruses often seems to us to be rather ineffective. This is due to the enormous variety of these viruses. Infection with or vaccination against a specific strain often confers good immunity against subsequent infections with the same strain. But a very adequate supply of new strains is arising apparently by mutation of other strains. In fact, adults are less prone to catch colds than are children. Adults often will be immune because of already having been infected with a strain to which they may be exposed, whereas a child will not have been previously exposed and will have no immunity.

Passive Immunity.   When a child is born, such immunity as he or she has is almost entirely *passive* immunity. The child will have a variety of plasma antibodies, but most come from the mother's circulation. The major portion of plasma antibodies, the *gamma globulins*, are small enough to cross the placenta into the fetal circulation. At birth the child has approximately the immune capacity of the mother. After about three months, the antibodies from the mother have largely disappeared, and the child must develop *active immunity* by producing its own antibodies. This can be a critical time in development. The immune system begins to function during the fetal period, but exposures to disease-associated antigens occurs primarily after birth.

Immune Tolerance.   One important feature of immunity is the ability to respond only to foreign antigens and not to the myriad of antigens that each of us possesses as part of our own being. It is necessary for the immune system to distinguish between "self" and "nonself." Furthermore, the fetus should not make antibodies against maternal antigens that are transferred to it during the fetal period. Neither of these unwelcome events normally occurs because of the phenomenon of *immune tolerance.* In brief, any antigen to which a fetus is exposed is permanently remembered as "self," and subsequent exposure to that antigen even in later life elicits no antibodies. The mechanism of immune tolerance is not well understood, although it can readily be demonstrated with a variety of antigens in experimental animals. It is as if any cell capable of responding to an antigen present during the fetal period were permanently suppressed. The system is not perfect, however, since there are a number of *autoimmune* diseases, such as *lupus erythematosus*, in which antibodies are made to one's own body constituents.

The Cellular Basis of Immunity

A number of different cell types are important in immunity. The leukocytes (white blood cells) are especially important. The origins and kinds of blood cells are summarized in Box 15.1. Lymphocytes, a subclass of white blood cells, are responsible for recognizing antigens and producing the antibodies specific for them. Two types of lymphocytes are formed in the body, B lymphocytes and T lymphocytes. When a B lymphocyte encounters the antigen for which it is programmed to make antibodies, it is stimulated to replicate and differentiate into a clone of plasma cells that secrete the antibody into the plasma. These clones are located in the lymph nodes and spleen (Fig. 15.1). This is why an infection causes the lymph nodes that drain an infected area to swell and be painful. T lymphocytes are processed in the thymus to form several different functional cell types. Some produce surface-bound antibodies that can bind the lymphocyte to an antigenic cell and destroy it. Others help in B cell differentiation. Others—memory cells— are long lived and keep the body ready to respond rapidly to any antigen to which it has been previously exposed.

FIGURE 15.1
Diagram of the body showing some of the organs and tissues important in immunity.

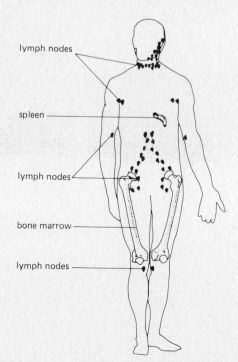

lymph nodes

spleen

lymph nodes

bone marrow

lymph nodes

Box **15.1**

Cells of Circulating Blood

All blood cells descend from cells in the bone marrow. These *stem* cells are multipotent; that is, each has the ability to differentiate into any of the blood cell types. During each cycle of replication of a stem cell, one product becomes committed to the differentiated pathway. After futher maturation, which involves several steps as well as influence from various circulating factors, the fully differentiated cells are released into the blood.

The nuclei of immature red cells disintegrate and are extruded during maturation, so that normal human red cells lack nuclei. The red cells of birds and reptiles retain their nuclei, however. Platelets are formed when the cytoplasm of the megakaryocytes breaks up into small anucleate bodies. Platelets are rich in enzymes, but neither platelets nor red cells have the ability to divide or to synthesize new messenger RNA.

The various types of granulocytes (my-

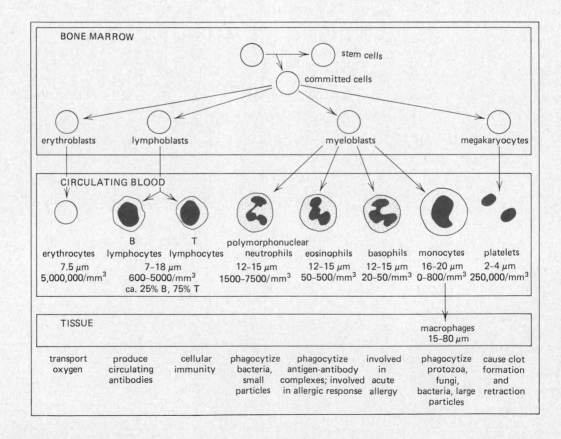

Box **15.1**

(*continued*)

elocytes), descendant from myeloblasts, are especially concerned with phagocytosis (engulfing) of foreign or other undesirable substances, including red cells that are coated with anitbodies. The macrophages are located in various tissues, such as the lungs, where they scavenge foreign particles.

The lymphocytes must go through additional maturational steps before they are completely functional immunologically. This process occurs in the thymus gland for T cells. The location for B cell processing is not known in humans; in chickens it is the organ known as the bursa of Fabricius, often called "bursa" for short, from whence the name "B" cell comes.

The actions and interactions of the various cells involved in immunity are complex and not yet fully understood. Proper functioning of the immune system is critical to survival, however. Without it, we would be subject to overwhelming infections. Indeed, a number of inherited defects in the immune system illustrate this point very well. *Severe combined immunodeficiency disease* (SCID) is an autosomal recessive disorder whose primary defect in many cases is absence of the enzyme adenosine deaminase. In others it is absence of an enzyme nucleoside phosphorylase. Patients with SCID are unable to produce functional T or B lymphocytes. The defect would thus appear to occur early in stem cell differentiation, before commitment to the T or B cell lineage. Other blood cells function normally, however. In other inherited defects of immunity, either cellular or humoral immunity may be primarily affected.

The Genetics of Antibodies

Antibodies have been a major puzzle of understanding how genes act. On the one hand, they are produced in response to exposure to environmental agents, including many "unnatural" chemicals that animals could never have encountered during their evolution until the advent of organic chemists. Any one individual may make thousands of different antibodies. This suggested that somehow the antigen conferred to the antibody-producing cells the information on how to make the corresponding antibodies. On the other hand, a large body of evidence had accummulated that protein structure is a function only of the DNA sequence of the corresponding gene, and no evidence had appeared in support of additional mechanisms of protein synthesis. But if this were true, there would have to be thousands of genes for immunoglobulins, each gene specifying the structure of an antibody that might someday be required.

The solution to the problem is closer to the second possibility; there are lots of immunoglobulin genes. But some unusual mechanisms of gene interaction also were found to operate in antibody production. To understand these it will be useful first to review antibody structure.

The Structure of Antibodies

Antibodies belong to the class of plasma proteins called *globulins* (not globins, which are part of hemoglobins). Many are part of the subgroup called gamma globulins, so-named because of their characteristic mobility in electric fields. The injection of gamma globulins after exposure to an infectious agent, such as hepatitis, is based on the premise that there will be antibodies against the agent in gamma globulins pooled from many donors. The term in general use for antibodies is *immunoglobulins*, symbolized by *Ig*.

Plasma has five major classes of immunoglobulins: IgG, IgA, IgM, IgD, and IgE. All are composed of rather similar subunits (Table 15.1). Each consists of two identical small polypeptide chains, called light chains or L chains, and two identical large polypeptide chains, called heavy chains or H chains. The formula for an Ig molecule can thus be written L_2H_2. IgM is much larger, consisting of five of these tetrameric units joined together by a polypeptide J: $(L_2H_2)_5J$. Figure 15.2 shows the arrangement of the L and H chains.

Any one person produces many different kinds of L and H chains. There are two major groups of L chains, kappa (κ) and lambda (λ). Although these chains are similar in size and other properties, the genes that produce κ chains are located on

FIGURE 15.2
Diagram of an antibody molecule, showing arrangement of the H (heavy) and L (light) chains and the antigen-binding sites. The two L chains are shown in light gray; the H chains in dark gray and white. The dark structures are carbohydrate chains. (From E. W. Silverton, M. A. Navia, and D. R. Davies, *Proc. Natl. Acad. Sci. U.S.A.* 74: 5140, 1977.)

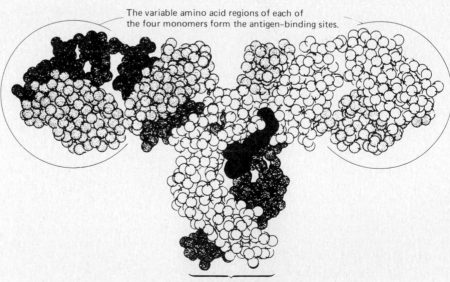

The variable amino acid regions of each of the four monomers form the antigen-binding sites.

In some immunoglobulin molecules, the carboxyl ends are extended by hydrophobic regions that bind to membranes.

There are five major classes of immunoglobulins, each defined by the type of heavy chain. The κ and λ light chains are found in all Ig classes and have molecular weights of 22,000 daltons.

TABLE 15.1
Plasma
Immunoglobulins

Ig Group	Plasma Concentration, mg/ml	Molecular Formula	Molecular Weight of Heavy Chain
IgG	8–16	$\kappa_2\gamma_2$; $\lambda_2\gamma_2$	53,000
IgA	2–4	$\kappa_2\alpha_2$; $\lambda_2\alpha_2$	60,000
IgM	0.5–2	$(\kappa_2\mu_2)_5J$; $(\lambda_2\mu_2)_5J$	70,000
IgD	0.5	$\kappa_2\delta_2$; $\lambda_2\delta_2$	70,000
IgE	0.1	$\kappa_2\epsilon_2$; $\lambda_2\epsilon_2$	80,000

chromosome 2, whereas those for λ chains are on chromosome 22. Each of these loci is complex, consisting of genes for a number of different κ and λ chains. Both κ and λ L chains are found in all the Ig classes.

The H chains are produced by a gene complex on chromosome 14. There are five major classes of H chain, each found in one of the five major classes of immunoglobulins (Table 15.1). Thus the type of H chain determines what class of immunoglobulin is produced.

The IgG isolated from the plasma of a normal person is a mixture of many similar but nevertheless different molecules, because the IgG contains hundreds, perhaps thousands of different antibodies, each with its unique structure. Chemical separation and analysis of such a mixture are virtually impossible. Much has been learned though from persons with multiple myeloma, a malignant disease in which one of the enormous number of plasma cells escapes regulation and forms a large clone of cells that secrete a single type of immunoglobulin. The immunoglobulin is a major component of the plasma proteins in such persons (Fig. 15.3) and can be isolated in pure form with relative ease. A comparison of pure immunoglobulins from a number of such patients shows interesting patterns. The "tips" of the Y-shaped molecule vary greatly from one person to another. These regions are known to be the antigen-binding sites. The base of the Y exists in only a few different forms. This region binds to the cell surface and is responsible for the properties that cause a particular Ig molecule to fall into a particular major class: IgG, IgM, etc.

Since the amino-terminal regions of the polypeptide chains are at the variable part of the molecule and the carboxyl terminals are at the base of the Y, the amino-terminal amino acid sequences for both the L and H chains are quite different in each molecule. Conversely the carboxyl-terminal regions are limited to a few different sequences. This difference presented a paradox at first, because a single polypeptide chain was thought always to be determined by a single stretch of DNA. Conversely, a particular stretch of DNA was thought to determine the sequence of no more than one polypeptide.

The immunoglobulins have proved to be the single exception to this model. A diagram of the H-chain gene complex is shown in Figure 15.4. Through mechanisms not fully understood, one of a large variety of DNA sequences that code for the variable (V) amino-terminal region of the H chain joins with one of the smaller number of DNA sequences that code for the constant region. No other DNA

FIGURE 15.3
Electrophoretic pattern of plasma from a person with multiple myeloma compared to the pattern from a normal person. The myeloma plasma has a high concentration of a single type of immuno-globulin that is produced by the clone of myeloma cells. The immunoglobu-lins in normal plasma form a broad band indica-tive of a mixture of molecular forms, no one of which occurs in high concentration.

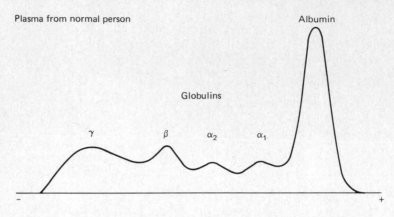

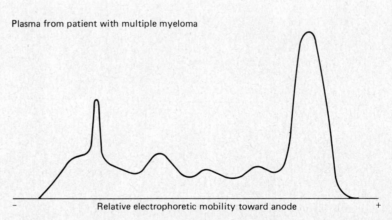

sequences are expressed in the H-chain complex, either on the active chromosome or on the homologous chromosome. Since there are so many possible V and C region combinations, a very large number of different kinds of polypeptide chains could be made. And since each Ig molecule consists of two different kinds of polypeptides, L and H, both of which have variable and constant regions, the number of L and H combinations would be the product of the number of L genes and the number of H genes. For example, if there were 100 L genes and 100 H genes, 10,000 combinations would be possible, probably more than enough to cope with most of the antigens to which we are exposed.

To summarize this rather complex situation, we may make the following generalizations.

1. The information for the great diversity of antibodies is contained entirely within the DNA of the antibody-producing cells, the lymphocytes.
2. Any one cell manufactures only one type of antibody, consisting of one type of L chain and one type of H chain. All other DNA sequences that code for antibody structure are suppressed, including the alleles of the expressed genes on the homologous chromosome.

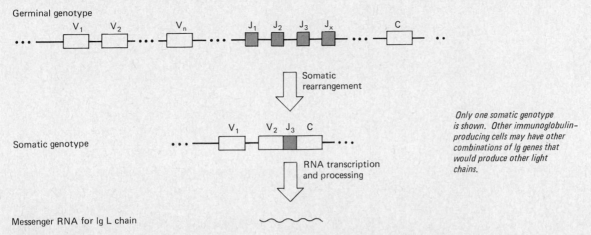

FIGURE 15.4
Diagram of an immunoglobulin light-chain gene complex. The immunoglobulins appear to be unique in that segments of DNA are eliminated in somatic cells to form a great variety of functional immunoglobulin genes.

3. When antigen is introduced into the body, those pre-existing cells that can make antibodies that bind to the antigen are stimulated to grow and divide to form a clone of cells identical in the kind of antibody that they produce. This accounts in part for the increased ability to respond to subsequent exposure to the antigen.

Antigenic Variation

The discovery of antigenic variation within the human species occurred in 1900, when K. Landsteiner observed that blood samples could be classified into four groups, based on the patterns of agglutination that occur when blood from two persons is mixed. Agglutination means that the red cells form large clumps. If this occurs during transfusion, death may occur. Only *compatible* blood, that is, blood that will not agglutinate, may be safely transfused. *Incompatible* blood must be avoided. Studies by Landsteiner and subsequent investigators demonstrated that agglutination is caused by antibodies in the serum (that is, the liquid part of the blood after a clot has formed and contracted). These antibodies react with antigens on the surface of the red blood cells and cause them to stick together into clumps. Persons belonging to the four groups have different combinations of antigens on their red cells as well as different antibodies in their sera.

   Not only do red cells vary in their antigenic properties, but other tissues also show a great variety of antigens. This accounts for the inability to transplant organs freely among different persons. In this section we will review the nature and inheritance of blood groups and tissue antigens and consider some of the applications of these systems to medicine, law, anthropology, and agriculture.

Blood Groups

The ABO Groups. Landsteiner found that red cells and sera from the four groups behave as shown in Table 15.2. In modern terminology, these four groups are designated A, B, AB, and O. Since the incompatibility is due to serum antibodies reacting with red cells, the cells and sera have been separated in Table 15.2. The results in the table can be explained by two kinds of antigens, A and B, with the corresponding antibodies, anti-A and anti-B. Types A, B, and AB cells have the antigens indicated in their designations. Type O cells lack both A and B antigens. The serum contains antibodies to whatever antigens are missing. Thus a person with type A cells would have antibodies to B cells, and a person with B cells would have anti-A antibodies. AB persons have neither type antibody, and O persons have both.

When incompatible blood is transfused, the donor's red cells are destroyed by the antibody in the recipient's blood. However, the donor's antibodies are too diluted in the transfusion to cause problems with the recipient's red cells. In theory, type O cells could be transfused into any other type person, and AB persons could accept cells of any type. In practice, this use of unmatched transfusions is made only in unusual or emergency situations.

The antigenic differences among ABO groups have now been well defined chemically (Fig. 15.5). The antigenic groups are carbohydrate polymers attached to large protein or fat molecules. The carbohydrate chains are built up by adding various sugar residues in a stepwise fashion, each addition being catalyzed by a specific enzyme (and therefore controlled by a corresponding specific gene). The A antigen is formed when the sugar N-acetylgalactosamine (GalNAc) is added to a precursor carbohydrate, the H substance, a reaction that is catalyzed by the A enzyme. The B enzyme catalyzes addition of a different sugar, galactose (Gal). If both enzymes are present, both A and B antigens are formed. If neither is present, the H substance is unchanged and can be detected with anti-H antibodies. However, H antibodies do not occur naturally, since low levels of H substance are always present, even in A, B, and AB persons. It is therefore "self." The presence of H is not routinely tested in blood typing, since all but a handful of persons are positive.

TABLE 15.2 Reactions of Red Blood Cells According to ABO Group

A plus indicates agglutination; minus indicates lack of agglutination.

| | | Red Cells of Group | | | |
		A	B	AB	O
Serum	A	−	+	+	−
from	B	+	−	+	−
	AB	−	−	−	−
Group	O	+	+	+	−

FIGURE 15.5
The relationship of alleles of the ABO locus to the ABO antigens found on red blood cells. The four common alleles each produce a protein that (a) acts as an enzyme to produce either A or B antigen or (b) is enzymatically inactive.

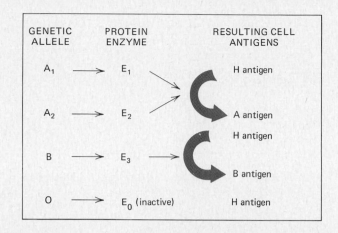

The inheritance of the ABO groups follows simple Mendelian rules. A single locus on human chromosome 9 determines the presence of the A and B enzymes. These enzymes are in fact almost identical to each other structurally with only a slight difference that results in one being active for galactose, the other for the closely related N-acetylgalactosamine. The genes that code for these enzymes are alleles, usually symbolized I^A and I^B, or just A and B. A third allele, I^O, codes for a defective form of the enzyme that is unable to catalyze the addition of a sugar group to the H substance. It is therefore what is often called a *null* allele.

The genotype–phenotype relationships are summarized in Table 15.3. There are six possible combinations of the three common alleles. Since no distinction can be made in the red cell phenotypes of AA and AO persons, at least by the standard tests, A is dominant to O. Similarly, B is dominant to O also. However, the presence of both A and B can be detected in AB heterozygotes, and these alleles are said to be *codominant* to each other. O is recessive to A and B.

One aspect of the ABO blood groups is unique. That is the presence of naturally occurring antibodies in the absence of known exposure to them. It is as if all persons were programmed to make anti-A and anti-B, but actual synthesis occurs only if the corresponding A or B antigens are absent. This inhibition would be expected on the basis of immune tolerance. Other ideas have been advanced, but no evidence strongly supports any one explanation.

Actually four common alleles are at the ABO locus. Type A can be separated into two subgroups, A_1 and A_2, based apparently on the variety of molecules to which the carbohydrate chain can be attached. This distinction is not made in routine typing, since it has no medical significance. The A_2 enzyme appears to be more limited in its activity. The more versatile A_1 enzyme obscures the presence of the A_2 enzyme. Phenotypically, A_1 is dominant to A_2. Therefore the A_2 type can only be detected in the absence of A_1. In the United States white population, 75% of the ABO alleles are A_1 and 25% are A_2. In the black population, the corresponding figures are 66% and 34%.

The production of the H substance is mediated by a specific enzyme that is coded by a corresponding gene. One would expect mutation at the H locus to produce rare persons who cannot carry out this step, provided, of course, that

**TABLE 15.3
Inheritance of ABO
Blood Groups and
Frequencies in Selected
Populations**

		Frequencies in U.S. Populations				
Genotypes	Phenotypes	White	Japanese (Hawaii)	Black	Mexican-American	Indian[a] (Pima)
AA, AO	A	.413	.400	.265	.283	.140
BB, BO	B	.099	.205	.201	.090	.000
AB	AB	.035	.103	.043	.024	.000
OO	O	.453	.293	.491	.602	.860

[a]Many Indian tribes in the Americas have only type O. Where the B allele is found, it can usually be explained by European admixture. The same is often true of the A allele, although certain North American tribes have a substantial frequency of the A allele that is not of European origin.

Frequencies from A. E. Mourant, A. C. Kopeć, and K. Domaniewska-Sobczak, *The Distribution of the Human Blood Groups*, 2nd ed., London: Oxford Univ. Press, 1976.

such a mutation is not lethal. Such persons have been found in India, where the mutant gene *h* has a somewhat higher frequency then elsewhere, although it is uncommon even in India. Persons homozygous for the *h* allele produce no H substance; therefore, even though they have active A and B enzymes, they cannot produce A or B antigens either. Their blood type, designated *Bombay* for the location of the first family described, can be distinguished from type O both by the absence of H antigen and by the presence of naturally occurring anti-H antibodies. In one of the first families studied, a type O woman had an AB daughter. This is not possible considering only the alleles at the *ABO* locus. Further investigation showed the woman to be *BO* at the *ABO* locus but homozygous *hh* at the *H* locus. This is an example of *epistasis*, in which variation at the *H* locus suppresses the phenotypic expression of variation at the *ABO* locus.

The Rh Blood Groups. In 1939, Philip Levine and R. E. Stetson reported an unusual case that revolutionized our understanding of a major cause of fetal and infant death. A woman had given birth to a stillborn infant. She was in need of a blood transfusion. Her husband's blood was compatible for ABO type, so he donated a pint of blood. She immediately suffered a severe hemolytic reaction; that is, the injected red cells were destroyed. When Levine and Stetson investigated the case, they found that the woman's serum contained antibodies against her husband's red cells, even though the blood had been matched for ABO type. Furthermore, her serum agglutinated approximately 80% of other persons who were ABO-compatible. Levine and Stetson proposed that the husband's red cells possessed an antigen that had been inherited by the fetus but that was not present on the woman's cells. Furthermore, some of the fetal red cells must have entered the mother's circulation, where they were identified as "foreign." Her immune system responded by making antibodies to the foreign antigen, attacking the red cells in the fetus, causing it to be stillborn, and also attacking the transfused red cells of her husband.

At about the same time, Landsteiner and A. S. Wiener were studying the antigens of rhesus monkeys by injecting rhesus red cells into rabbits. They produced antibodies that reacted with rhesus red cells and that also reacted with red cells from 80% of humans. The factor, named the *Rhesus* or *Rh* factor, was subsequently shown to be the same as the factor first detected by Levine and Stetson.

In the initial studies, the inheritance of the Rh factor followed simple Mendelian rules. The presence of the factor, designated Rh-positive or Rh(+), was dominant to its absence, Rh-negative or Rh(−). Thus persons were genotypically *RR*, *Rr*, or *rr*, where *R* is the dominant allele for Rh(+) and *r* the recessive allele for Rh(−). Additional studies have shown the *Rh* locus to be much more complex than a simple two-allele system. There are a number of alleles each specifying some combination of antigenic factors. The antigenic factor that is most important is labeled D, and in most families Rh(+) can be equated with the presence of the D factor. For our purposes, we will assume that this is true, although the student should remember that exceptions to occur.

If an Rh(−) woman marries an Rh(+) man, the matings may be of two types: male *RR* × female *rr* and male *Rr* × female *rr*. In either case, the father possesses an antigen D, coded by the *R* allele, that is lacking in the woman. Unlike the A and B antigens, antibodies do not occur against D, unless Rh(+) cells have been introduced into the circulation of a person who lacks D. This can occur by transfusion, although the Rh type of transfused blood is now always matched to the recipient. Most importantly, it also occurs when cells from an Rh(+) fetus get into the circulation of an Rh(−) mother. This may happen at any time during pregnancy, but it happens most often at the time of delivery. When it does, the Rh(−) mother may produce antibodies against the Rh(+) factor (Fig. 15.6). The consequence is that, although the first incompatible pregnancy usually escapes ill effects, subsequent incompatible pregnancies are very likely to get into difficulty. The first pregnancy escapes because immunization of the mother does not occur

When a child is born, some of the fetal red blood cells escape into the mother's circulation.

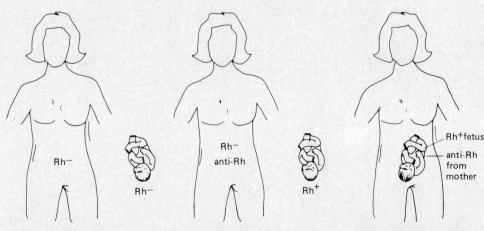

FIGURE 15.6
Hemolytic disease of the newborn (HDN) due to Rh incompatibility in humans.

If both mother and child are Rh− , no antibodies are formed.

If the mother is Rh− but the child is Rh+, antibodies to the Rh+ cells may form.

In subsequent pregnancies, the antibodies from the mother cross the placenta. If the fetus is Rh+, the antibodies combine with red cells in the fetus, causing them to be destroyed.

until termination of the pregnancy. But when a second incompatible pregnancy occurs, the mother is already immunized, and antibodies against the Rh(+) cells of the fetus are produced rapidly. The antibodies cross from the maternal into the fetal circulation and destroy the fetal cells, producing the condition *hemolytic disease of the newborn* (HDN).

Children with HDN may be only mildly affected with jaundice or they may be severely affected and require exchange transfusions. In especially severe cases, the fetus is affected and often aborted. Since the problem depends largely on the transfer of fetal cells to the mother at the time of birth, a rather effective prevention is achieved by injecting antibodies to Rh(+) cells into the mother shortly after each birth. The injected antibodies coat the fetal cells, causing them to be removed rapidly from the circulation before they have had time to stimulate the mother's immune system to produce antibodies. Such a treatment is effective only if the mother has never been immunized against Rh(+) cells.

Other Blood Groups. The second human blood group discovered was described in 1927 by Landsteiner and Levine. They injected samples of human blood into rabbits and found that some samples of blood then reacted with certain of the rabbit antibodies but others failed to react. The reactive factor was designated M. Shortly thereafter, a related factor N was also discovered, and M and N were recognized as codominant alleles. People are either blood type M, MN, or N, corresponding to the genotypes *MM*, *MN*, and *NN*. The *MN* locus has become more complex as new antigenic factors are discovered to be part of it.

The MN groups illustrate an important aspect of blood groups and of immunology in general. Injection of a foreign antigen does not necessarily mean that antibodies will be formed. Thousands of blood transfusions are performed every day with the donor and recipient not matched for MN groups. But such "incompatible" transfusions almost never lead to a clinical problem. To the immunologist, M and N are not "good" antigens, that is, they are not good in inducing antibodies because they are so similar that both are identified as "self." Rarely, a person with multiple transfusions or a woman with multiple pregnancies (or some combination of the two) will produce antibodies to M, N, or any of a large number of other red cell antigens that usually can be safely ignored. It is these rare persons who provide the test antibodies used to classify red cells.

Xg is a red-cell surface antigen whose locus is on the X chromosome. The first known production of antibodies to Xg occurred in 1962 in a man who had received multiple transfusions, many of them containing Xg-positive red cells while his red cells were negative. For several years, this man in Grand Rapids, Michigan, was the only source of serum that could be used to test for Xg. Yet, since the introduction of blood transfusion, countless Xg(−) persons had been transfused with Xg(+) red cells. By contrast, the D factor of the Rh system is a very good antigen and is likely to stimulate antibody production whenever Rh(+) cells are introduced into an Rh(−) person.

These rare antisera have made it possible to identify many different blood groups. For the most part, these groups have little medical significance, but they can be very helpful to geneticists, anthropologists, and lawyers.

Blood Groups in Domestic Animals. Every domestic animal tested has been shown to have inherited variation at a number of blood group loci. Interest in blood groups has been especially strong for horses and cattle, since blood groups are the primary means of documenting breeding records. A milliliter of semen from a prize bull is worth thousands of dollars, but only if the offspring produced from it can be shown by blood group test to have been sired by that bull.

Tissue Antigens

Red cells are the most convenient human tissue to study and to transfer to other human recipients. Furthermore, red cells have a normal life span of only 120 days. Hence, transfused red cells would be destroyed within 3–4 months by normal ageing unless the recipient has already been immunized against one or more of the red-cell antigens carried by the transfused cells, in which case the transfused cells would be destroyed earlier by antibodies to them.

Other tissues also have antigens, generally different from those of the red cells, although some are shared in common. For example, ABO antigens occur on a variety of other cell types but not on all. The existence of these other tissue antigens is readily demonstrated by transplantation of tissues or organs, which are virtually always "rejected" unless the exchange is between identical twins or the immune system is suppressed by therapeutic drugs. Tissue antigens are readily demonstrated by small skin transplants. If the recipient has not been previously immunized against the donor antigens, the transplant may appear to take for a few weeks. Eventually, however, the host antibodies destroy the transplanted cells and the transplant is rejected.

The antigens expressed by tissues are called *histocompatibility* (*histo-* = tissue) antigens. There must be hundreds of surface structures on cells capable of acting as antigens, but most are not very effective in inducing antibodies. The most important antigens with respect to transplantation are produced at a complex locus called the *major histocompatibility complex* (MHC). It is also known in humans as the *HLA* locus, because the locus was first studied by means of the antigens on human leukocytes rather than on transplanted organs. The same locus, known in mice as the H-2 locus, has been extensively studied in that species by means of skin transplants.

Several loci in the MHC region are separable by crossing over (Fig. 15.7). The entire region is approximately two map units long, which is large enough to accommodate many additional structural genes. The four loci that produce detectable antigens have a very large number of alleles, each recognized by a

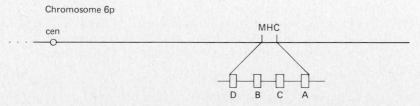

FIGURE 15.7
Diagram of the major histocompatibility complex (MHC) in humans. This complex is located on the short arm of chromosome 6. Although the region appears quite small in the diagram, there is sufficient DNA for hundreds of additional genes.

different antibody. This great richness of alleles has a very direct consequence for transplantation.

1. The great majority of the population is heterozygous for these loci, since the likelihood of a person inheriting identical chromosomes from both parents is small.
2. The likelihood that an unrelated person would have the same genotype is very small.

Since the loci of the MHC are closely linked, they are usually transmitted in meiosis as a unit, even though crossing over may rarely occur within the MHC region. A particular combination of alleles at the several loci is called a *haplotype*. For convenience, we can designate various haplotype combinations as *a*, *b*, *c*, and so forth. A typical mating would then be *ab* × *cd*, with both parents heterozygous. This gives four possible offspring combinations, *ac*, *ad*, *bc*, and *bd*. If a person needs a kidney transplant, for example, there is a 25% chance that any particular sib has the same genotype. That likelihood is very much larger than for a parent and especially for unrelated persons. For heart transplants, one cannot ordinarily look to one's sib or to any specific person. Rather, the need for a particular histocompatibility type is made known, and potential donors, typically accident victims, are matched, sometimes with someone very distant geographically. This is the reason persons who need a kidney or heart transplant often are reported in the press as waiting for a suitable donor.

Other antigenic differences exist among people but seem less important for long-term transplantation. It is generally possible to manage some antigenic incompatibilities by suppressing the immune system with drugs. This must be done with great care, since response to infectious bacteria and viruses also is suppressed. Therefore persons who have received organ transplants are at great risk of infection during the period after surgery and must be kept isolated. Kidney transplant patients generally do very well, and later rejection of the transplant kidney is uncommon if the genotypes of the donor and the recipient are well matched. The prognosis for other organ recipients is much less optimistic.

Applications of Inherited Antigens to Human Problems

The great number of blood groups and tissue antigens has made these inherited traits useful for purposes other than the medical situations already discussed. Chief among these has been their use in settling questions of biological relationships for legal purposes, for the most part, in paternity suits.

Blood Groups and Paternity

Courts now accept a variety of genetic evidence in paternity suits, although the primary evidence is red-cell antigens and HLA types. In a typical example, a man is accused of being the biological father of a child. The blood types of the mother, the child, and the putative father are tested to see if the man indeed could be the father or if the blood groups rule out that possibility. Looking only at the ABO locus, let us assume that the mother is type O, the child type A, and the man type B. The mother must be homozygous *OO* and could only transmit an *O* allele to the child. Therefore the child must be genotypically *AO*, and the *A* allele must have come from the father. However, the genotype of the accused man is either *BB* or *BO*. He could not be the father of the child.

If the man's blood type had been A, he could have transmitted an *A* allele to the child. In that case, he could be the father. On the other hand, so could any type A or AB man, which would include a large portion of the population. Blood groups can rule out relationships, but they cannot prove them. One can only establish that such relationships are compatible or incompatible with the genetic evidence.

If only the ABO system were available, the probability of exonerating a falsely accused man is approximately 20%. If additional genetic systems are added to the battery of tests, the probability increases. A man who is not excluded by the ABO system may be by the MN or the Rh. With the addition of HLA types to the list, the probability of exclusion approaches 99%. One must add to the battery several other genetic "markers," such as haptoglobin tests, a plasma protein that shows simply inherited variants. Faced with the results of such studies, many litigants now choose to settle out of court rather than go through a trial whose outcome is predictable.

Genetic Markers and Human Evolution

Another use of blood group information has been by anthropologists and geneticists interested in the origins of and relationships among human populations. This application will be discussed more extensively in Chapter 20. For the present, we will cite only one example. The Diego blood group system has two alleles, Di^a and Di^b, that are codominant. This system was first detected in Venezuela associated with a case of hemolytic disease of the newborn. Señora Diego produced antibodies against an antigen, designated Diego (a) or Di(a), on red cells from her baby. The antigen was also found on red cells of her husband and several members of his family. The cells of 11% of Chippewa Indians and about 10% of Japanese are Di(a)-positive, but persons without Asian or American Indian ancestry are uniformly negative. These results support the theory that American Indians and Asians are more closely related to each other genetically than they are to other ethnic groups, a theory that initially was proposed on the basis of quite different evidence.

Review Questions

1. A skin graft from person A to unrelated person B is almost certain to be rejected. Why?
2. The injection of a foreign protein into our body elicits what is called an immune response. Describe this response.
3. What is vaccination?

4. A newborn child is generally immune to common disease organisms for several months after birth. Why? Why does the immunity disappear?
5. What function does an immunoglobulin perform? How specific is this function? What confers the specificity to the immunoglobulin?
6. In what significant ways do structural genes for immunoglobulins differ from other structural genes?
7. Joyce has blood type A,MN and a child of type O,MN. Grace has blood type AB,MN and a child of type B,MN. Hope has blood type O,N and a child of type O,N. Greg is type A,MN; John is type AB,MN; William is type O,M. Assuming no illegitimacy and that each woman is married to one of the three men, deduce who the three couples are.
8. What is so important about the HLA locus?
9. What significant differences between the ABO group and the MN group have an important bearing on blood transfusion?
10. Under what circumstances does hemolytic disease of the newborn occur?
11. How can hemolytic disease of the newborn be prevented?

References and Further Reading

Fudenberg, H. H., J. R. L. Pink, A.-C. Wang, and S. D. Douglas. 1978. *Basic Immunogenetics*, 2nd ed. Oxford Univ. Press, New York. 262 pp. An elementary text for persons with a strong background. Emphasis is on gene function and malfunction, especially in humans.

Hood, L. E., I. Weissman, and W. B. Wood. 1978. *Immunology*. Benjamin/Cummings, Menlo Park, Calif. 467 pp. A textbook for beginners. A clear exposition of the major concepts in this field.

Leder, P. 1982. The genetics of antibody diversity. *Scientific American* 246(May):102–115. A brief explanation of the theory current in 1982 by a leading researcher in the field.

Race, R. R., and R. Sanger. 1975. *Blood Groups in Man*, 6th ed. Blackwell Scientific Publ., Oxford. 659 pp. Through its many editions, "Race and Sanger" has been the prime source of information on human blood groups and on many related topics as well.

One of the principal concerns about increased levels of radiation in the environment, chemical additives to food, and pollution in general has been the possibility that these agents might increase the human mutation rate and add to the number of persons born with genetic defects. In the previous chapters, we have seen how the structure of the DNA can change and produce mutations, which in turn often produce defective activity of genes. In this chapter, we will discuss how mutations are measured in human beings and in other species, and we will consider what is known about some of the environmental agents that influence mutation rates.

The Measurement of Mutation Rates

A mutation was defined earlier as any heritable change in the genetic material. Traditionally, heritable has meant that the change must be transmitted to offspring. With the development of methods of culturing mammalian cells, heritable also includes any change that is stable and is transmitted to daughter cells and their descendants. Any change in DNA sequence, such as base-pair substitutions, deletions, additions, or rearrangements, is a mutation, as are changes in chromosome structure and number.

With the exception of large chromosome rearrangements and deletions, one cannot readily detect mutations by examination of DNA or chromosomes. Rather, one must look for effects on the phenotype, from which the genotype changes can be inferred. The sudden appearance of an inherited trait in a family may be due to mutation. One well-documented example of a new X-linked mutation is hemophilia in the royal families of Europe (Fig. 2.6).

Any phenotypic effect used to detect mutations will reveal only certain types of genetic changes. For example, mutant forms of hemoglobin and other proteins are often detected by changes in the electrical charge of the hemoglobin due to substitution of an amino acid for one with a different charge. This would be due to a base-pair substitution. But most base-pair substitutions do not lead to substitution of an amino acid with a different charge. A mutation at the hemophilia locus might be expressed when the normal gene product, factor VIII in the blood clotting system, is not functional. This might result from a variety of changes at the DNA level. But many changes would lead to altered but still functional factor VIII and would not be detected as hemophilia. The measured frequency of mutation will therefore vary depending on the phenotypic effect observed.

In order to assess the influence of environmental agents on the induction of mutations, it is necessary to be able to measure the frequency of mutation. The mutation rate is most often expressed as the number of mutations per locus per generation (or per gamete, which is the same thing). For example, achondroplasia is a form of dwarfism (Fig. 16.1). In a study of achondroplasia in Northern Ireland, A. C. Stevenson found 37 cases who had no affected sibs and who were born to parents, neither of whom was affected. Since achondroplasia is an autosomal dominant trait, this means that each person in this group had one new mutant

16

Mutation and the Environment

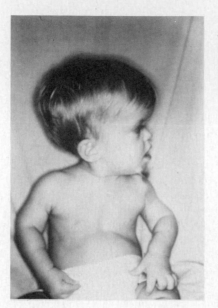

FIGURE 16.1
A child with the autosomal dominant condition achondroplasia. This rare disorder has been used in several studies of mutation rates. (Photograph courtesy of Dr. J. Friedman, The University of Texas Health Science Center at Dallas.)

gene for it. The total population of Northern Ireland at that time was 1,387,000 persons, each of whom represents two gametes with the potential for mutation. Therefore, the mutation rate for achondroplasia is $37/(2 \times 1,387,000)$ or 1.3×10^{-5}. The molecular nature of these mutant genes is not known, and each is very likely different from the others, although presumably all represent changes at the same locus. But all cause a defect in gene function that eventually is expressed as achondroplasia. This particular calculation assumes that persons with achondroplasia are equal to normal persons in viability. Somewhat different strategies may be required to measure mutation rates for a gene that leads to early death.

Many different phenotypic endpoints can be used to measure mutation rates, depending on the organism. The different endpoints will give quite different numerical mutation rates, even though the underlying mutational process is uniform. It is important therefore that mutation rates be compared only when based on comparable observations.

Measurement of Mutation in Experimental Organisms

Many genetic systems can be used to detect mutation. Some are most useful in studying the basic mechanisms of mutation, some were developed especially for testing mutagenicity of specific chemical agents to which we are exposed, and others were developed primarily to measure the genetic effect on humans of exposure to various agents.

The first system was that developed by H. J. Muller, who demonstrated that X rays can cause mutations in the fruit fly *Drosophila* (Box 16.1). Muller's approach was to count all the recessive lethal mutations induced on the X chromosome, mutations that are lethal when present in males with only one X chromosome, but that are recessive and therefore viable when combined with a normal X chromosome in females. Using this system, Muller and others were able to show that the number of mutations produced is directly proportional to the amount of radiation received.

Microbial Test Systems. Mutations can be measured most readily in microbial systems, such as bacteria, molds, and yeasts, where the life cycle is short and millions of cells can be grown in a single test tube or petri dish. Much of our knowledge of mutational mechanisms comes from studies of microbial systems. They are easy to manipulate, and the results are obtained in days rather than years.

However, for purposes of estimating human risk, microbial systems have the disadvantage that they are not, after all, miniature humans. There are many differences in metabolism, some of which are important in the way environmental chemicals are processed. They do not differentiate into cells with specialized functions as do higher organisms. Nevertheless, microbial systems can tell us whether a chemical being tested does react with DNA or with other cell constituents that in turn react with DNA to produce mutations. This information, combined with the results of other test systems, helps us reach a decision on the risks of human exposure to that chemical.

Box 16.1

The Use of Drosophila *to Test for X-linked Lethal Mutations*

The value of an experimental organism such as *Drosophila* is apparent in the early work of H. J. Muller on mutation. Not only does *Drosophila* have a short life cycle, it also has only four chromosomes with many known genes. Sex determination is similar to that in humans—XX females and XY males—although there are differences in the mechanisms by which these chromosome complements cause sex differentiation (Chapter 12).

Muller constructed stocks of *Drosophila melanogaster* that made detection of one class of mutations, X-linked recessive lethals, very efficient. Various versions of the test have been used, the Muller-5 test being a common example. In this test, females are homozygous for the dominant Bar-eye (*B*) mutation and for the recessive gene *apricot* (w^a) that changes the eye color from the usual wild-type red to orange. Both loci are on the X chromosome. The *Bar-apricot* chromosome, usually designated as (*In*)*Basc*, also has some chromosomal inversions that suppress crossing over, thus the (*In*) included in the designation of this chromosome.

To test a mutagen, wild-type males are treated with the mutagen and then mated with the females, as illustrated. The F_1 females are heterozygous for the (*In*)*Basc* X chromosome and a treated wild-type X chromosome. F_1 males have the (*In*)*Basc* chromosome. The F_1 males and females are then allowed to inter-

Box **16.1**

(*continued*)

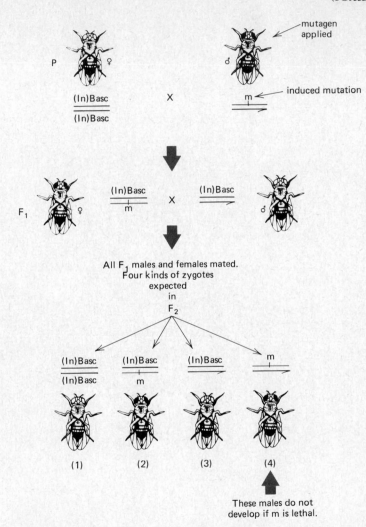

mutagen applied

induced mutation

P ♀ X ♂

$$\frac{(In)Basc}{(In)Basc}$$

$$\underset{m}{\rule{1cm}{0.4pt}}$$

F₁ ♀ X ♂

$$\frac{(In)Basc}{m}$$

$$(In)Basc$$

All F₁ males and females mated.
Four kinds of zygotes
expected
in
F₂

$$\frac{(In)Basc}{(In)Basc}$$ $$\frac{(In)Basc}{m}$$ $$(In)Basc$$ $$\underset{m}{\rule{1cm}{0.4pt}}$$

(1)   (2)   (3)   (4)

These males do not
develop if m is lethal.

breed. They will produce four types of F₂ offspring:

1. females homozygous for (*In*)*Basc* recognizable by the recessive eye color;
2. females heterozygous for (*In*)*Basc* and a treated wild-type X chromosome, recognizable by the wild-type eye color;

3. males with the (*In*)*Basc* chromosome, recognizable by expression of both apricot and Bar;
4. males with the treated wild-type X chromosome.

Any recessive lethal mutation that was induced on the X chromosome during treatment will be expressed in the fourth

Box **16.1**

(*continued*)

group. Therefore, if there are no wild type males in the F_2, such a mutation must have occurred.

The frequency of X-linked lethals induced would be simply the frequency of F_1 matings that produce no wild-type males in the F_2. It doesn't make any difference at which locus on the X chromosome the mutation occurs so long as it is recessive in females and lethal in males.

This means that one can deal with events that are fairly common (any X-linked lethal) rather than looking for very rare mutations at a specific locus.

Drosophila continues to have an important role in mutagenesis testing, both because of the efficiency of the detection systems and because it is a eukaryote with chromosomal mechanisms similar to those of humans.

The most widely used microbial test is the Ames test (Fig. 16.2). The bacterium used is *Salmonella typhimurium*, which is easily grown under laboratory conditions, provided that the essential nutrients are present in the culture medium. Wild-type *S. typhimurium* manufactures most of its own requirements, including the amino acid histidine needed for protein synthesis. Mutant forms that cannnot make histidine can grow only if it is added to the medium. These *his*⁻ strains are called *auxotrophs* because they require an auxiliary nutrient, as opposed to the nonmutant forms, which are called *prototrophs*. As is true of many but not all mutant forms, some can mutate from *his*⁻ to *his*⁺; that is, they can undergo reverse or back mutation to the original prototrophic form and can grow without histidine supplements.

In the simplest form of the Ames test, a petri dish containing culture medium in a gel form (agar) is heavily seeded with *his*⁻ bacteria and kept at 37°C for a day or two. If no histidine supplement is added, the vast majority of the cells cannot grow. But a few colonies will appear, scattered about the surface of the medium, because the single cell from which each colony arose had mutated spontaneously from *his*⁻ to *his*⁺. If, at the beginning of the experiment, a mutagen is added to the culture medium, the number of *his*⁺ revertants is increased. One way to do this, which shows the effect very dramatically (but not very quantitatively), is to place the mutagen in a well cut into the medium and allow it to diffuse radially. The high concentration of mutagen near the well causes a much higher frequency of mutants near the well.

A number of special strains of *Salmonella* have been developed that are more sensitive to mutagens and that respond differently to different classes of mutagens. An additional component often added to the medium is an extract of mouse or other mammalian liver. Because of the enzymes present in it, this extract can "metabolize" the test mutagen in much the same way as in the mammal from which it was derived. This may convert some mutagens into nonmutagenic substances, and it also can convert some nonmutagens into mutagens. The results are thus more relevant to human beings.

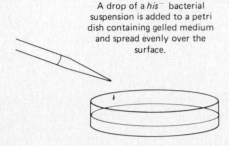

The bacterium *Salmonella typhimurium* ordinarily can grow without added histidine because it has the genes (*his*⁺) necessary to synthesize what is needed.

Certain mutant strains of *S. typhimurium* (*his*⁻) cannot synthesize histidine. These will grow only if the histidine is added to the medium. By *reverse* mutation, cells of these strains can reacquire the ability to synthesize histidine.

The induction of such reverse mutations can be used to measure the mutagenicity of substances added to the medium.

A drop of a *his*⁻ bacterial suspension is added to a petri dish containing gelled medium and spread evenly over the surface.

After 2–3 days incubation, the petri dish is examined for growth of bacterial colonies.

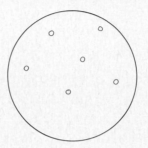

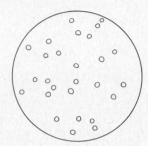

If no histidine is in the medium, only those bacteria can grow that have mutated from *his*⁻ to *his*⁺.

If a chemical mutagen is added to the medium, many more *his*⁺ colonies appear.

FIGURE 16.2
The Ames test for mutagenicity of chemical compounds. The only colonies that can grow are those that have mutated from histidine dependence to histidine independence. Therefore many more colonies are observed if a mutagen is added to the medium.

When a drug or other substance is being considered for commercial development, it has become common to submit it to an Ames test. If the test is strongly positive, the company will consider whether the potential benefits of the drug, perhaps as a treatment for cancer, justify its continued development. If the Ames test is negative, additional tests with mammalian systems are required before the substance can be considered nonmutagenic for humans. Unfortunately requirements for mutagenicity testing are recent, and most of the substances to which we are exposed have not been tested.

Drosophila.   The fruit fly is not only the first organism in which artificially induced mutations were detected; it also remains one of the more useful for testing chemicals for mutagenicity. The life cycle is short, only about two weeks, and large numbers can be used in a single test. More importantly, it is a eukaryote, as are mammals. Therefore, if the presence of a nucleus is important in sensitivity to mutagens, then *Drosophila* should tell us more about human risk than *Salmonella*. Many modifications of the test procedures are currently in use with *Drosophila*. Most resemble the original test of Muller in principle.

Mice.   Of the many organisms that have been used to study mutation, mice are especially interesting since they are mammals and should be most relevant to humans. The loci most often studied are recessive coat-color genes. In a typical experiment, males homozygous for the dominant coat-color genes are treated with the suspected mutagen and then mated with females homozygous for recessive alleles at several specific loci that influence coat color (Fig. 16.3). If a mutation has occurred at one of these loci in the male, the offspring from the mutant sperm will have the recessive coat characteristic of animals homozygous at that locus. If no mutation has occurred at any of these loci, the offspring will be heterozygous and will have the dominant coat pattern.

Mice have been used especially to study the effects of radiation but also are being used for other potential mutagens. The difficulty with mice is the very large number—and associated cost—required for a test. For example, in the strain used at Oak Ridge, seven loci are tested simultaneously. Mutation rates of untreated animals are on the order of 0.5×10^{-5} per locus per gamete. That means that only $7 \times 0.5 = 3.5$ mutants would be expected in 100,000 progeny tested. If treatment doubled the mutation rate compared to the spontaneous rate in the untreated

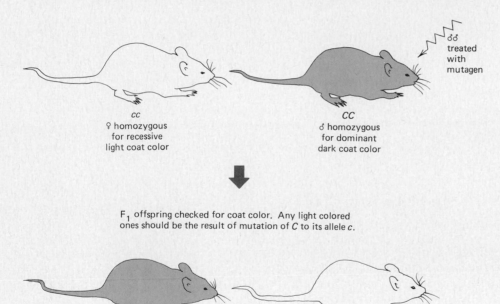

cc
♀ homozygous
for recessive
light coat color

CC
♂ homozygous
for dominant
dark coat color

♂♂
treated
with
mutagen

F$_1$ offspring checked for coat color. Any light colored
ones should be the result of mutation of C to its allele c.

Cc

cc
mutant

FIGURE 16.3

Diagram of mouse test used to detect mutations. The female parent is homozygous for a recessive mutation that affects coat color. The male parent is homozygous for the dominant allele that produces a dark coat. In the test, the male is treated and then mated to the homozygous recessive female. If no mutation has occurred at that locus as a result of the treatment, the F$_1$ offspring should be heterozygous and should resemble the male parent in coat color. Any offspring produced by a sperm that is mutant at the locus will resemble the female parent. In actual practice, the female is homozygous for seven different recessive alleles that affect coat color. Therefore a mutation at any one of these seven loci can be detected in the F$_1$.

control group, only seven mutants would be expected per 100,000 progeny. Each group would have to contain 200,000–300,000 mice in order to produce statistically valid results, and that number would not permit study of the dose response, that is, the number of mutants produced per unit of exposure. Small wonder then that geneticists speak of megamouse experiments! The large financial requirements for such experiments and the limited facilities available assure that only studies indicated by other observations to be especially important to humans can be contemplated.

Cultured Mammalian Cells.   In Chapter 10, the techniques for growing mammalian cells in the laboratory were discussed. Such cells express the genotypes of the individuals from which they are derived and often the particular differentiated functions of the tissue of origin as well. When mutations occur, they can often be recognized by the altered phenotype of the mutant cells and their clonal descendants.

One of the mutant traits often used in cell culture is azaguanine resistance. Azaguanine is similar to guanine and can replace guanine to some extent in metabolism. However, it does not function normally, and cells that are fooled into using it are killed. The first step in using azaguanine is catalyzed by an enzyme, hypoxanthine phosphoribosyl transferase or HPRT, that converts azaguanine into a nucleotide by addition of ribose phosphate (Fig. 16.4). The gene that codes for HPRT is on the X chromosome. Many individuals have mutant forms of the HPRT gene. If the mutant gene has lost its catalytic activity, the persons have Lesch–Nyhan syndrome, a condition that involves mental retardation and self-mutilation.

If azaguanine is added to a culture of cells from a normal person, the cells are killed except for rare mutant cells that are said to be azaguanine-resistant. These Ag^r cells can grow quite well in the presence of azaguanine because the HPRT gene has mutated from $HPRT^+$ to $HPRT^-$, that is, from a form in which normal enzyme is coded to a form in which normal enzyme is not produced. Therefore azaguanine cannot be incorporated into the cell and kill it. The frequency of AG^r cells is low, on the order of one per million, but the number of mutants can be increased by various mutagens.

The HPRT locus has been especially useful also because reverse mutations can be detected. As shown in Figure 16.4, a special selection medium called HAT (**h**ypoxanthine–**a**minopterin–**t**hymidine) does not support growth of $HPRT^-$ cells. Mutant $HPRT^+$ cells can grow in HAT and be counted, isolated, or otherwise studied.

Chromosome alterations also can be studied in cultured cells. The types of alterations are those that involve chromosome or chromatid breaks with abnormal rejoining. Some, such as balanced translocations, produce balanced daughter cells with a full complement of genes and therefore will persist both *in vivo* and *in vitro*. They are said to be *stable* rearrangements. Other rearrangements, such as deletions and dicentrics, are *unstable*, since there is no mechanism to assure that the daughter cells receive a full complement of genes. Cells with unstable rearrangements would rarely survive the first cell division.

The cultured cells studied may be short-term cultures of lymphocytes from a person or animal exposed to a potential mutagen, or they may be established cell lines exposed during growth to a mutagen. If the exposure occurs *in vivo*, the effect should reflect the particular metabolism of the species and individual and should

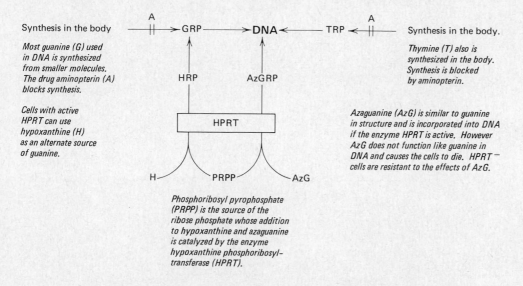

Synthesis in the body

Most guanine (G) used in DNA is synthesized from smaller molecules. The drug aminopterin (A) blocks synthesis.

Cells with active HPRT can use hypoxanthine (H) as an alternate source of guanine.

Thymine (T) also is synthesized in the body. Synthesis is blocked by aminopterin.

Azaguanine (AzG) is similar to guanine in structure and is incorporated into DNA if the enzyme HPRT is active. However AzG does not function like guanine in DNA and causes the cells to die. HPRT⁻ cells are resistant to the effects of AzG.

Phosphoribosyl pyrophosphate (PRPP) is the source of the ribose phosphate whose addition to hypoxanthine and azaguanine is catalyzed by the enzyme hypoxanthine phosphoribosyl-transferase (HPRT).

HAT medium contains hypoxanthine (H), aminopterin (A), and thymidine (T). Only HPRT⁺ cells can grow in HAT medium, since such cells must use hypoxanthine to make guanine for DNA.

FIGURE 16.4

The system used for detection of mutations at the HPRT (hypoxanthine phosphoribosyl transferase) locus in cultured cells. Normal cells are killed by azaguanine unless the HPRT becomes nonfunctional through mutation. Such HPRT⁻ cells can grow quite well so long as they are allowed to make their own purines. However if purine synthesis is blocked with the drug aminopterin (which also blocks thymidine synthesis), only cells that have mutated back to HPRT⁺ can grow in HAT medium, since they must depend on conversion of hypoxanthine for a source of purines.

therefore be a reliable predictor of exposure. Cell lines have many of the metabolic features of intact animals, but they also lack many differentiated functions. For example, a fibroblast culture does not express the many enzymes of liver important to metabolic conversion of potential mutagens. In some senses then, mammalian cell cultures may be almost like bacterial cultures. Since all cultured cells are somatic rather than germinal, the mutation rates observed are not necessarily representative of germinal mutations.

Mutation Rates in Human Populations

Point Mutations.   In the case of human beings, one cannot make the necessary crosses to demonstrate lethal mutations as one often does with *Drosophila*. Rather, one must look for traits that were not transmitted from the parents. Unfortunately, very few human genes are suitable for measuring mutation rate. The great majority of mutations are recessive and could only be detected if they happen to be combined with a similar mutant gene in homozygous combination. But homozygous recessives typically do not have parents or other ancestors also affected. Therefore there is no way to distinguish between a homozygous affected person, both of whose alleles were transmitted from prior generations, and a person one of whose alleles represents a new mutation. Instead, one must rely primarily on dominant traits to measure mutation. Indeed, relatively few dominant traits are very useful for this purpose. Traits must be easy to recognize, and they must be very rare so that most affected persons are new mutants. Only about a dozen genes meet these requirements. The population to be studied must be very large, at least a million or more persons in size.

Measurement of the mutation rate of achondroplasia was discussed earlier in the chapter. Other examples of dominant and X-linked mutations that have been used to study human mutation rates are given in Table 16.1. It is important to remember that each mutation rate refers to a single locus. But there are a very large number of loci, probably 20,000 to 50,000. Therefore the likelihood that a particular gamete will have a new mutation at *some* locus is not at all small. If there are 20,000 per genome, each with a likelihood of mutating with a frequency of 1×10^{-5}, then each gamete would have roughly an 18% chance of carrying at least one new mutation somewhere among its 20,000 loci, and some will have more than one.

Chromosome Mutations.   The frequency of aberrant karyotypes among newborns is 0.5%, or 1 in 200 live births. Most of these are trisomies and monosomies, due to nondisjunction. A few involve structural rearrangements that must involve breakage and abnormal rejoining. Most of the newborns with karyotype abnormalities do not become fertile adults. Hence the larger fraction of the abnormalities are new mutations rather than having been transmitted from a similarly affected parent.

The abnormalities in newborns is just the tip of the iceberg. Approximately 15% of recognized conceptions result in spontaneous abortion. Of these, some 50% have karyotype abnormalities (Table 9.1). As with newborns, the most common

TABLE 16.1
A Partial List of Human Genes Whose Mutation Rates Have Been Measured

Trait	Population	Mutations Per Million Gametes
Achondroplasia	Denmark	10
Aniridia	Denmark, Michigan	2.9, 2.6
Retinoblastoma	England, Michigan, Switzerland, Germany	6–7
Acrocephalosyndactyly	England, Germany	3, 4
Neurofibromatosis	Michigan, USSR	100, 46
Intestinal polyposis	Michigan	13
Polycystic disease of the kidneys	Denmark	65–120
vonHippel–Lindau syndrome	Germany	0.18
Hemophilia A	Germany, Finland	57, 32
Hemophilia B	Germany, Finland	3, 2
Duchenne muscular dystrophy	Various	43–105

From F. Vogel and R. Rathenberg, Spontaneous mutation in man, *Advances in Human Genetics* Vol. 5, Plenum, New York, 1975, pp. 223–318.

abnormalities are due to nondisjunction. But 20 of the 1000 in Table 9.1 had structural rearrangements that could well be increased by agents, such as radiation, that break chromosomes. The high frequency of triploidy has been shown to be due in part to fertilization by two sperm.

Aberrant chromosomes may occur in somatic cells as well as in germ cells. The persons in whom they occur (all of us) need not be abnormal as a consequence, since most of the somatic cells would still be normal. The typical observation is based on the culture of white blood cells. Most of the cells examined will have a normal complement of chromosomes, but occasional cells will have an abnormal karyotype. Reference has already been made to the use of lymphocyte cultures in mutagen testing. Abnormal karyotypes also arise spontaneously but at rather low frequency. Furthermore, and most importantly, the frequency of abnormal karyotypes is very dependent on the way the blood sample is handled and on the conditions of culture. It is difficult therefore to select a particular frequency of aberrant karyotypes as representing the background or spontaneous rate. It is possible, however, to compare the frequencies of aberrations in different groups of persons if great care is taken to assure comparable handling of samples. One may thus recognize that one group has had greater exposure to agents that produce aberrations. Examples of this use of somatic chromosome aberrations are given in later sections.

Radiation and Mutation

All molecules above a temperature of absolute zero ($-273°C$) are unstable due to the thermal energy that they contain. They continually collide with each other and they vibrate as if the covalent bonds holding them together were made of springs. Occasionally these collisions and vibrations cause a bond to break, changing the chemical nature of the molecule in which it happens. Many molecules are very stable under ordinary conditions of temperature and environment. Both proteins and nuclei acids are stable, as are most of the structures important in biological systems. This must be so if we are to measure life spans in years rather than milliseconds. But every molecule, no matter how stable, has a likelihood of undergoing spontaneous chemical change due to the thermal energy in the molecule. When such changes occur in DNA, they may well become permanent in the form of mutations. One can never expect to reduce the mutation rate to zero except at absolute zero.

All molecules also are subject to change by interacting with other molecules or with physical agents in the environment. One would expect that mutation, which is ultimately a chemical reaction, would be influenced by chemical and physical agents as are other chemical reactions. The first agent found to influence mutation rate was ionizing radiation. This discovery was made by H. J. Muller, who developed for this purpose the X-linked lethal test in *Drosophila* described earlier. Radiation remains of great concern because of its potential detrimental effect on human genes.

The Nature of Radiation

Many different kinds of radiation are of concern to biologists. The most familiar is electromagnetic radiation (Box 16.2). The longer wavelength electromagnetic radiation—radio waves, infrared and visible light—has energy too low to break covalent bonds. When an infrared photon strikes a molecule and is absorbed, it increases the thermal vibration of the molecule, making it "hotter" but not otherwise altering it. When a particle of gamma radiation strikes a molecule, the energy imparted is sufficient to alter the distribution of electrons in the shells around the atomic nuclei. This often causes the covalent bonds in which these electrons participate to disrupt, forming free radicals or ion pairs. For example, if a molecule of water absorbs an X ray, the reactions that may occur are shown in Figure 16.5. Radiation of this amount of energy is often referred to as ionizing radiation because of these reactions.

Free radicals are very unstable chemical forms, since they have unshared electrons, and they will react further with other molecules in the vicinity. These reactions that are illustrated with water can occur with many other cell components, including DNA, although it must be remembered that a high proportion of cell contents is water. It has been shown by W. S. Stone and co-workers that bacteria added to previously irradiated media have increased mutations, proving that the alteration in DNA can occur as a secondary reaction with the primary products of irradiated cell constituents.

Ultraviolet (uv) light lacks sufficient energy to cause ionization. Nevertheless, it can cause mutations. Many molecules absorb uv very efficiently. Some especially sensitive molecules can undergo chemical change with relatively less energy.

Box 16.2

The Nature of Electromagnetic and Other Radiation

When a pulse of electric current flows through a wire, a magnetic wave is generated that radiates from the wire. When an atom of cobalt-60 disintegrates, it emits a gamma ray that can be detected by its track on a photographic film. These seemingly different kinds of radiation are actually two examples of electromagnetic radiation. The energy levels are very different as are the effects on chemical and biological systems.

Electromagnetic radiation is both wave-like and particle-like. At low energy levels, such as radio waves, the wave characteristic can be used most effectively to explain its behavior. At high energy levels, such as X rays and gamma rays, radiation is more effectively treated as particles. The energy level is related to the frequency by the equation $\varepsilon = h\nu$, where ε is the energy, h is a constant (Planck's constant), and ν is the fre-

Box 16.2

(*continued*)

quency. Since electromagnetic rays travel at the speed of light $c = 3 \times 10^{10}$ cm/s, the wavelength $\lambda = 3 \times 10^{10}/\nu$. These relationships were developed by Max Planck in 1900 as part of the quantum theory, and the particles of radiation are called *photons*, each with a *quantum* of energy.

The various kinds of electromagnetic

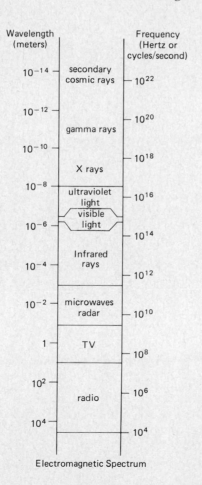

Electromagnetic Spectrum

rays are shown in the diagram. It must be emphasized that there is no qualitative difference as one goes from one "kind" of radiation to another. Rather, the energy of the photons change, and this may cause different responses by molecules struck by the photons. Radar and infrared rays cause molecules that absorb the energy to move faster and vibrate faster, thereby becoming hotter. Visible light has enough energy to cause shifts in the electrons of many molecules, including eye pigments, and thereby cause us to have vision. Ultraviolet rays cause electron shifts in virtually all biological molecules but cannot break covalent bonds. X rays and gamma rays have sufficient energy to break bonds, producing free radicals and ion pairs.

Not all radiation is electromagnetic. When atoms disintegrate, several types of subatomic particles may be emitted capable of causing ionization of molecules that they may strike. The electron is the familiar negatively charged particle in the shells of atoms and in electric currents. Protons are positively charged particles (hydrogen nuclei) with a mass 1859 times greater than that of an electron. Neutrons are uncharged particles of the same mass as protons and are found in all nuclei except the predominant form of hydrogen. Positrons have the same small mass as electrons but have a positive charge. Alpha particles are helium nuclei, consisting of two protons and two neutrons. They are therefore about four times as massive as a proton. Because of their mass, high-speed alpha particles can have very high energy and are very effective at producing ionization.

FIGURE 16.5
The effect of X rays on water. Since most of the cell volume is water, these reactions are very important in radiation damage to cells.

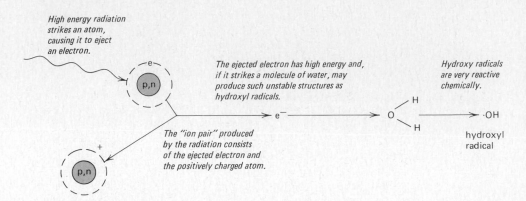

High energy radiation strikes an atom, causing it to eject an electron.

The ejected electron has high energy and, if it strikes a molecule of water, may produce such unstable structures as hydroxyl radicals.

Hydroxy radicals are very reactive chemically.

hydroxyl radical

The "ion pair" produced by the radiation consists of the ejected electron and the positively charged atom.

Absorption of uv by these molecules increases the mutation frequency. Because of the efficient absorption of uv, it cannot penetrate more than a millimeter or two into the skin. Therefore, mutations can only be induced in surface cells or, in the case of microorganisms, in cells at the surface of the medium.

Measurement of Radiation. The traditional unit of radiation is the *roentgen*, symbolized R, named after W. C. Roentgen, who discovered X rays in 1895. The roentgen is a physical unit, defined as the amount of radiation that produces ionization equal to 2.58×10^{-4} coulomb/kg air. A related unit is the *rad* (**r**oentgen **a**bsorbed **d**ose), equal to 100 ergs absorbed/g. The roentgen is a measure of exposure, whereas the rad measures only the fraction of radiation actually absorbed by the medium through which it passes. For biological systems, the two units are not very different, but the rad is more often used, since unabsorbed radiation would have no effect. A third unit is the *rem*, or **r**oentgen-**e**quivalent-**m**an. The rem takes into account not only the amount of radiation absorbed but also the biological effectiveness of the particular kind of radiation. For electromagnetic radiation (X rays, gamma rays, cosmic rays), the rem and rad are approximately equal and are often used interchangeably. For heavier particles such as neutrons and alpha particles, the relative biological effectiveness (RBE) is much greater per unit of energy absorbed as compared to electromagnetic radiation, and the rem and rad are not equivalent.

It is often necessary to deal with small fractions of a rem or rad when discussing human exposures to ionizing radiation. It is convenient to use units that are 1/1000 as large, the millirad (mrad) and millirem (mrem). One must be careful to keep the distinction clear. A whole body exposure of 500 rem would induce severe illness and often death, while 500 mrem is the amount most of us receive as background normally over a 5-year period.

Sources of Radiation. We are never free from exposure to radiation. It surrounds us and penetrates us at all times. Indeed it emanates from us. The principal source of radiation for most persons is the radioactive elements of which we and our surroundings are composed. Many of the common chemical elements are a mixture of isotopes (Chapter 3), some of which are radioactive. For example, the element hydrogen consists primarily of one proton and one electron, a very stable arrangement. An additional form or isotope has a neutron in the nucleus. This "heavy hydrogen" or deuterium is also stable (not radioactive). It can be used experimentally to replace the usual hydrogen in a molecule and provide a tracer for that molecule in biological or chemical experiments. A third form of hydrogen, tritium, has two neutrons in the nucleus. This isotype is unstable and disintegrates according to the reaction

$$\, {}_{1}^{3}\mathrm{H} \rightarrow {}_{2}^{2}\mathrm{He} + \beta^{-}$$

The β^{-} emitted as a product is an electron that is readily absorbed by materials. Hence it is useful experimentally to trace the movement of organic compounds that contain hydrogen (which virtually all do) through biological systems.

Table 16.2 shows the major sources of radiation plus a number of minor sources to which we are exposed. The three major sources of nonmedical radiation are (1) terrestrial, primarily the rocks on which we walk and with which we build; (2) cosmic, the very high energy rays from outer space, some of which are absorbed by the atmosphere; and (3) isotopes within the body, primarily ^{40}K (potassium-40). These sources add up to approximately 100 mrem/yr. Medical X rays add an additional 100 mrem average per person per year. Except for occupational exposures, other sources of radiation are small by comparison.

In considering radiation dose, one must distinguish between whole body

TABLE 16.2   Sources of Exposure of U.S. Population to Ionizing Radiation
Only a selection of the minor sources is given.

Terrestrial		
Atlantic and Gulf Coast States	23 mrem/yr average	
Colorado plateau area	90 mrem/yr average	
Remainder of U.S.	46 mrem/yr average	
U.S. average, weighted by population		40 mrem/yr
Cosmic rays		
Sea level	26 mrem/yr	
10,000 ft. altitude	74 mrem/yr	
U.S. average, weighted by population		31 mrem/yr
Internal radionuclides		
Potassium-40, gonadal dose	19 mrem/yr	
U.S. average, all radionuclides		28 mrem/yr
Medical X rays, bone marrow dose		103 mrad/yr
Atmospheric nuclear tests prior to 1971		4.4 mrem/yr
Naval nuclear ships, exposure to workers only (1975)		123 mrem/yr
Air travel, cosmic radiation to passengers		0.2 mrem/h
Luminous wristwatches, gonadal dose to wearer		1–3 mrem/yr
Fossil fuels, lung dose, U.S. average		
Coal		0.05–10 mrem/yr
Oil		0.004 mrem/yr
Gas		7 mrem
Tobacco products, bronchial epithelium dose of users		8,000 mrem/yr
Television receivers		0.5 mrem/yr

From The Effects on Populations of Exposure to Low Levels of Ionizing Radiation, National Academy of Sciences, Washington, D.C., 1980.

radiation and radiation to a particular organ. For example, when X rays are used therapeutically—that is, to treat a particular tissue such as a cancerous growth—rather than diagnostically, the X-ray beam is aimed at the target organ only. Other parts of the body are shielded with metal, usually lead, that absorbs X rays. The amount of radiation reaching the bone marrow, a very sensitive tissue, or the gonads may be quite small. The high exposure from smoking is to the lungs only and not to the rest of the body.

Even with whole body radiation, there is attenuation of the exposure for organs on the interior of the body compared with the skin. The attenuation is not large, only about 20% for typical X rays. It is necessary though to correct for body absorption for some purposes. For example, for genetic effects, the important exposure is that of the gonads. No other exposure has any relevance. The amount of radiation that reaches the gonads is influenced by whether one is male or female, the direction of the radiation, and the ability of the radiation to penetrate tissue. The genetic importance of the radiation that raches the gonads is also a function of the likelihood of having offspring. These variables have been combined into a concept of the *genetically significant dose* (GSD). The GSD is the exposure of the gonads multiplied by the probability of subsequent offspring. Radiation of a person who will have no more children has no genetic significance, although there may be somatic effects important to that person.

Experimental Studies of Radiation

The first demonstration of the genetic effects of radiation were reported by Muller. (Refer to Box 16.1.) In his work and that of many others with *Drosophila*, it was demonstrated that the recovery of mutants is directly proportional to the dose of radiation administered (Fig. 16.6). These studies suggest that the straight line extends all the way to zero radiation, meaning that there is no threshold of exposure below which there is no effect. One cannot verify this easily because of the very small effect at such exposures.

Studies of mice have been much more limited because of the problem of rearing mice as compared to *Drosophila.* The major studies have been carried out at two centers—Oak Ridge National Laboratories in the United States and Harwell in England. Historically the studies were based primarily on coat-color mutations. Many of the mutants detected are probably deletions, and it is possible that base-pair substitutions are not usually detected. More recently, dominant skeletal mutations have proved very useful. Much has been learned about the quantitative effects of radiation on mutation, especially as related to spermatogenesis and oogenesis. Figure 16.7 shows the recovery of mutants versus radiation applied at various parts of the maturation cycle of sperm. Mature mouse oocytes, which are arrested in the early stages of meiosis, are very resistant to radiation. Whether a comparable situation exists in humans has not been established.

Radiation studies of mice and *Drosophila* show clearly the differences in sensitivity to radiation among species. *Drosophila* is relatively resistant to radiation. The LD_{50} (the dose required to kill 50% of the population) is approximately 100,000 rad. The LD_{50} for mice is only 600 rad. While no human studies have ever

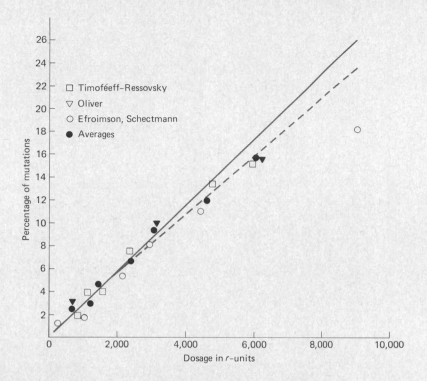

FIGURE 16.6

The relationship between induction of X-linked lethal mutations in *Drosophila* and exposure to X rays. *Drosophila* is relatively insensitive to X rays as compared to mammals. Therefore the doses are very high compared to human exposures. The dashed line is drawn from the experimental data. The solid line takes into consideration the fact that many mutations at higher exposures occur on chromosomes that already have another lethal mutation. A fly can die only once! (From A. H. Sturtevant and G. W. Beadle, *An Introduction to Genetics*, W. B. Saunders, 1939; after N. W. Timoféeff-Ressovsky, K. G. Zimmer, and M. Delbrück.)

been carried out to study this point, studies of persons accidentally exposed indicate that the human LD_{50} is about 450 rad.

Radiation Effects on Humans

No one has ever been exposed deliberately to radiation in order to measure the damage produced. Some of the early pioneers of radiation physics did expose themselves to radiation, sometimes largely by accident but also by design, to see what the effects would be. Pierre Curie exposed his arm to radium and was burned. Henri Becquerel accomplished much the same result accidentally by carrying some radium in his vest pocket for several days. But much of what we know of the

FIGURE 16.7
The recovery of X-ray in-
duced translocations in
mice as a function of the
period during sper-
matogenesis when expo-
sure occurred. In the
experiment illustrated,
male mice were subjected
to 300 rad X rays and
then mated once each
week to nonirradiated
females. The progeny
produced from successive
matings arise from germ
cells at earlier and earlier
stages of development at
the time of the treatment.
The cells are much more
sensitive at some periods
than at others, with spermatids being the most
sensitive. Other measures
of mutation and other
species give generally
similar results, although
the details vary. (Data
from A. Léonard and
G. Deknudt, *Canadian J.
Genet. Cytol.* 10: 495,
1968.)

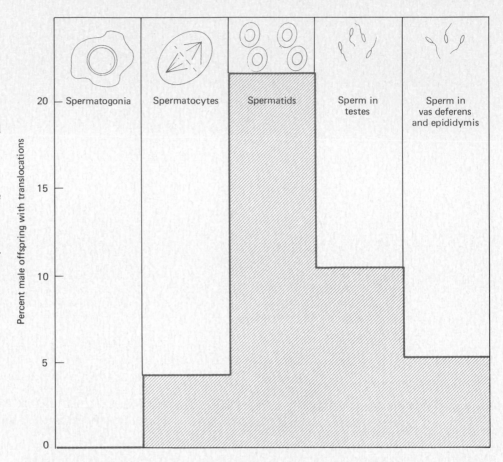

damaging effects of radiation comes from accidental exposures or from exposures
for other purposes, such as the atomic bombs and medical therapy.

Atomic Bomb Studies.   The atomic bombs dropped on Hiroshima and
Nagasaki in 1945 killed over 100,000 civilians and an unknown number of military
personnel, some by the heat and trauma of the blasts, others by the heavy dose of
radiation. Many others survived but received substantial amounts of radiation,
primarily gamma radiation. By reconstruction of the amounts of radiation each
person received, based on location at the time of the blast, shielding at that
particular position, and presence of symptoms of radiation sickness, it has been
possible to study the subsequent biological effects based on dose. It should be
noted that the atomic radiation was an acute dose, that is, most of the radiation
was received at the time of the detonation or shortly thereafter from the "black
rain" fallout. Chronic radiation is received at low rates over a long period. Much
higher total doses of chronic radiation can be tolerated without causing radiation
sickness or death, but there need not be any such difference for mutations.

After the end of World War II, a study was set up jointly by the United States and
Japan to measure the effects of the radiation on the atomic bomb survivors. This

study is still underway and has made important contributions to our knowledge of human radiation effects. A number of approaches have been used to detect and measure the effects. These studies can be divided into those that measure somatic effects, that is, effects on tissues and organs of the persons who were themselves irradiated, and genetic effects or effects on offspring conceived after their parents were irradiated. Relatively few survivors received more than 500 rad, as might be expected since this dose is near the mean lethal dose for humans. In addition, such heavily irradiated persons were closest to the point of detonation and were most likely to have been killed or injured by heat and trauma from the blast. The most numerous irradiated survivors were those who received moderate or low levels of radiation. The magnitude of expected effects would be correspondingly smaller in this group.

Of the various somatic effects observed, the most interesting from a genetic point of view was the induction of cancer, especially leukemia. It was already known that leukemia could be induced by ionizing radiation. Within 5 years after exposure, the atomic bomb survivors had a substantially increased incidence of leukemia compared to nonirradiated controls (Fig. 16.8). This increased risk has diminished with time and has returned essentially to the background level of nonirradiated populations. In the meantime, other tumors that were not increased initially have since become a much higher risk. There is substantial evidence that one or more of the steps in carcinogenesis is mutation. The latent period—the time between exposure to a carcinogen and appearance of a tumor—varies with the type of cancer and is shorter for leukemia than for other cancers. The Japanese studies also show that the risk of leukemia is higher in young children who were irradiated than in adults, and the greater the amount of radiation, the shorter the latent period.

A second somatic effect was observed among those who were irradiated *in utero* during the first three months of gestation. There was substantial reduction in head

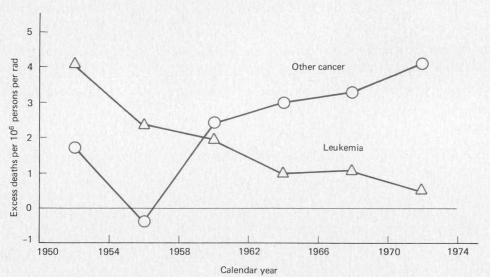

FIGURE 16.8
The increased risk of leukemia and other malignant tumors in irradiated survivors of the atomic bombs in Hiroshima and Nagasaki. (Redrawn from G. W. Beebe, H. Kato, and C. E. Land, *Radiation Research* 75: 138–201, 1978.)

circumference, often associated with mild mental retardation. This apparently occurs only with substantial radiation, and the effect on intelligence seems restricted to the first part of the gestation period. The effect was initially observed in England in pregnant women who had received medical X rays. It is now a generally accepted effect that argues against any prenatal radiation that is not absolutely necessary. The effect is not a genetic effect, of course, but rather a developmental defect.

One additional somatic effect that can still be seen in atomic bomb survivors is the presence of structural rearrangements in chromosomes of peripheral lymphocytes. One might expect most of these to be eliminated over time, since some, such as dicentrics, cannot divide properly at anaphase. To be sure, most of those still detected tend to have stable rearrangements. But a few unstable rearrangements can still be observed. It may well be that some of the small lymphocytes that are stimulated to grow in culture survive for dozens of years *in vivo* without dividing. A chromosome spread from a survivor of the atomic bomb at Hiroshima is shown in Figure 16.9.

Genetic effects should be detectable, if at all, in the offspring conceived after their parents were irradiated. A number of traits have been examined for mutation: growth and development, intelligence and various intellectual skills, karyotypes, and point mutations that affect enzyme functions. To date no significant effect of mutation on any of these traits has been noted. This does not mean that mutations did not occur. It is very well established that ionizing radiation does cause mutations. It means rather that the number of mutations produced by the amount of radiation received was too small to be distinguished from the background of genetic and environmental variations that are present both in the exposed and unexposed populations. Perhaps a sample twice as large would have been adequate to show an effect, but that is an experiment we can hope to avoid.

Medical Exposures.   Most of us are familiar with X rays used as a diagnostic tool. Generally the exposure from a diagnostic X ray is small. The usual chest X ray is about 45 mrad per exposure to the skin, but values for other types of X rays may be much higher than this. The X-ray dose to the skin is attenuated as the beam passes through tissue, so that the exposure of the gonads is substantially less. Also, only a part of the body is exposed during diagnostic X rays. Therefore, the high levels of radiation reported for one particular organ are not at all comparable to whole body radiation of the same intensity.

Radiation is also used as a therapeutic agent. Currently the principal use is in the treatment of cancer, where the potential benefit clearly outweighs the risks. Radiation has been used also in many other situations where the benefit/risk ratio is much less certain. For example, it was used at one time to treat ringworm of the scalp, with skin doses of 600 rad or more, brain doses of greater than 100 rad, and bone marrow doses of nearly 100 rad. Small wonder that some patients showed nausea and vomiting, symptoms of radiation sickness! These patients now show an increase in tumors in various parts of the body.

Ankylosing spondylitis is a disease involving fusion of the vertebrae, for which radiation of the spine has often been used as a treatment. A series of patients in England have been observed for some years for evidence of radiation damage. A typical course of ten treatments exposed the spinal region to some 500 rad, a very high dose were it given all at once. These patients now show a substantial increase

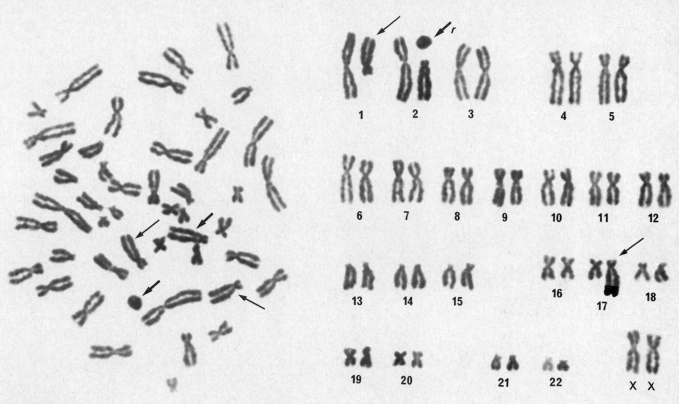

FIGURE 16.9
A chromosome spread and karyotype from a female exposed to atomic radiation in Hiroshima when she was 13 years old. The chromosome preparation was made 29 years after the exposure. The arrows indicate unusual aberrations that still can be found in greatly increased frequencies in white blood cells of atomic bomb survivors. (Supplied by Dr. A. A. Awa, Radiation Effects Research Foundation, Hiroshima.)

in various types of cancer. Such patients are unlikely to have additional offspring after treatment, and no reports are available on the genetic consequences in this group.

Internally Deposited Radionuclides. The previous discussions have dealt with external radiation, that is, radiation that strikes the body from some external source. Radioactive substances may also occur inside the body. The example of potassium-40 was given earlier. Potassium is uniformly distributed in the body so that the radiation from it is also evenly distributed. Many other radionuclides are deposited in very localized regions and cause intense radiation to that region but not to others.

Iodine-131 occurs in fallout from some nuclear weapons, such as the hydrogen bombs. Iodine also concentrates in the thyroid gland. For persons exposed to nuclear fallout, one of the major concerns is ^{131}I, which can cause heavy irradiation to the thyroid, destroying many of the cells and causing thyroid malfunction and deficiency. In the 1954 nuclear tests at Bikini atoll, the winds changed unexpectedly, causing several hundred Marshall Islanders to be in the fallout path. Some had symptoms of radiation sickness, but none died as a direct consequence of the irradiation. The only long-term effect detected is a high frequency of thyroid disorders. ^{131}I has a very short half-life so that the danger from ^{131}I in fallout drops off rapidly.

Uranium miners incur a special risk of lung damage from locally deposited particles of uranium ore. Such particles give intense radiation to the area immediately surrounding the deposited particle. Since the ore particles are insoluble, they persist in the lungs and continue to irradiate over a long period. The cells are heavily damaged or killed, and they have a high risk of becoming cancerous.

One of the more bizarre exposures to radiation occurred in the 1920s and 1930s among women who painted the dials of luminous watches with paint that contained radium. No precautions were exercised to prevent exposure. Indeed, in order to make sure that the brush had a good point, a woman would often moisten the brush in her mouth. Substantial amounts of radium were thus ingested. Radium is deposited internally in bone, where it persists for years. The first recognition of the health risk to radium painters was the appearance of tumors of the jaw bone, a type of tumor that is otherwise exceedingly rare. A number of women have died as a result of these tumors. Some 4000 are still alive and are being monitored for other evidence of health problems. As yet, they do not seem to be at high risk for other types of defect, and the high risk of bone tumors seems largely to have passed.

Radiation and Human Sensitivity.   The studies summarized above along with a number of others that have been carried out provide little solid information on the genetic sensitivity of humans to radiation. They do suggest that humans are not very different from mice in the magnitude of their response to a given amount of radiation. Some of the studies, especially the studies of atomic bomb survivors, permit conclusions about the upper limit of risk of genetic effects. And we can be reassured that humans are not more sensitive than are mice.

The measurement of genetic effects is often expressed as the *doubling dose*, which is the amount of radiation necessary to double the frequency of mutations over the spontaneous rate. Existing data required to calculate a doubling dose have large potential error. Therefore the calculated doubling dose has a large error. A recent estimate of the doubling dose is 50–250 rem, a fivefold range. The annual total for background and medical exposures is approximately 200 mrem. If we take 30 years as the average human generation time, the cumulative exposure for parents would be 6 rem, or ⅛ the minimum estimate of the doubling dose. Otherwise stated, only about ⅛ of the present "spontaneous" mutation rate can be attributed to radiation at a maximum, and the actual fraction could be as low as ¹⁄₄₀.

One point of concern in basing risk figures on radiation sensitivity of mice is that inbred strains of mice do not have the genetic variability of human populations. Some people may be more sensitive than others. Indeed, cells from persons with the recessive disease ataxia telangiectasia are known to be more sensitive to killing by radiation. They may well be more sensitive to mutation also. There is no reason to believe that the majority of humans is substantially more sensitive than the mice studied, but present knowledge would not exclude perhaps 5–10% of the population being more sensitive.

Chemical Mutagenesis

Since DNA is a chemical substance capable of reacting with many other chemicals, it is hardly surprising that many chemicals can cause mutations. The first report of a mutagenic chemical was made by Charlotte Auerbach, who found that formaldehyde can cause mutations in *Drosophila*. The list of mutagenic chemicals is now very long indeed.

Radiation has been recognized as a mutagen for half a century, and, in spite of our gaps in knowledge, more is known about it than any other environmental agent. By comparison, there are tens of thousands of chemicals in our environments, and we have only a few scant measures on a small sample of them. Whereas radiation has been with us throughout our evolution, many of the chemicals are new creations. We can take no comfort in the argument that humans have adapted to current levels of exposure. There has been no time for such adaptation.

Table 16.3 lists some of the substances that have been documented as mutagens. It should be remembered that none of these has been directly tested in humans. In order to test them in mice, a very large number is required. Therefore, most have not been extensively tested in mice either. Where testing has occurred, the exposures have been at very much higher levels than would occur naturally. It is necessary to do that in order to get a statistically meaningful result. Application of the results to humans requires the assumption that reducing the exposure to 0.1% reduces the risk by a corresponding amount, an assumption that may not always be correct.

We cannot remove chemicals from our environment any more than we can eliminate radiation. We and all our food are made of chemicals. We can ask, however, whether mutagenic chemicals are necessary or whether the benefits of their use outweight the risks. Measurement of the risks is much less satisfactory than for radiation, so any benefit/risk ratio will have a large error. But sometimes the benefits are so near zero that the decision is easy. The mutagens in hair dyes can be (and were) replaced by nonmutagenic dyes. Chemotherapeutic agents used

Substance	Comments
Nitrosamine	Formed in stomach from nitrite used in preservation of bacon and sausage.
Aflatoxin	Produced by certain molds on peanuts and other foods. Very carcinogenic.
Caffeine	Caffeine is probably not a mutagen itself, but it enhances the mutagenicity of other mutagens by interfering with repair mechanisms.
Vinyl chloride	A common industrial agent that is very carcinogenic and has been identified as the agent responsible for one outbreak of cancer near a manufacturing plant.
Ethylene oxide	Used to sterilize equipment in hospitals.
Ethylene dibromide	Used to fumigate fruits, vegetables, and grains.

TABLE 16.3
A Selected List of Chemical Mutagens

to treat cancer are probably active because they are mutagenic. In any event, the potential benefit easily justifies the risk.

The use of nitrite to preserve bacon and sausage is a more difficult decision. The nitrite is converted into very mutagenic and carcinogenic nitrosamine in the stomach. To what extent this may contribute to stomach cancer is unknown. If the nitrite were left out of the meat, the bacterium *Clostridium botulinum* could grow, producing botulism and death. No one knows where the scales would balance. What is needed is a replacement for nitrite, but one is not yet available.

The situation with nitrite will undoubtedly prove true for many other chemicals as well. Considering the tens of thousands of chemicals in our environment, chemicals that have not been and rarely can be adequately tested for mutagenicity, we will have to make many decisions about relative risks on the basis of very limited experimental knowledge. Many of the decisions will be based instead on general principles relating structure of the chemical to mutagenicity with minimal testing in microbial and other systems. It is imperative then that we learn as much as possible about basic mechanisms, while keeping a wary eye on genetic indicators that might indicate increased mutation in the human population.

Review Questions

1. Our most accurate determinations of the mutation rates at specific loci in the human population are from dominant mutations. Why?
2. Aniridia is a simple inherited dominant condition involving absence of the iris of the eyes. In Michigan, 4,664,799 persons were reported to have been born between 1919 and 1959. Among these, 41 children with aniridia were born to parents with normal irises. What is the probable mutation rate at this locus?
3. Most of our data on spontaneous mutation rates of specific genes are obtained with bacteria and fruit flies. Do the results agree well with the data obtained for mice and men?
4. The rate of spontaneous mutation at a specific locus is low, generally about 10^{-5} to 10^{-6}. But if we assume that a cell has about 20,000 genes per haploid set, what would be the total number of new mutations in a zygote?
5. Distinguish among roentgens, rads, and rems.
6. What is the difference between a stable isotope and a radioactive isotope?
7. What radioactive isotope is a normal constituent of our bodies?
8. The mutational doubling dose for X rays is about 100 rem according to one observer. What does this mean?
9. There may be considerably more danger in our exposure to the many different manmade chemicals than to the current levels of ionizing radiations. Why?
10. It is almost certain that even "normal" persons vary widely in their susceptibility to mutagenic agents in the environment. What explanations can be advanced for these differences?

References and Further Reading

Evans, H. J., and D. C. Lloyd (eds.). 1978. *Mutagen-Induced Chromosome Damage in Man.* Yale Univ. Press, New Haven. A number of authors present information on recent studies of chromosomal mutations.

Fishbein, L., W. G. Flamm, and H. L. Falk. 1970. *Chemical Mutagens. Environmental Effects on Biological Systems.* Academic Press, New York, 364 pp. The emphasis is on the potential chemical hazards. The book was written before many of the current test systems became available, but the information on chemicals is still very good.

National Research Council: Advisory Committee on the Biological Effects of Ionizing Radiations. 1980. *The Effects on Populations of Exposure to Low Levels of Ionizing Radiation.* National Academy of Sciences, Washington, D.C. 638 pp. The so-called BEIR III report resulting from many years of intensive study of the problems by a committee of the National Academy of Sciences drawing on all available data.

Sinclair, W. K. 1981. Effects of low-level radiation and comparative risk. *Radiology* 138:1–9. A technical but authoritative summary by an outstanding worker in the field.

Upton, A. C. 1982. The biological effects of low-level ionizing radiations. *Scientific American* 246 (February):41–49. Another evaluation by a highly qualified observer, written at a nontechnical level.

The publication by Charles Darwin of his book *On the Origin of Species by Means of Natural Selection* in 1859 brought as great a revolution into human thought as any other idea before or since. Darwin's theory proposed first that all organisms, incuding humans, arose by evolution from earlier forms of life. He further proposed that the mechanism by which this occurred was *natural selection*. Leaving aside for now the evidence relative to human and other evolution, let us examine what is meant by natural selection and how it operates.

Darwin viewed natural selection as "survival of the fittest." This view, in marked contrast to the view that the meek shall inherit the earth, led to intense, acrimonious debate (see Box 17.1). "Nature red in tooth and claw" became a label attached to the Darwinian theory. The relative absence of red teeth and claws in the gentle English countryside made the idea of natural selection seem remote and irrelevant to modern times.

Biological Fitness.   Darwinian fitness, however, is not measured by clawing or pommeling one's fellows into submission or premature death. It is measured solely by the number of fertile offspring that one produces. In humans, the red claw may be no more fearsome than a family that has three children compared to the one next door that has only two. The first family is 50% more fit than the second because it will be represented by 50% more persons in the next generation. Both are more fit than the olympic athlete who lives a long vigorous life but dies at age 80 without leaving children, or the soldier who is killed in war before he has a chance to produce children. Natural selection is blind to the qualities that we as humans might consider biologically desirable. The only thing that is important is fertility.

If everyone were the same genetically, the fact that some might produce more or fewer children would be of no consequence for selection or evolution. Since natural populations consist of diverse individuals, the opportunity exists for individuals of certain genotypes to produce more offspring and so to alter the character of the next generation.

Darwin was unaware of the mechanisms of heredity as we now know them. Indeed, he accepted the common belief, associated with the name of the great French biologist Lamarck, that acquired traits could be transmitted. This view is now thoroughly discredited. Centuries of circumcision have not shortened the foreskin of newborn Jews. But Darwin's theory was not based on a specific mechanism of heredity. So long as offspring inherit the characteristics of their parents, differential fertility and selection may occur.

Mendelian Heredity and Selection

Understanding the Mendelian transmission of inherited traits also makes it possible to understand how selection operates to change the nature of a popula-

Natural Selection

Box 17.1

The Discovery of Natural Selection and the Darwinian Controversy

The discovery of natural selection as the force of evolution is associated with the names of two great intellects of Victorian England—Charles Darwin (1809–1882) and Alfred Russel Wallace (1823–1913). History accords both these men the honor of having discovered the principles of natural selection independently of each other. Darwin gained his appreciation of evolution and adaptation during his early years as naturalist aboard the H.M.S. *Beagle* during a five year trip (1831–1836) whose main purpose was to survey the coastal regions of South America and the South Pacific. The many local variations and adaptations of species convinced him that those individuals that are best adapted in a population are the ones most likely to serve as parents of succeeding generations. The theory of evolution had been around a long time, but Darwin gave it life by providing the mechanism, natural selection.

Alfred Russel Wallace had much the same experience as Darwin. After an expedition to the Amazon (1848–1850), he spent a number of years surveying the biology of the Malay archipelago (1854–1862). In February, 1858, it occurred to Wallace that natural selection could provide the means for evolution. Three days later, he sent Darwin, whom he had not met, a copy of an essay containing his theory. Darwin had been preparing a long manuscript on the same idea since 1856, based on sketches dating back to 1842. He received Wallace's letter in June, 1858, and was startled to find his own theory fully set forth in it. After consulting with several scientific friends, Darwin prepared an abstract of his own work, which was read together with Wallace's paper on July 1, 1858, before the Linnaean Society. They were published the same year in the *Journal of the Linnaean Society*. Darwin's full manuscript, *On the Origin of Species by Means of Natural Selection, or the Preservation of Favoured Races in the Struggle for Life* was published on November 24, 1859, with all 1250 copies printed being sold on that day.

Evolution was no threat to the creationists so long as it lacked a mechanism. But Darwin and Wallace changed that. An intense controversy developed that involved the greatest intellects of England and included scholars from Germany and France as well. Darwin's views were championed by a very talented biologist and physician, Thomas Henry Huxley (1825–1895), who also had served as ship's surgeon and naturalist on the H.M.S. *Rattlesnake* during its voyages in the Southwest Pacific (1847–1850). Darwin was often too ill to attend meetings to defend his own views, having contracted a disease in South America, now thought to have been Chagas disease. His surrogate, Huxley, was probably much more effective at such public confrontations than the reticent Darwin would have been. Huxley seems to have enjoyed immensely his role as Darwin's champion and was a formidable adversary for those trying to combat reason and evidence with obscurantism.

The climax of the controversy occurred on June 30, 1860, at Oxford at a meeting of the British Association for the

Box **17.1**

(*continued*)

Advancement of Science. Of the various distinguished persons on the platform, the principal antagonists were Huxley and Samuel Wilberforce, Bishop of Oxford, a churchman noted for his conservatism, his oratorical abilities, and his charm. The last earned for him the nickname "Soapy Sam" among his detractors. The Bishop launched a strong attack against Darwinism and finally addressed a question directly to Huxley, asking whether it was "through his grandfather or his grandmother that he claimed his descent from a monkey?" Huxley responded by presenting Darwin's theory quietly and rationally. He then delivered the coup de grace by stating that he would not be ashamed to have a monkey for an ancestor but he would be "ashamed to be connected with a man who used great gifts to obscure the truth." In the excitement that followed this response, one woman is said to have fainted, a reaction rarely seen at recent meetings of the British Association.

Following that episode, the confrontation between evolutionists and creationists lost much of its intensity. Perhaps Huxley lacked an adequate adversary. In part, the change was due also to new doctrines adopted by many of the major religious groups, doctrines that no longer required literal interpretations of Biblical accounts of creation. The controversy between evolutionists and creationists still continues, but the flavor has changed.

tion or indeed to stabilize a population and buffer it against change. First we must consider the genetic structure of populations and how this structure can change.

The Hardy–Weinberg Law

Gene Frequencies and Genotype Frequencies. Up to this point, we have discussed genes only as they occur in individuals. It is useful sometimes to treat the genes of an entire population as if they constitute a *pool of genes*. Let us suppose that there are a thousand persons in a particular population. At any one autosomal locus, each person would have two genes, for a total of 2000 genes in the population. Considering the MN blood group locus, 210 might be type M, 480 type MN, and 310 type N. Among the 2000 genes at this locus, there would be $(2 \times 210) + 480 = 900$ M alleles and $(2 \times 310) + 480 = 1100$ N alleles. This constitutes the pool of M and N alleles for that population. The frequencies of these alleles, expressed as a fraction, would be $900/2000 = 0.45$ M; $1100/2000 = 0.55$ N. Since there are no other kinds of alleles, $0.45 + 0.55 = 1.00$.

Rather than talking about the number of M, MN, and N persons, we could describe the population as having 45% M alleles and 55% N alleles. From these figures, we could predict how many M, MN, and N persons there are, provided the alleles are combined two at a time at random to represent people. We can think of this in terms of drawing beans from a jar that contains 45% black beans and 55% white. Each bean that we draw is an independent event; that is, the results are not influenced by other draws. If the beans are drawn two at a time, there is a 45% chance that each will be black and, since the draws are independent, the chance that both will be black is the product of the individual events, $(.45)^2 = 0.2025$ or approximately 20%. Similarly, the chance of two white beans will be $(.55)^2 = 0.3025 \cong 30\%$. There are two ways to obtain a white and a black bean. The first bean could be black (.45) and the second white (.55) for a combined probability of $(.45)(.55) = 0.2475$. Or the first bean could be white and the second black for a combined probability also of 0.2475. If we do not care in which sequence the beans were drawn, we could add these two together for a total of 0.4950. Thus nearly half the pairs of beans would be black–white combinations.

We can substitute M and N alleles for black and white beans. Each person can be treated as a sample of two alleles taken from the overall gene pool. There should thus be $(.45)^2 = 0.2025$ type M persons, $2(.45)(.55) = 0.4950$ type MN persons, and $(.55)^2 = 0.3025$ type N persons. The three values $0.2025 + 0.4950 + 0.3025 = 1.0000$, indicating as expected that all the combinations are accounted for.

The combination of alleles can be generalized as shown in Table 17.1. With p and q representing the frequencies of the M and N alleles, the various genotypes occur in the frequencies $p^2(MM)$, $2pq(MN)$, and $q^2(NN)$. This is usually written

$$p^2 + 2pq + q^2 = 1$$

where 1 is the sum of all possible genotypes. This is the expansion of the binomial $(p + q)^n$, for $n = 2$.

With this formula, the expected frequencies of genotypes can be calculated for any values of p and q. If there are only two alleles, then $p + q = 1$. If there are three or more alleles, the general formula is

$$(p + q + r + \ldots + z)^2 = 1$$

TABLE 17.1
The Frequencies of Various Combinations of M and N Alleles Based on Random Pairing

The frequency p *of* M = 0.45 *and the frequency* q *of* N = 0.55. *The frequencies of* MM, MN, *and* NN *are equal to* p², 2pq, *and* q² *respectively.*

	M $p = 0.45$	N $q = 0.55$
M $p = 0.45$	MM $p^2 = 0.2025$	MN $pq = 0.2475$
N $q = 0.55$	MN $pq = 0.2475$	NN $q^2 = 0.3025$

The usefulness of the binomial for estimating genotype frequencies was recognized independently by the British mathematician G. H. Hardy and the German physician W. Weinberg. This application is thus known as the Hardy–Weinberg law. It applies to populations in which mating is at random; that is, mates are chosen independently of the genotype under study. There are many such traits, and the Hardy–Weinberg law has many applications in genetics.

Frequencies of Recessive Genes: Cystic Fibrosis.   Many detrimental traits are due to homozygosity for alleles at an autosomal locus. Except for red-green colorblindness, the most common inherited defect in European-derived populations is cystic fibrosis, found in approximately 1 in 2500 births in the white population of the United States. The primary defect in cystic fibrosis is not known, but there is a general dysfunction of glands such as the pancreas. Death usually occurs by the midteens, although better management of patients in recent years has extended the expected life span. Nevertheless, affected persons rarely survive into adulthood, and those who do have greatly reduced fertility. In a Darwinian sense, persons with cystic fibrosis are very unfit, since they do not leave their genes to the next generation.

The frequency of homozygotes for the recessive *cf* allele is 1/2500 or 0.0004. If we let p be the frequency of the normal *Cf* allele and q the frequency of *cf*, then $q^2 = 0.0004$ and $q = 0.02$. Since $p + q = 1, p = 1 - q = 1 - 0.02 = 0.98$. This tells us that of each 100 genes at the *Cf* locus, 98 are *Cf* and 2 are *cf*. We can further calculate the frequency of heterozygous "carriers" of cystic fibrosis with genotype *Cfcf* by the middle term of the binomial expansion, $2pq = 2(.98)(.02) = 0.0392$. In other words, nearly 4% of the population, 1 in 25 persons, carries the *cf* allele in heterozygous combination. The remainder of the population, 96%, is homozygous for the normal *Cf* allele.

Sickle Cell Anemia.   A similar situation exists for sickle cell anemia in the American black population. One difference is that persons heterozygous for sickle cell anemia can be identified by the presence of both hemoglobins A and S in their red cells, whereas cystic fibrosis heterozygotes cannot be distinguished from normal except by difficult and often unreliable tests. The frequency of the Hb^S allele has been found to be approximately 4% among blacks, with a frequency of the normal allele Hb^A of 96%. From these figures, we can calculate the expected frequency of genotypes at the *Hb* locus:

$$
\begin{array}{llll}
\text{homozygous } Hb^A\, Hb^A = p^2 & = (.96)^2 & = 0.9216 \\
\text{heterozygous } Hb^A\, Hb^S = 2pq & = 2(.96)(.04) & = 0.0768 \\
\text{homozygous } Hb^S\, Hb^S = q^2 & = (.04)^2 & = \underline{0.0016} \\
& & \text{Total} & 1.0000
\end{array}
$$

Nearly 8% of the normal population is heterozygous for Hb^S and 1.6/1000 has sickle cell anemia.

Gene Frequencies of Dominant Traits: Huntington Disease.   Dominant traits are handled very much like recessives except that the portion of the population with the dominant trait is the sum of the two terms, $p^2 + 2pq$. For a rare dominant trait, there will be very few homozygous dominant individuals. For example, if the frequency of the dominant allele D is 0.001, the frequencies of the three genotypes would be

$$0.000\,001\ DD + 0.001\,998\ DR + 0.998\,001\ RR = 1$$

or 2/thousand persons would be heterozygous but only 1/million would be homozygous. For rare dominant traits, then, homozygotes are exceedingly rare and often are ignored in calculations. For example, Huntington disease is a rare dominant trait associated with degeneration of the central nervous system. The disease ordinarily doesn't appear until the affected person is 30 years old or more, thus providing ample opportunity to transmit the gene to additional generations. The frequency of affected persons is 1 in 10,000. Since these are essentially all heterozygotes, we can consider the population to consist of two genotypes: Hh and hh. The frequency of H would then be half the frequency of Hh persons, half of whose alleles are H. This would give a frequency of 0.000 05 for H. The expected frequency of persons homozygous for H would be $(0.000\,05)^2$, or 25/10 billion persons. It is small wonder that no homozygote has ever been recorded. Indeed, the HH phenotype may be very different from Hh and may be very severe or lethal.

Gene Frequencies of X-Linked Traits: Colorblindness. In the case of X-linked traits, there are two kinds of people, females and males. In females, with two X chromosomes, gene frequencies are handled just as for autosomal traits. Males have only one X chromosome, however, so the frequency of alleles in males is simply the enumeration of the trait frequencies. Expressed algebraically, the relationship is

$$(p + q + r + \ldots + z)^1 = p + q + r + \ldots + z$$

For a trait such as red-green colorblindness, for example, some 8% of white males are affected. Thus the frequency of the recessive allele c is 0.08 and the frequency of the dominant allele C for normal color vision is $1 - 0.08 = 0.92$. The genotype frequencies in females would be

$$
\begin{array}{lll}
CC: & (.92)^2 & = 0.846 \\
Cc: & 2(.92)(.08) & = 0.147 \\
cc: & (.08)^2 & = \underline{0.006} \\
& & 0.999
\end{array}
$$
normal color vision

colorblind

The ratio of affected males to affected females is $q/q^2 = 1/q$. For $q = 0.08$, $1/q = 12.5$. As q decreases, the frequency of affected males decreases proportionately, but the ratio of affected males to females increases enormously.

The Hardy–Weinberg Law and Selection

In the absence of selection, the allele frequencies will not ordinarily change from one generation to the next. If the population is very small, they may fluctuate randomly, but that does not occur for the large populations found over most of the world today. This stability of allele frequencies depends on each genotype being equally fit, that is, producing as many viable and fertile offspring on the average as the other genotypes. If one or more genotypes is more or less fertile than the remainder, then the gene frequencies may change and the distribution of phenotypes will differ in the next generation.

Selection for Favorable Dominant Mutations. Let us consider how this might happen with a rare dominant trait that makes a person more fit. Such a trait might be increased resistance to a disease, such as falciparum malaria, a debilitating and often fatal tropical condition. A resistance allele could arise by mutation and would be exceedingly rare at first. Indeed, the most likely thing to happen is that it will be lost from the population by chance. But occasionally, also by chance, the mutant allele will be transmitted to a small number of persons, who will have an advantage over those who lack the mutant allele.

Let us suppose that this beneficial allele, which we designate M^R for malaria resistance, conveys a 100% advantage over its malaria-sensitive allele, M^S. This means that the average number of children produced by an $M^R M^S$ heterozygote is twice the average number produced by an $M^S M^S$ homozygote. This advantage could be due to increased survival through the reproductive period. Or it could be due to increased likelihood of having a child because of superior fertility or less frequent periods of illness with correspondingly greater interest in the opposite sex. Or it could be due to some combination of these and other differences. The results are the same, whatever the mechanism. Each $M^R M^S$ person will leave twice as many offspring on the average as $M^S M^S$ persons.

The consequences of this are shown in Table 17.2. In only 20 generations, the frequency of the beneficial allele would increase from 1/million to 0.45, and over half the popoulation would have the M^R allele, either in homozygous or heterozygous combination. In a few more generations, almost everyone in the population would have the M^R allele. With respect to this locus, the population will have evolved very rapidly. Such marked reproductive advantage must be rare in new mutations, most of which are detrimental. But smaller advantages will lead to the same result over more generations.

Favorable Recessive Mutations. Recessive mutations that are advantageous have a more difficult time getting started, since only homozygotes would have the advantage. This might happen in a small tribal population where high inbreeding occurs but would not be likely to happen in a large outbred population. During most of human evolution, however, the small tribal group has been the predominant population unit. Once a substantial number of homozygous recessive individuals is established, the deleterious dominant allele would be selected against quite efficiently.

Selection Against Deleterious Recessive Genes. To speak of the reproductive advantage of an allele is to speak also of the disadvantage of its homolog. The frequency of the malaria-sensitive M^S allele in the earlier example would decrease until sensitive persons became rare. The rate at which this would happen is shown in Figure 17.1 for several situations, including one in which homozygotes do not reproduce at all. Although selection occurs rapidly when the recessive allele frequency is fairly high, it becomes very inefficient at low frequencies. This is because at very low frequencies nearly all of the alleles are in heterozygous combination rather than homozygous, and selection would operate only against the small fraction of alleles in homozygotes.

**TABLE 17.2
Increase in Frequency
of a Hypothetical
Dominant Allele That
Confers a 100%
Selective Advantage in
Resistance to Malaria
Compared to the
Normal $M^S M^S$
Homozygotes**

Generation	Initial MR Frequency	Frequency of Zygotes			Frequency of Gametes Produced	
		$M^R M^R$	$M^R M^S$	$M^S M^S$	M^R	M^S
0	.000 001	10^{-12}[a]	.000 002	.999 998	.000 002	.999 998
1	.000 002	10^{-12}	.000 004	.999 996	.000 004	.999 996
2	.000 004	10^{-11}	.000 008	.999 992	.000 008	.999 992
3	.000 008	10^{-11}	.000 016	.999 984	.000 016	.999 984
4	.000 016	10^{-10}	.000 032	.999 968	.000 032	.999 968
5	.000 032	10^{-9}	.000 064	.999 936	.000 064	.999 936
6	.000 064	10^{-9}	.000 128	.999 872	.000 128	.999 872
7	.000 128	10^{-8}	.000 256	.999 744	.000 256	.999 744
8	.000 256	10^{-8}	.000 512	.999 488	.000 512	.999 488
9	.000 512	10^{-7}	.001 024	.998 976	.001 023	.998 997
10	.001 023	.000 001	.002 044	.997 955	.002 042	.997 958
11	.002 042	.000 004	.004 076	.995 920	.004 067	.995 933
12	.004 067	.000 017	.008 101	.991 882	.008 069	.991 931
13	.008 069	.000 065	.016 008	.983 927	.015 883	.984 117
14	.015 883	.000 252	.031 261	.968 486	.030 795	.969 205
15	.030 795	.000 948	.059 693	.939 358	.058 068	.941 932
16	.058 068	.003 372	.109 392	.887 236	.104 367	.895 633
17	.104 367	.010 892	.186 949	.802 158	.174 258	.825 742
18	.174 258	.030 366	.287 784	.681 850	.264 398	.735 602
19	.264 398	.069 906	.388 983	.541 110	.362 464	.637 536
20	.362 464	.131 380	.462 168	.406 452	.454 914	.545 086

[a]Frequencies less than one per million are expressed as orders of magnitude.

Balanced Polymorphisms. One additional model of selection, known as *balanced polymorphism*, should be considered at this point. The term *polymorphic* is applied to loci for which at least two alleles are found in frequencies greater than 1%. The ABO and Rh blood groups are familiar polymorphic traits. But many others are known also. The question arises as to how two or more common alleles can coexist. One would expect that it is rare for two alleles to be exactly equivalent under all circumstances and that the superior allele would slowly displace the selectively inferior allele. The existence of *neutral* alleles, that is, alleles that are equal with respect to selective advantage, has been a much debated topic. They almost certainly do exist, but it is very difficult to prove that alleles are equivalent. On the other hand, it is sometimes possible to prove that one is detrimental.

In some instances though, the heterozygote may be superior to either homozygote. The best known example is sickle cell anemia. Persons homozygous for sickle cell homoglobin, $Hb^S Hb^S$ (which, for convenience, we shall shorten to *SS*), do not ordinarily survive to the reproductive period, although such homozygotes do occasionally survive and reproduce. On a population basis, the *S* allele is virtually completely lethal when homozygous. Heterozygous *AS* persons are quite normal with respect to red cell function. Their red cells can be made to sickle in the laboratory, but that does not occur in nature. In one respect, sickle cell heterozygotes are superior to homozygous *AA* persons. They are more resistant to fal-

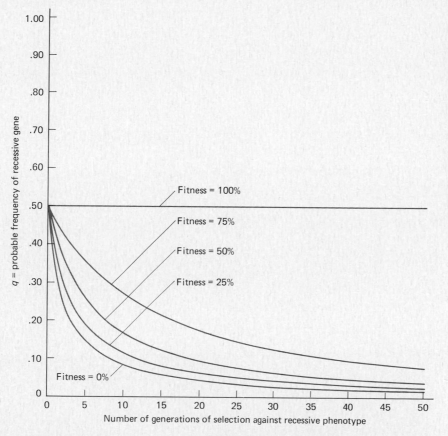

FIGURE 17.1
The change in gene frequencies with selection against homozygous recessive genotypes. In the lower curve, homozygous recessives do not reproduce, that is, their fitness is zero. The upper curves show the change in frequencies with other values of fitness. For 100% fitness, the gene frequencies do not change.

ciparum malaria. The mechanism for the increased resistance is not known definitely, but it seems to have something to do with the ability of the malaria parasites to invade and multiply in the red blood cells, an obligatory part of their life cycle.

In areas where falciparum malaria occurs, *AA* persons are therefore at a disadvantage with respect to *AS* heterozygotes. The heterozygotes are thus superior to either homozygote, *AA* or *SS*. This heterozygote superiority has the effect of maintaining both *A* and, more importantly, *S* in the population. If the frequency of either allele becomes too high, there is increased elimination of that allele relative to the other allele through loss of homozygotes. Such a polymorphism is said to be balanced because selection operates to maintain both alleles in the population.

The frequency of the alleles at the equilibrium point is determined by the relative fitness of the two homozygotes (Fig. 17.2). Let us consider the fitness of *SS* homozygotes to be zero and the fitness of *AA* homozygotes to be 80%. That is, *AA* persons leave 80% as many offspring as *AS* heterozygotes, to whom we assign a fitness of 100%. For each *SS* person who fails to reproduce, two *S* alleles are lost. For each *AA* person with a 20% reduction in fertility, $2 \times 0.2 = 0.4$ *A* alleles are lost.

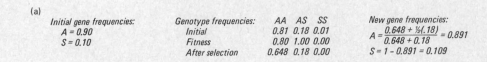

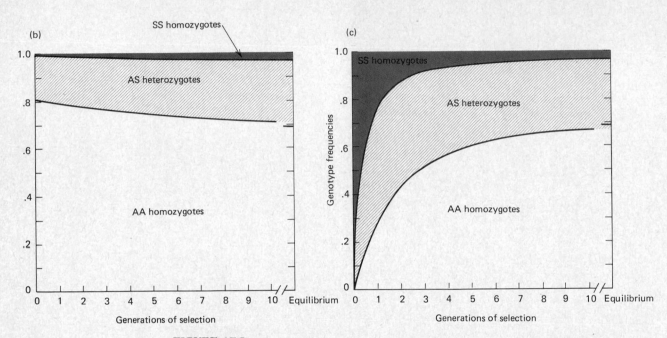

FIGURE 17.2
Diagram showing the return of the allele frequencies to equilibrium values in a balanced polymorphism. The final equilibrium values depend entirely on the relative fitness of the two homozygous genotypes. (a) Calculation of allele frequencies of *A* and *S* after one generation of selection, starting with frequencies of 0.90 and 0.10, respectively, and assuming the fitness of *AA* to be .80 that of *AS*, while the fitness of *SS* is zero. (b) A graph of allele frequencies of *A* and *S*, starting with the conditions in (a) and carried to 10 generations. (c) A similar graph but starting with an *A* frequency of 0.10 and an *S* frequency of 0.90. After ten generations of selection, the same equilibrium values have nearly been attained.

Intuitively, one would expect the *A* allele to be more frequent than *S*, since *AA* homozygotes are more favored than *SS*. It is possible to show algebraically, by the Hardy–Weinberg relationship, that the equilibrium frequencies are a function of the relative fitness of the homozygotes. The calculations can be handled more easily in terms of selection coefficients. These are defined as the reduction in fitness of a particular genotype compared to the most fit genotype. Expressed algebraically, the selection coefficient $s = 1 - w$, where w is the fitness. Specifically, if s_1 is the selection coefficient for AA persons ($s_1 = 0.2$), where s is defined as the reduction in fitness compared to the *AS* heterozygotes ($s_2 = 0$) and s_3 is the selection coefficient for SS persons $(s_3 = 1.0)$, then

$$q_e = s_1/(s_1 + s_3) = 0.167$$

where q_e is the equilibrium value of the S allele.

Sickle cell anemia also illustrates the great dependence of selection on the environment. The heterozygote advantage occurs only where falciparum malaria is endemic. If malaria is removed from the environment or if the people are removed from the environment, as happened when ancestors of the American blacks were taken from Africa, the selective disadvantage of AA persons no longer exists. The SS persons will still be strongly selected against, and the frequency of the S allele will slowly decline as will any detrimental recessive (Fig. 17.1).

Not all polymorphisms are necessarily balanced. For most polymorphic loci, we have no evidence that selection is operating to keep two or more alleles in the population. Some may well be completely neutral and exist only because of the continuous generation of new alleles, with chance increases (*drift*) causing some to become common. Other polymorphisms may be subject to selection but may not have attained the final frequencies. Such *transient* polymorphisms would be very difficult to recognize as such, since we can only look at gene frequencies at a single point on the evolutionary time scale. We do not know, for example, whether the B allele of the ABO locus is increasing, is decreasing, or is stable. Perhaps in another 10,000 years, the answer can be obtained. We must be content to search for factors that might operate in selection. Especially when we see a detrimental gene such as Hb^S at high frequency, we suspect that it may be part of a balanced polymorphic system that is maintained by selection against the "normal" homozygote.

Natural Selection in Contemporary Populations

From time to time, one hears alarm expressed that modern society with its improved medical practices has destroyed the natural selection that caused humans to evolve to their present state. Another concern is that the introduction of birth control has meant that the more intelligent have fewer children than do those less well endowed genetically. A corollary of these arguments is that *Homo sapiens* is accumulating deleterious genes that are contributing to the overall decline in biological quality of the species. Is there in fact any evidence that these assertions are true? To what extent does natural selection operate in modern human populations?

To answer the first question, we need to establish whether differential reproduction can in fact be associated with detrimental genes. There is ample evidence that a high proportion of fetal loss is attributable to genetic conditions. Approximately half of spontaneous abortions have a recognizable chromosomal abnormality. There are suggestions also that the loss of early embryos—embryos that fail to implant in the uterus or that are aborted during the first weeks of pregnancy—may be large. Many of these undoubtedly are due to genetic conditions. Their impact would vary with the social structure. In a society where family planning is practiced, they would be replaced with other zygotes. The only effect would be a slightly increased period between pregnancies, which is a selective disadvantage, however. The following are situations where natural selection is an important factor, at least under some circumstances.

Reduced Fertility

Mention has been made of the many inherited traits associated with lack of or reduced fertility. Selection in these cases is obvious. Persons with PKU, sickle cell anemia, cystic fibrosis, and hundreds of other well-documented Mendelian traits do not survive to the reproductive period or, if they do, they are unlikely to have normal numbers of offspring. In some instances, reduced fertility has a biological basis. In others, the person may be less likely to marry. The result is the same in either case.

With some diseases, of which PKU and hemophilia are examples, the frequency of the genes and of affected persons is so low as to suggest that there is strong selection against them with no compensating or favorable selection. PKU occurs with a frequency of approximately 1 in 10,000 births, corresponding to a frequency of the PKU gene of 0.01 ($q^2 = 0.0001$; $q = 0.01$). Hemophilia also occurs in approximately 1 per 10,000 male births. Since it is an X-linked recessive trait, the gene frequency is simply the disease frequency in males.

Other detrimental diseases occur at much higher frequencies, which suggests that they are part of a balanced polymorphic system, similar to sickle cell anemia. Cystic fibrosis (CF) is the most frequent Mendelian hereditary disease in European children, occurring approximately once per 2500 births, corresponding to a gene frequency of 0.02. This seems much too high for a lethal disease. It has been suggested, therefore that cystic fibrosis is part of a balanced polymorphic system. There is no direct evidence on the nature of the advantage that CF heterozygotes might have over homozygous normal persons, but there is evidence that they may be slightly more fertile.

Infectious Diseases

We think of infectious diseases as being entirely environmental, since the disease is dependent on presence of the disease agent in the environment. However, people are not equally susceptible to infectious agents, some being more resistant, others being very prone to invasion by a particular germ or virus or, in rare cases, to all infectious agents. In the case of typhoid fever, for example, some persons become very ill, others appear to become infected and recover without ever having been noticeably ill, and a few otherwise healthy persons harbor the typhoid bacillus permanently in the gastrointestinal tract. The last group are carriers and a continuous threat to the health of others. The differences in these responses seem to be in the constitution—presumably the genetic constitution—of the host. Many bacteria and viruses can infect only one or a few species of animals, another example of inherited differences in susceptibility.

With the development of antibiotics, the importance of infectious agents has been greatly diminished, both in health problems and in natural selection. Before this era, however, infectious agents are thought to have been one of the most important forces in natural selection. Epidemics such as the black plague, which killed a third of the population of Europe in the fourteenth century, used to sweep entire continents. Smallpox was a major problem for centuries until its recent eradication. The American Indians were decimated by infectious agents brought to the New World by Europeans. It is usually stated that these were agents to which the Indians had developed no resistance. It is more nearly correct to say that susceptible genotypes had not been eliminated by previous exposures to the agents. If certain genotypes were more resistant to infection or associated with less severe symtoms, they might have strong selective advantage.

Malaria is another agent that has clearly played an important role in selection. Malaria is due to infestation by microscopic parasites, of which there are several common species. *Plasmodium vivax* causes the relatively benign malaria of the temperate zones. *P. falciparum* occurs only in the tropics and causes a much more severe, sometimes fatal, disease. Malaria parasites have a complex life cycle, part of which must be in mosquitos and part in red blood cells (Fig. 17.3). Several human genes show distribution patterns similar to that of falciparum malaria (Fig. 17.4). These all affect red cell properties, and it is thought that these alleles offer some protection from malaria. Sickle cell anemia has already been discussed as the best-documented example. Children heterozygous for Hb S appear less likely to become infested and, if infested, have lower parasite counts and lower mortality than children homozygous for Hb A. The high frequencies of Hb C in Africa, of Hb E and Hb Constant Spring in Asia, of thalassemia in Africa, the Mediterranean area, and Asia, and of glucose-6-phosphate dehydrogenase deficiency in Africa and the Mediterranean area are all thought to be due to the increased resistance to malaria conferred by these genes. In each case, the gene has some deficit in function, which may be severe in homozygotes.

One red cell antigen also is involved in malaria. The Duffy blood group was first detected when antibodies were produced in a man who had received multiple transfusions. In European populations, two common alleles, Fy^a and Fy^b, produce characteristic antigens. In African black populations, a third allele, Fy, is virtually the only allele found at this locus. No antigen corresponding to this allele has been identified as yet. It has been established, however, that the site of the Duffy antigen on the surface of the red cell is also the receptor to which the *P. vivax* malaria parasites bind in order to invade a cell. Apparently both Fy^a and Fy^b alleles produce receptors for malaria, but Fy does not, or it may produce a receptor that can be recognized neither by antibodies nor by parasites. The prevalence of malaria in Africa undoubtedly gave selective advantage to homozygous $FyFy$ persons. The normal function of the Duffy receptors is not known. $FyFy$ persons have no obvious defect attributable to their genotype at this locus.

Intelligence

Perhaps no topic has generated more controversy than the question of whether intelligence is inherited and whether the low IQ people are outproducing the high

FIGURE 17.3
A typical life cycle of a malaria parasite of the genus *Plasmodium*.

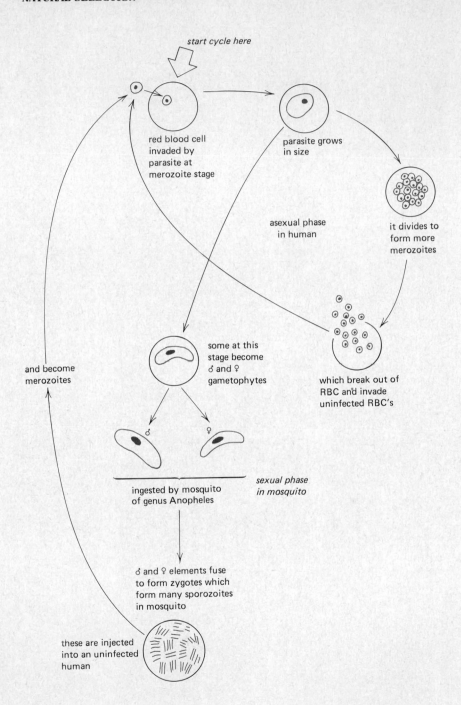

start cycle here

red blood cell
invaded by
parasite at
merozoite stage

parasite grows
in size

it divides to
form more
merozoites

asexual phase
in human

which break out of
RBC and invade
uninfected RBC's

some at this
stage become
♂ and ♀
gametophytes

and become
merozoites

*sexual phase
in mosquito*

ingested by mosquito
of genus Anopheles

♂ and ♀ elements fuse
to form zygotes which
form many sporozoites
in mosquito

these are injected
into an uninfected
human

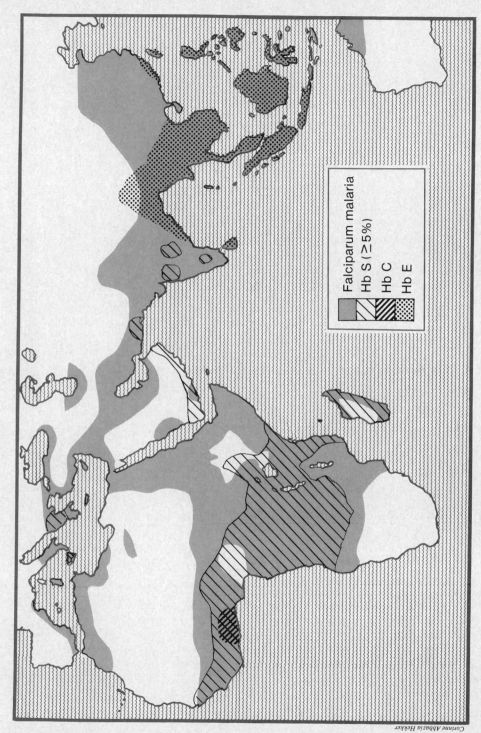

FIGURE 17.4 Geographic distribution of falciparum malaria and of several abnormal hemoglobins. In addition, the various forms of thalassemia occur throughout the tropical regions in which falciparum malaria was once prevalent. The gene for glucose-6-phosphate dehydrogenase deficiency, an X-linked trait, is found in high frequency in the Mediterranean area and in tropical Africa.

Corinne Abbazia Hekker

331

IQ people, to the detriment of future populations. In Chapter 18, we will discuss the evidence that a substantial part of the *variation* in intelligence can be attributed to genetic variation. If that is so, then differential fertility of low or high IQ persons will lead to an increase in genes associated with low or high IQ. The principle is the same used in agricultural breeding to select cows that have high milk yields or chickens that lay lots of eggs. But what is the evidence for differential reproduction related to intelligence?

The study that has been cited extensively showing differential fertility was of Wellesley graduates, who had fewer children during their child-bearing years compared to women who did not go to college. Such a finding may well be true but not necessarily informative. For example, a comparison of high school graduates with college graduates examines only the high end of the IQ distribution. Further-more, graduation from high school versus college depends on many factors in addition to intelligence. A full comparison should include *all* the persons born into a defined population during a specific time. Some of these will not be able to meet the academic requirements of high school or elementary school. Many will be retarded and not attend school. Some will be placed in institutions for severely retarded persons, perhaps outside the geographic area of the study. But these persons are just as important to our study as the few who obtain PhDs. And all these must be studied after their child-bearing years! Another complication is the change in patterns of childbearing at different times and in different cultures. A study done on one population may tell us relatively little about others.

One study that attempted to overcome some of these problems was carried out in Grand Rapids, Michigan. C. J. Bajema located 979 persons who were tested for intelligence in 1916 and 1917 when they were in the sixth grade and whose later reproductive history could be obtained. As can be seen in Figure 17.5, the persons with the highest IQ were actually more fertile than average, while the group with the lowest IQ was least fertile. Contributing to this difference is the fact that over twice as many persons in the lowest IQ category had no children compared to those in the highest IQ group. One must also remember that persons at the lowest end of the IQ range would never have made it to the sixth grade. Many could have been in homes for the mentally retarded, where their opportunities to reproduce would be very limited, even if they were biologically fertile. Selection in this population would be strongly in favor of persons within the normal IQ range, a finding that should not surprise us. In order to reproduce, one must be biologically capable and intelligent enough to attract a mate. The intelligence required may seem minimal, but it must be sufficient for survival outside an institution.

As noted earlier, one must be careful in extending these results to other populations. The ideal family size changes with time. During the depression of the 1930s, emphasis was on small families. After World War II and during the 1950s, it was popular to have large families. When the threat of the population explosion became evident in the 1960s, small families became the vogue again. Persons in different parts of the IQ range very likely respond differently to such trends. It is no longer rare for couples, especially when both are professionals, to marry with the understanding that they do not plan to have children. Selection with respect to intelligence is thus very complex, but it seems safe to assume that there still is

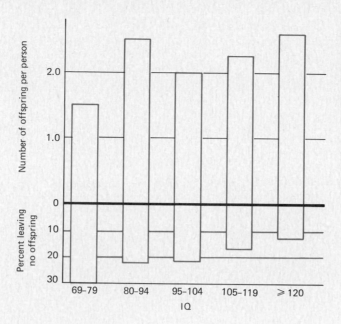

FIGURE 17.5
Number of offspring per person and percent of persons with no offspring versus IQ. The study populations were in the sixth grade at the time the IQ tests were administered. (Data from C. J. Bajema, *Eugenics Quarterly*, 10: 175–187, 1963.)

strong selection against very low intelligence, as there must have been during human evolution.

Generation Time

Fertility is commonly discussed in terms of discrete generations. That is both a matter of convenience and because observations typically are made only on the fertility of one generation. A very important variable not measured by the fertility of individuals is the average generation time. Let us consider two species that double in number each generation. Species A has two generations per year; species B has only one. Starting with the same number of individuals in each species, it is obvious that at the end of one year, there will be twice as many of species A as species B. At the end of two years, there will be four times as many.

The same principle applies to human populations. Let us consider two populations, which may be ethnic groups, religious groups, or persons distinguishable by genotype. Let us further consider that the average number of children born per adult is 1.5 (i.e., each family has three children). But in population A, the average age of parents at the birth of their children is 25 years. In population B, it is 30 years. At the end of a thousand years, population A would have completed 40 generations; population B would have completed 33.3. Each new generation of offspring would have 50% more individuals than the parental generation. Therefore, we can calculate the number of persons at the end of a thousand years, starting with 100 in each population, as follows.

$$\text{Population A:} \quad 100 \times 1.5^{40} \quad = 1,105,733,230$$
$$\text{Population B:} \quad 100 \times 1.5^{33.3} = \quad\; 73,086,796$$

There would now be over fifteen times as many persons in population A as in population B, even though the average family size has been the same. In actual practice, of course, parents who marry younger are likely to have more children, especially in cultures in which family planning is not practiced, and also to have them sooner on the average.

We do not know of genes that favor shorter reproductive cycles. However, one can imagine several situations in which selection would operate based on the length of the reproductive cycle. Consider a population in which the ideal number of offspring is three. The most fit families would be those that had their three children early, spaced as close together as possible. Those with longer spacing, due to miscarriages, low conception rates, or any other reason, would be less fit. To the extent that any of these reasons is associated with certain genotypes, the genes responsible would be slightly less fit.

Reproductive Compensation

When a child dies, parents often "replace" that child with another that they would not otherwise have had. This is called *reproductive compensation*. It is equivalent to a couple deciding that they will have a certain number of children, regardless of the number of pregnancies required to attain that number. The genetic consequences of compensation are to favor detrimental recessive genes that contribute to fetal loss or to childhood death.

Consider a mating between two persons who are heterozygous for an autosomal recessive trait, such as Tay–Sachs disease. The expected zygotes are

$$Tt \times Tt \rightarrow \tfrac{1}{4}TT + \tfrac{1}{2}Tt + \tfrac{1}{4}tt$$

One-half of all zygotes would be heterozygous for Tay–Sachs disease. But among *nonaffected* offspring, $\tfrac{2}{3}$ would be heterozygous. The gene frequencies in the parents are $0.5\,T$ and $0.5\,t$. Among all zygotes, the gene frequencies are $0.5\,T$ and $0.5\,t$, as in the parents. If the tt offspring were not "replaced," the frequency of t would be reduced by half in the offspring of these parents. For example, among 100 zygotes from such matings, 25 would be lost as tt homozygotes. Fifty t alleles would be transmitted through heterozygotes, and 100 T alleles would be transmitted through TT and Tt children. But there would be only $\tfrac{3}{4}$ as many children because of the loss of the tt children. Therefore, of the 200 alleles in the original 100 zygotes, 100 T and 50 t would be transmitted to normal offspring. If the 25 tt zygotes are replaced with TT and Tt zygotes, an additional $\tfrac{1}{3} \times 50 = 16.67\ t$ alleles would be transmitted, for a total of 66.67 t alleles rather than 50. The selection against t alleles is still present, but it is diminished by compensation. Should the parents have "extra" children to compensate (overcompensate) for the lost children, the selection against *t* would be still further diminished.

Parental Investment

If matings of two heterozygous parents, such as *Aa* × *Aa*, produces a defective *aa* offspring, the parents may decide not to have more children, whatever their original plans may have been. This provides strong selection against *a*, since subsequent *Aa* as well as *aa* children are eliminated. This decision often takes the form of recognizing that the homozygous recessive *aa* offspring will require a great amount of care and money and that additional children, even though normal, would be penalized by the family situation. The profoundly defective zygote that miscarries or dies as a newborn has much less impact in this regard than a zygote that produces a moderately affected person who requires extensive care. The latter is much more likely to reduce the number of subsequent children than the unrecognized early abortion. Persons homozygous for a mildly detrimental recessive allele may therefore cause greater selection against that allele through voluntary avoidance of subsequent pregnancies than if the allele caused only fetal wastage.

Late Onset Diseases

Some of the diseases about which we are most concerned are ordinarily of late onset. These include the major diseases, such as heart disease and cancer, that take so many lives. To what extent are these diseases subject to selection? To arrive at the answer, we must remember that selection is measured only by the number of offspring. Diseases that occur after the child-bearing and rearing years should have little selective disadvantage. Furthermore, if one considers that few persons survived to old age during the millenia of evolution of primitive *Homo sapiens*, dying instead of trauma or infection, the degenerative diseases of our present long-lived society must have been of little evolutionary significance. From our present perspective, these late onset diseases may be very important, and there is ample evidence of genetic predisposition to them. But unless the diseases are shown to influence the number of surviving children, they are irrelevant to selection. It is possible, of course, that the same genes that predispose to heart disease later in life are beneficial in younger people and are selected for rather than against. Until more is known about specific genes, such speculations must remain just that.

From the above examples, it should be clear that natural selection is still very much with us. It occurs not just in terms of death or sterility but also as part of the changing patterns of social practices. We influence selection every time we decide how many children to have and when to have them. With the advent of family planning, selection is often mediated by philosophies and mental attitudes as well as by purely biological variables.

Review Questions

1. Many persons have difficulty in coping with the idea of evolution of living organisms because they fail to distinguish between fact, theory, and mechanism. Can you distinguish these in a few words?

2. Darwin's theory of natural selection has as its chief element the "survival of the fittest." What is meant by "fittest?"

3. What is the central idea in the concept currently called Lamarckianism?

4. According to the Hardy–Weinberg law, a population with a certain pair of alleles, a and a^+, presently at frequencies of 0.10 and 0.90, respectively, would maintain these frequencies for an indefinite number of generations, provided the fitnesses of the three possible genotypes, aa, aa^+, and a^+a^+, remain identical. What other conditions would have to be met for this to happen?

5. Children homozygous for the gene for phenylketonuria (pk/pk) occur about once in every 15,000 births. Calculate from this the frequency of the pk allele and the frequency of heterozygotes (pk^+/pk). Taking the population of the U.S. as 220,000,000, how many persons in the U.S. are heterozygous for this disease?

6. Why are there more colorblind men than women?

7. Deleterious alleles that have low or zero fitness when homozygous do indeed have a low frequency. But why are they not eliminated entirely by natural selection?

8. In a population of 11,565 people isolated on an island, 9,626 have blood group A. The rest are either AB or B. None has type 0. Using these data and assuming that A, AB, and B persons are equally fit, calculate (a) the gene frequencies for A and B; (b) the expected number of persons with blood type AB and B.

9. What can be the result when a heterozygote is more fit than either homozygote? Would this still be true even though the fitness of one of the homozygotes is zero?

10. Discuss the importance of the following to natural selection: (a) infectious disease, (b) generation time, (c) reproductive compensation, (d) parental investment.

11. How important would you expect cancer to be as a force in natural selection?

References and Further Reading

Ayala, F. J., and J. W. Valentine. 1979. *Evolving, The Theory and Process of Organic Evolution.* Benjamin/Cummings, Menlo Park, Calif. An elementary introduction to the topic of evolution.

Cavalli-Sforza, L.L., and W.F. Bodmer. 1971. *The Genetics of Human Populations.* Freeman, San Francisco. 965 pp. A general textbook of human genetics with especially good chapters on selection in human populations.

Darwin, C. 1859. *On the Origin of Species by Means of Natural Selection, or Preservation of Favoured Races in the Struggle for Life.* Murray, London. The original classic now available in the first and sixth editions. This is probably the most important book written in the nineteenth century.

Darwin C. 1871. *The Descent of Man and Selection in Relation to Sex,* 2 vols. Murray, London. The sequel to the "Origin." Frequently referred to but rarely read.

Dobzhansky, T. 1970. *Genetics of the Evolutionary Process.* Columbia Univ. Press, New York. 505 pp. A classic textbook by one of the outstanding students of evolution. The treatment is largely nonmathematical and very clearly written.

Friedman, M. J., and W. Trager. 1981. The biochemistry of resistance to malaria. *Scientific American* 244 (March): 154–164. An elementary explanation of the resistance of persons with sickle cell trait and thalassemia to malaria.

Hardy, G.H. 1908. Mendelian proportions in a mixed population. *Science* 28:49–50. In a short, easy paper, a mathematician explains why gene frequencies don't change in the absence of selection. One of the important historical papers in genetics.

Motulsky, A. G. 1964. Hereditary red cell traits and malaria. *Amer. J. Trop. Med & Hyg.* 13:147–158. A review by one of the major participants in the study of natural selection in human populations.

Scientific American 239 (September): 1978. The entire issue is devoted to evolution, from prebiotic molecules to behavior.

Few human or animal traits are more fascinating than behavior. By behavior, we mean the entire universe of intellectual skills, personality, special abilities, perception and processing of visual, aural, and other stimuli, communication skills, etc. In other words, we shall use the broadest definition of behavior.

Much of behavior is learned. Yet, we know from studies of inbred strains of dogs, mice, and other animals that many "personality" characteristics in these species are inborn, therefore inherited. Is it not possible that in humans many of these learned traits are also strongly influenced by heredity? Prospero thought so when he described Caliban as

> A devil, a born devil, on whose nature
> Nurture can never stick; on whom my pains,
> Humanely taken, all, all lost, quite lost;
>
> Shakespeare (*The Tempest*, IV, i)

The fact that a trait is strongly influenced by heredity need not make it immune to environmental influence, although heredity may set the limits of what can be accomplished.

To what extent are we, like Caliban, limited in our abilities to learn, to perform, and to perceive? In this chapter, we will consider some of the observations and inherited conditions that help answer that question.

Single Gene Effects on Behavior

The influence of heredity on behavior is particularly evident in some of the simple Mendelian traits. Even though these may be individually rare, they raise the possibility that much of our behavioral variation is conditioned by the combinations of genes we inherit. A few examples will illustrate these points.

Phenylketonuria. One of the best studied examples of inherited mental retardation is phenylketonuria (PKU). This autosomal recessive disorder occurs approximately once in 10,000 births and is due to the inability to convert the amino acid phenylalanine to tyrosine (Fig. 9.5). Phenylalanine is an essential building block for proteins, but much more is consumed in the diet than is required for protein synthesis. The excess is disposed of via tyrosine, with recovery of the energy from metabolic conversion to CO_2, H_2O, and urea. Persons homozygous for the PKU gene lack the active form of the enzyme phenylalanine hydroxylase that catalyzes the addition of oxygen to phenylalanine. Since phenylalanine cannot be metabolized, it accumulates in the blood and tissues. Normal persons typically have about 1 mg of phenylalanine per 100 ml blood. Phenylketonurics may have up to 100 times that much.

The name *phenylketonuria* means "phenylketones in the urine." The phenylalanine that accumulates undergoes a variety of conversions to other substances,

18

Behavioral Genetics

including phenylketones, which are excreted. Some of the unchanged phenylalanine also is excreted in the urine.

(a)

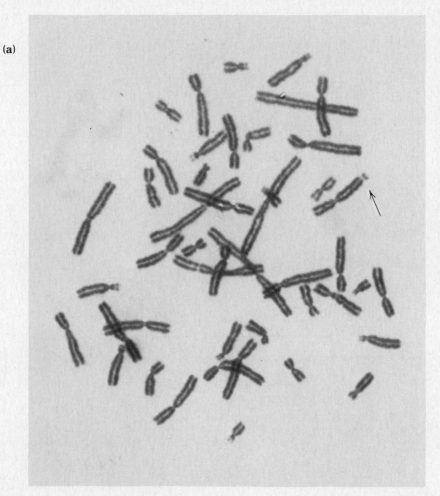

FIGURE 18.1
X-linked mental retardation due to the fragile-X syndrome. (a) A chromosome spread from an affected male. The fragile site is marked with an arrow. (b) A pedigree of fragile-X syndrome, showing affected males related to each other through normal heterozygous females. (Photograph provided by Dr. Patricia N. Howard-Peebles, The University of Texas Health Science Center at Dallas. The pedigree is from H. A. Lubs, *American Journal of Human Genetics* 21: 231, 1969.)

(b)

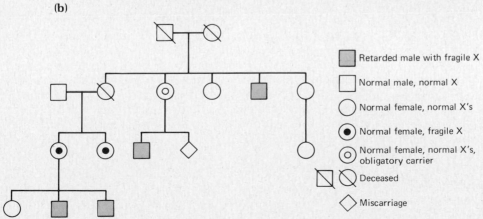

■ Retarded male with fragile X

□ Normal male, normal X

○ Normal female, normal X's

◉ Normal female, fragile X

⊙ Normal female, normal X's, obligatory carrier

⊠ ⊘ Deceased

◇ Miscarriage

chromosomes bearing the fragile site can usually be recognized. A number of pedigrees of this particular form of X-linked mental retardation are known, but many other pedigrees of X-linked recessive mental retardation clearly have a different primary defect.

Acute Intermittent Porphyria.   Although acute intermittent porphyria (AIP) is less well understood in terms of genetic mechanisms than PKU, it is of special interest to Americans, since King George III of England is thought to have had it. It is an autosomal dominant disorder. Affected persons often have episodes of psychotic behavior interspersed with periods of complete sanity. The biochemical defect is in the synthesis of heme, the red pigment of hemoglobin. In this instance, synthesis is excessive, and the metabolic products, porphyrins, are excreted in the urine. Porphyria exists in several forms, some dominant and some recessive. Acute intermittent porphyria is very episodic and is triggered by a variety of environmental agents. Not all affected persons express psychological disorder, but some do have episodes of hallucination or other forms of insanity. While one may view the psychological effects as a by-product of the basic metabolic disease, it must be remembered that King George was viewed only as psychotic, since the attending physicians didn't know about porphyria. How many of our present day psychoses are unrecognized metabolic disorders?

Lesch-Nyhan Syndrome.   Perhaps no example of deviant behavior is more bizarre than that in Lesch-Nyhan syndrome. This is a rare X-linked recessive disorder. Therefore only males are affected. The defect is well-understood metabolically. The enzyme hypoxanthine phosphoribosyl transferase (HPRT), an enzyme important in purine metabolism, is nonfunctional. The behavioral effect is mental retardation and self-mutilation. Boys with Lesch-Nyhan syndrome bite their lips and fingers off. This is not due to lack of pain receptors. It is rather a compulsive behavior that is extremely damaging. They must be restrained, but even so, substantial mutilation occurs.

The interest in this disorder lies in part in the specificity of the behavioral abnormality. No other disorder is known that produces this behavior pattern. It is an overwhelmingly destructive and painful behavior. Just how the biochemical defect causes this particular effect is not known, but again it points out the behavioral consequences that may be associated with a single mutant gene.

Affective Disorders.   The affective disorders, known also as manic-depressive psychosis, have been strongly suggested as inherited defects. Affected persons tend to cluster in families, which could be due to either genetic or environmental effects. Some investigators believe that the major genetic effect is due to a gene on the X chromosome, a gene that is largely dominant, that is, it is expressed in most heterozygous females. In other studies, the results are less clear-cut. Of the various common major psychiatric disorders, this has the strongest evidence for a single mutant gene. It may well be that affective disorder can result from several different genotypes and possibly from environmental causes as well. Continued research will be necessary to resolve the issue.

Perceptual Deficiencies

Colorblindness.   The inability to distinguish certain colors has long been known as an inherited trait. There are in fact several kinds of colorblindness, at least two of which are due to mutant genes on the X chromosome, genes that are close to each other but that can be separated by crossing over. The most common is red-green colorblindness, in which there is a deficiency in the ability to discriminate between red and green. Not all colorblindness is absolute. There may well be alleles with different degrees of defect. Total colorblindness is very rare but occurs as an autosomal recessive trait.

Females heterozygous for X-linked colorblindness should have two populations of retinal cells, in accordance with the hypothesis of X-chromosome inactivation (Box 12.1). This in fact is observed. If special instruments are used to test the color vision of individual cells, the retina is found to be a mosaic of cells with normal color vision and cells that are deficient. The information from the normal cells is used by the brain to give an overall impression of normal color vision. Rarely, a heterozygous female will have a preponderance of retinal cells in which the X chromosome with normal color vision genes was inactivated. Such a female would be essentially colorblind, or more interestingly she might be essentially colorblind in one eye with normal color vision in the other.

PTC Taste Sensitivity.   Phenylthiocarbamide (PTC), also called phenylthiourea, was first synthesized many decades ago by organic chemists, who noted its extremely bitter taste. Many years later, another organic chemist synthesized what should have been PTC but found that the product was tasteless. Convinced that he should have PTC from the procedure he used, he asked a colleague to taste it. The colleague confirmed that the supposed PTC did indeed taste very bitter. This became the first and still the best example of "taste blindness."

Most persons, approximately 70% of caucasians, can taste extremely small quantities of PTC. The remaining 30% often cannot taste levels hundreds of times larger. The ability to taste PTC is a dominant trait, tasters being *TT* or *Tt*. Differences in taste sensitivity to PTC do not extend to other kinds of substances. Compounds structurally related to PTC are found in certain plants and are known to cause goiters. It is interesting therefore that among persons with certain types of goiter, the frequency of nontasters of PTC is greater than in the general population.

Deafness.   Deafness has many forms, some due to simple genetic mechanisms, some to trauma or other environmental causes, and many of unknown cause. Most involve a neurological deficiency in the ability to sense sound waves or to transmit the nerve impulses from the ear to the brain. Of more interest in many ways are deficiencies in the ability to process the information received by the brain.

One example described recently is *tune deafness* or *amusia*. H. Kalmus found that about 4% of Europeans are unable to recognize mistakes deliberately introduced into familiar songs, mistakes that seem ludicrously wrong to most persons. An example is given in Figure 18.2. The defect seems to be in the memory

FIGURE 18.2
An example of a common melody used to test persons with tune deafness. Which version is correct? (From the test developed by H. Kalmus. Used by permission.)

of tunes rather than in the neurological ability to hear the notes correctly. This interesting trait has been studied genetically only to a very limited extent. The preliminary results indicate that it may be inherited as an autosomal dominant trait. Perhaps an example is the unfortunate man in the old limerick cited by Kalmus.

> There was an old fellow of Sheen
> Whose musical sense was not keen.
> He said, "It is odd
> I never tell God
> Save the Weasel from Pop Goes the Queen."

The ability to remember tunes and to discriminate between similar tunes must surely be a requisite for success as a musician or as a student in a music appreciation course.

Examples of Behavior with a Complex Genetic Basis

Intelligence

Few areas of research have generated as much controversy as intelligence. Even though most of us can identify persons who are very intelligent compared to those with subnormal intelligence, it is not easy to describe with precision the behavior that is the basis for such judgments. Intelligent people are able to solve problems that are difficult and solve them faster than others might, they are able to extract selected information from an enormous number of stimuli and make meaningful associations, they can remember details, and they can express their ideas verbally with facility. Not every intelligent person is equally adept in each of these areas, but most are adept in several. Those with below normal intelligence are deficient in these areas.

The world is not divided neatly into intelligent and stupid people, however. Rather, there seem to be people with all gradations of intelligence. A large portion fall somewhere in the middle, with smaller and smaller numbers as one moves to the very bright or very dull. These statements imply that intelligence is a continuous variable, which carries an additional implication that it can be measured. But is intelligence a single ability, as some psychologists think, or is it a composite of several, perhaps many abilities, as others believe? If it is a composite, how meaningful is an overall measure, especially from the point of view of genetic analysis?

These questions cannot be answered rigorously. But even if intelligence is a composite of many abilities, it is subject to genetic analysis just as is a complex trait such as stature. A number of tests have been developed to measure various aspects of intelligence, of which the Stanford-Binet and related tests are the best known (see Box 18.1). These yield an IQ or *intelligence quotient* as an overall measure, and most genetic studies have been based on IQ.

Comparison of pairs	No. of pairs	Expected correlation*	0.0	0.1	0.2	0.3	0.4	0.5	0.6	0.7	0.8	0.9	1.0
MZ twins reared together	4672	1.00											
MZ twins reared apart	65	1.00											
DZ twins reared together	5546	0.50											
Siblings reared together	26,473	0.50											
Siblings reared apart	203	0.50											
Parent-offspring reared together	8,433	0.50											
Parent-offspring reared apart	814	0.50											
Adopted-natural sibs	345	0.00											
Adopted-adopted sibs	369	0.00											
Parent-adopted offspring	1397	0.00											

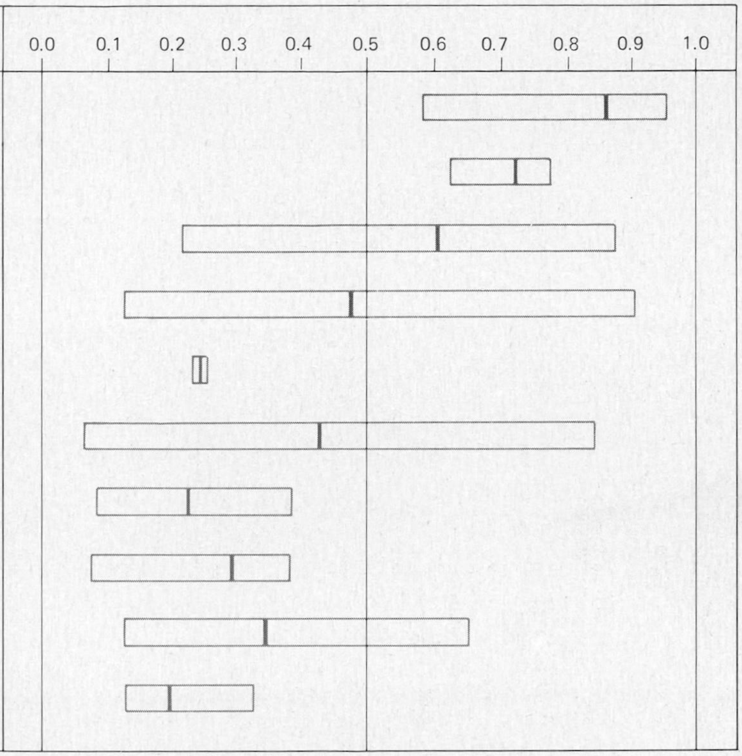

*The expected correlation is based on 100% genetic variation.

FIGURE 18.3
Summary of studies of correlations of intelligence between pairs of persons who are related biologically or by adoption. The heavy vertical bar is the weighted average of all the correlations from studies in which the particular comparison was made. The shaded horizontal bar is the range of correlations. (Data from T. J. Bouchard Jr. and M. McGue, *Science* 212: 1055–1059, 1981.)

Several different approaches can be used to study genetic factors in intelligence. The most often used are correlation coefficients of persons who share some portion of their genes (Box 11.1). If heredity were the *only* variable that influenced intelligence, we would expect correlation coefficients of $+1.0$ for monozygotic twins, $+0.50$ for dizygotic twins, sib pairs, and parent-offspring pairs. Unrelated persons, such as foster sibs and foster parent-child pairs, should give correlations of zero. A summary of a large number of such studies is given in Figure 18.3. It is obvious that, although the theoretical values for biologically related persons are not obtained, the observed values are strongly in accord with the predictions of major hereditary variation.

The studies of adoptive children are especially interesting. On the basis of genetic theory, adopted children should resemble their biological parents rather than their adoptive parents. Conversely, if intelligence is largely environmental, the adoptive children should resemble their foster families. The data for such compari-

Box **18.1**

How to Measure Intelligence

The use of tests to help us predict events is ancient and commonplace. We predict the ripeness of a banana by the appearance of the peel. We learn to do so by removing the peel and testing the ripeness directly. Our prediction is thus subject to validation. Other predictions, such as reading our future in tea leaves, cannot be validated so readily because the predicted events are too far removed in time and often too ambiguous for most of us to bother with. Tests can be developed to measure psychological attributes such as intelligence, so long as there is an independent expression of intelligence with which to validate the predictions.

The present-day intelligence tests had their origin in 1904 when authorities of the Paris school system asked the eminent psychologist Alfred Binet for advice on the education of retarded students. It was apparent to Binet that there was no objective way to predict which students would be successful with the usual classroom instruction and which would do better with special instruction. Binet set out to solve this problem by devising a series of questions that would assess how much information and understanding had been acquired by the students. In order to be applicable to all students, the questions had to be based on material and experiences that would be faced by most French children living at that time rather than on academic subject matter.

On the basis of the results, Binet was able to predict which children would do well in class and which would have difficulty with the academic material, as rated by the teachers. Norms were established for each age level, and a child was rated as performing at the eight-year-old level, say, although his chronological age may have been seven, eight, or ten years. A German psychologist, William Stern, noting the inconvenience of a constantly changing *mental age* as each child grew older, devised the *intelligence quotient* or *IQ*, obtained by dividing the mental age by the chronological age, and multiplying by 100 to convert the quotient to whole numbers. Thus a child whose mental age was 6.5 years and whose chronological age was also 6.5 years would have an IQ of 100.

Binet accomplished what he set out to do: predict performance in a French school. He was not attempting to measure intelligence as such, although most would agree that, whatever we mean by intelligence, it is an important ingredient in school success. The test was culturally biased, but that was not important for the purpose of the test. Professor Lewis Terman of Stanford University adapted the Binet test for use in America, and current modifications of the *Stanford-Binet* test are still widely used. Several other tests are also available for use with very young children, with adults, or with group testing.

A psychological test (or any other test) can be constructed to yield whatever distribution of scores is desirable. By selecting questions of various degrees of difficulty, one can cause the results to be distributed in a variety of ways. Intelligence tests are constructed so that the results yield a *normal distribution*, as shown below in the figure. The mean IQ is approximately 100 for the white population on which the test was standardized, and 50% of the population falls

Box **18.1**

(*continued*)

between 90 and 110. The test results are standardized by administering the test to very large numbers of persons, and questions that produce these results are used in the final version. It is important to remember that *the questions used are selected to produce a particular distribution in the population on which the test was standardized.*

Because of the way tests are developed and standardized, it is hazardous to compare test results from populations whose cultures are different. Intelligence is certainly a strong component in whatever it is that IQ tests measure. But persons with very different experiences may produce different scores, even though

there is no difference in intelligence. For this reason, one should only make comparisons among persons within a defined population. Even this will not remove all environmental bias, but the errors due to that source are presumably randomly distributed among the subjects and can be estimated.

There has been much criticism of studies in which IQ scores of U.S. blacks are compared with the white population. Blacks generally score lower on IQ tests, some 15 points lower on the average. From the above discussion, it should be apparent that we cannot attribute this difference between whites and blacks to differences in intelligence. The test questions were selected to yield certain results in white populations and, to the extent that there are cultural differences between blacks and whites, many of the questions may be less useful with blacks. It probably is true that blacks who are more intelligent perform better *in general* on IQ tests than do less intelligent blacks. But two equal scores, one from a black, the other from a white, do not necessarily mean that the two persons are equally intelligent. It may be, however, that such persons with equal scores would have about equal prospects for school performance.

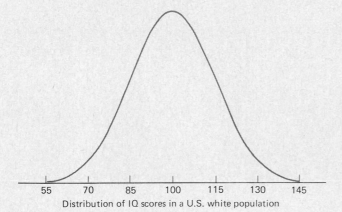

Distribution of IQ scores in a U.S. white population

sons have not often been available, but the results of one large study are shown in Table 18.1. Again the results are consistent with a large genetic component in intelligence. One criticism of such studies is that adoption agencies, perhaps unconsciously, tend to match biological and foster parents for socioeconomic level, education, and so on. Where it has been possible to study such effects, they have not altered the basic conclusion that heredity is very important in intelligence.

One must be careful though not to extend these results to situations where they

**TABLE 18.1
Correlations Between
Biological and Foster
Relatives for IQ**

Although there is a positive correlation between adopted children and their foster parents, it is not as large as the correlation between the adopted children and their biological mothers whom they have never seen. This supports the hypothesis that heredity is an important factor in intelligence (as measured by IQ tests), though heredity is not the ony factor.

Comparison	Expected Correlation[a]	Observed Correlation
Foster father/adopted child	0.00	0.14
Foster father/biological child	0.50	0.28
Foster mother/adopted child	0.00	0.17
Foster mother/biological child	0.50	0.20
Biological mother/adopted child	0.50	0.31
Foster father/foster mother	0.00	0.24
Foster father/biological mother	0.00	0.11
Foster mother/biological mother	0.00	0.14

[a]Based on variation in IQ attributable entirely to genetic variation. Also, it is assumed that there is no correlation between husband and wife of the adoptive family or between the biological mother and the foster parents. However, as the last three values indicate, husbands and wives do resemble each other in IQ, and the adoption agency also matched the biological mother with the foster parents to some extent.

From J. M. Horn, J. C. Loehlin, and L. Willerman, Intellectual resemblance among adoptive and biological relatives: The Texas Adoption Project. *Behavior Genetics* 9:177–207, 1979.)

are unwarranted. Let us suppose for the sake of argument that all of the variation in intelligence in the U.S. white population is due to genetic variation. One cannot conclude that the same would be true for other populations, although of course it might be. And, most importantly, *one cannot conclude that differences between populations are due to heredity.* A simple example will illustrate this point. Let us take a population from central Europe where some persons have very light skins, others have darker skins. Let us further assume that this variation in skin color is largely genetic. We divide the population into two similar groups, sending one to a cold overcast land where the inhabitants are never exposed to sun. The other group is sent to a sunny tropical isle where clothing is minimal. At the end of a year, we check our experiment. By now the two populations should be quite different from each other in the degree of skin pigmentation. That difference is entirely environmental. Within each population, there will continue to be persons with lighter or darker skins, due to heredity.

Until it is possible to measure innate intelligence with tests that are not culturally biased, it will not be possible to draw conclusions about the relative intelligence of one population compared to another. And we certainly cannot draw conclusions about the role of heredity in such differences. Individuals should, of course, be viewed entirely as individuals rather than as members of a group. Those who can take advantage of special opportunities should be given the chance, and those who require special help should receive it, regardless of race.

Personality

The measurement of personality is at least as difficult as the measurement of intelligence. There is even less basis for supposing that the characteristics that we

Separated at birth, the Mallifert twins meet accidentally.

(Drawing by Chas. Addams; © 1981 The New Yorker Magazine, Inc.)

find of interest correspond to underlying biological or psychological factors. Whereas intelligence has school performance to help define it, however imperfectly, personality lacks any such measure. Personality categories are whatever we choose to make them: calm, nervous, hyperactive, hostile, friendly, phlegmatic, feminine, masculine, compulsive, and so on. It would seem that all these are largely the result of training and experience. Yet those who have observed infants and young children growing up in similar environments recognize that elements of later personality differences can be expressed even in the very young.

The principal experimental approach to genetic studies of personality has been by comparison of monozygotic and dizygotic twins. Table 18.2 is a summary of two of these studies. In general there is no consistent finding from one study to another for individual test measures. This may be largely due to the imperfection of personality tests. Certainly also personality must be very complex. The generally higher correlations between MZ twins compared to DZ twins suggest an important contribution of genetic variation to the overall personality differences among people. One can imagine inheritance of an underlying stratum of temperament,

**TABLE 18.2
Comparisons of
Correlations Within
Monozygotic (MZ) Twin
Pairs and Dizygotic (DZ)
Twin Pairs for Various
Measures of Personality**

The personality scales are from the California Psychological Inventory (CPI).

CPI Scale	High School Boys[a]		Male Veterans[b]	
	MZ	DZ	MZ	DZ
Dominance	.58	.13	.44	.25
Capacity for status	.57	.36	.45	.18
Sense of well-being	.54	.33	.38	.13
Responsibility	.57	.29	.20	.28
Socialization	.53	.15	.44	.21
Good impression	.48	.32	.37	.11
Communality	.32	.24	.22	.06
Achievement via conformance	.49	.06	.36	− .05
Achievement via independence	.58	.36	.39	.05
Intellectual efficiency	.59	.27	.34	.21
Psychological mindedness	.47	.28	.39	.20
Flexibility	.43	.25	.49	.10
Femininity	.42	.27	.15	.17

[a]Cited from R. C. Nichols, The resemblance of twins in personality and interests. *National Merit Scholarship Corp. Res. Rep.* 2:1–23, 1966.
[b]Data of J. M. Horn et al., based on measures of twin veterans of World War II.
From J. M. Horn, R. Plomin, and R. Rosenman, Heritability of personality traits in adult male twins. *Behavior Genetics* 6:17–30, 1976.

but there must be a large overlay of culture that modifies these tendencies. We only know how to test the surface structure at present.

Schizophrenia

Schizophrenia is a form of mental illness expressed primarily as difficulty in reality testing and logical thinking. Many schizophrenics have hallucinations. A majority are able to function in society in spite of these problems, but many require hospitalization or close supervision. It is estimated that 1% of the population is schizophrenic, which makes it the major mental illness.

Virtually nothing is known about the causal mechanisms of schizophrenia. A variety of theories attribute schizophrenia to environmental experiences. Many of these have focused on early maternal-child relationships. Other theories have attributed schizophrenia to inherited changes in metabolism. Indeed a number of chemical changes occur in the blood of schizophrenics, but most seem to be the result of schizophrenia rather than the cause. The primary biochemical defect remains elusive. In spite of that, many scientists are convinced that schizophrenia is a biological defect that is inborn and that can be controlled but not "cured." As such, it could well be due entirely or largely to the person's genotype.

Early efforts to establish a genetic basis for schizophrenia were based largely on twin studies. If schizophrenia is inherited, then monozygotic twins should both be affected. If it is strongly influenced by heredity, then MZ twins should both be

The twin pairs were identified because at least one member had a diagnosis of schizophrenia. If the twins were MZ, the likelihood was much greater that the other twin would also have the diagnosis than if the twins were DZ. The frequencies differ substantially among the studies because of the way in which the affected twins were located. In the study of Allen et al., both twins had served in the armed services and therefore were not severely affected as young adults. The important comparison is between MZ and DZ twins within the same study.

**TABLE 18.3
Concordance for
Schizophrenia in
Monozygotic (MZ) and
Dizygotic (DZ) Twin
Pairs**

	Percent of Twin Pairs Both of Whom Have Schizophrenia		
	Kallman (1946)	**Inouye (1961)**	**Allen et al. (1972)**
MZ twin pairs	86	76	27
DZ twin pairs	15	22	5

Citations from M.G. Allen, S. Cohen, and W. Pollin, Schizophrenia in veteran twins: a diagnostic review. *American Journal of Psychiatry* 128:939–945, 1972.

affected more often than DZ twins. A summary of some of the twin studies is given in Table 18.3. The consistent increase in concordance of MZ twins as compared to DZ twins supports a strong genetic predisposition to schizophrenia.

Another approach to understanding the cause of schizophrenia is provided by adoption studies. Several such studies show that the risk of schizophrenia is greater in an adopted child if a biological parent also has been diagnosed as schizophrenic. One of the more rigorous efforts to separate genetic from potential environmental factors was carried out in Denmark. A number of children who had been adopted at birth and who subsequently became schizophrenic were located, along with a comparable number of adopted children who had not developed schizophrenia. The investigators then asked whether the frequency of schizophrenia was higher in the paternal half sibs of the schizophrenic adoptees than of the comparison group. Paternal half sibs were selected because such persons share only the genes received from the fathers' sperm. Since they had different mothers, even the intrauterine environments were different. Of 63 half sibs of schizophrenics, 14 had definite or probable schizophrenia. Of 64 half sibs of nonschizophrenics, only 3 had definite or probable schizophrenia. It is very difficult to interpret these results in any way except as indicating a genetic influence on risk of schizophrenia.

Schizophrenia is very likely a mixture of mental disorders that share some common elements. Possibly some forms are more influenced by heredity than are others. There is little evidence that any form of schizophrenia is transmitted as a simple Mendelian trait, although it would be very difficult to identify such families if they were only a small part of the total. The disorder may be both genetically complex (influenced by several loci) and genetically heterogeneous (due to several different genotypes). Clarification of the role of genes in schizophrenia will solve one of our most important health problems.

Chromosomes and Behavior

The association of aberrant karyotypes with mental retardation has been noted in earlier chapters. Here the emphasis will be on the *qualitative* differences that characterize some of the aneuploid states.

Down Syndrome. The best known behavioral effect of trisomy 21 is the reduction in intelligence. Some affected persons are mildly retarded and can learn to work productively in protected environments. Most are more severely retarded and can be trained to a more limited extent.

Persons with Down syndrome can be distinguished from others of the same intelligence by their personalities. They are pleasant people who respond to attention and affection. A group of adults with Down syndrome is like a gathering of overgrown happy children. Whatever the genic imbalance caused by the trisomy, it leads to a specific Down personality that is independent of environmental exposures.

XYY Syndrome. As noted in Chapter 12, approximately 1 male/1000 has an extra Y chromosome, with the karyotype formula 47,XYY. When a high frequency of XYY males was found among tall prisoners, this karyotype was unfortunately called by some popular writers a "criminal" karyotype. This is a misnomer, however. Most XYY males do not become criminals. There does seem to be a slightly elevated risk that XYY males will get into trouble. The significance of this has been argued. Since XYY males are taller on the average than XY males, it has been proposed that taller males are more likely to get into trouble and that this explains the excess of XYY among criminals. However, other taller males do not seem to get into trouble more than their shorter peers. The type of criminal activity is impulsive aggression. XYY males tend to destroy material objects, although assault of persons also occurs. There is no increased risk of white collar crime, such as embezzlement.

The most compelling evidence of the increased likelihood of antisocial acts associated with XYY comes from comparison of XYY and XY sibs. If one looks at sibs of imprisoned XYY and XY males, one finds fewer of the sibs of XYY males also in difficulty with the law. This is what one would expect if the environment had relatively less to do with the criminal activities of XYY males than with that of XY males. In other studies, the average age of the first offense was found to be younger in XYY males than in XY males who had committed crimes, 13 years compared to 18 years. Again, this suggests an inborn tendency.

The fact that a person charged with a crime has an XYY karyotype has been offered as evidence for acquittal due to diminished responsibility. This defense has generally been rejected by courts and is not accepted in the United States (see Chapter 23).

Turner Syndrome. In Chapter 12, it was pointed out that persons with Turner syndrome (45,XO) have difficulty with certain spatial imagery. The basis for this is not well understood, but there is evidence that it involves the memory for spatial relationships. Affected persons cannot draw geometric figures well from memory. One college student with Turner syndrome had difficulty remembering how to get to classes.

It has been suggested that there is a gene on the X chromosome that has to do with spatial imagery. Persons with Turner syndrome offer no support for such a gene. They have the same number of X chromosomes as males, who are not conspicuously different from XX females in spatial perception. The way in which the XO condition leads to this particular problem remains quite obscure.

Other Aneuploid Conditions.   The other karyotypes involving imbalance of autosomes are almost universally associated with mental retardation, for the most part very severe retardation. Trisomies 13 and 18, the most common autosomal trisomies next to trisomy 21, are too severely affected to test and rarely survive more than a few weeks. Many of the aneuploidies are known only by a few examples, too few to permit generalizations beyond the fact of severe retardation. Clearly the human nervous system is not able to tolerate substantial genic imbalance.

Sociobiology

The previous examples have dealt with intraspecies variations in behavior. Only these variations can be investigated by the techniques of genetics, since one can only make genetic crosses within a species. (One can make interspecies crosses in the form of cell hybrids, described in Chapter 10, but these are of no value yet in studying behavior.) Yet, one wonders to what extent the typical behavior of humans both as individuals and in groups is programmed in our genes. The instinctive behavior of animals is often very obvious. Each generation shows the same complex behavior, even though the generations do not overlap, as when newly hatched insects in the spring have no exposure to their parents who mated and left fertilized eggs the previous fall.

But humans do learn from their parents and from other humans, including those of past generations. Even primates kept in a zoo seem to have difficulty at times in carrying out sexual activities unless they have grown up in the presence of adults of their species, and monkeys reared artificially do not know how to be mothers. We therefore have little basis for making assumptions about the underlying "instincts" on which our patterns of life are constructed.

Society may be defined as a group of individuals that interact in an organized manner. Sociobiology is the study of the biological basis of societies. The members of a human society follow certain rules in dealing with each other. In different societies and at different times, the rules may be very different. Indeed, the extremes of behavior that characterize some societies support the idea that specific patterns of behavior can hardly be strongly influenced by heredity. Yet the very plasticity of humans in adapting to distinctive cultures is part of the human biological and hence genetic heritage.

There has been much discussion of the extent to which traits found in many humans are part of our genetic heritage—genetic as opposed to purely learned behavior. The ability to learn is inherited to be sure, but are we innately aggressive because of our genes, or are many of us aggressive because we have learned to be? If aggression is an inherited characteristic of all human populations, however much it may have become modified by learning, then natural selection must have favored it during human evolution. Is there an underlying genetic program favoring small family units rather than troops or herds? Or does the adaptive value of this common arrangement in human societies have purely a cultural basis?

At present one cannot say how much of the general behavior that is found in *Homo sapiens* is derived from the genes shared by most members of the species. Sound conclusions must be based on controlled observations, and there are no comparison groups against which to judge the whole of mankind. Even our closest primate relatives have had sufficiently different evolutionary history so that comparisons are hazardous. Nowhere is the uniqueness of humans so apparent as in social behavior. Progress in understanding human sociobiology is likely to be very slow, but the subject is important and must be pursued as opportunities occur.

Review Questions

1. What is the inherited metabolic defect in phenylketonuria? How did knowledge of this defect help design an effective therapy?
2. What other form of mental deficiency can be treated by diet?
3. There are several forms of X-linked mental retardation. Would you expect males or females to be affected more often? Do you know to what extent this expectation is in agreement with the populations found in institutions for the mentally retarded?
4. In the fragile-X syndrome, how are carriers identified?
5. What other abnormal behavioral variations have been suggested as resulting from mutant genes on the X chromosome?
6. How have studies of adopted children provided insight into the heritability of intelligence?
7. Why are behavioral traits such as schizophrenia very hard to study genetically?
8. Ordinarily we look for evidence of genetic variation by searching for similarities among close relatives. Yet only rarely is mental retardation due to trisomy 21 found in more than one person in a family. How do you resolve this discrepancy?
9. A couple is concerned about the risk of trisomy 21 in their offspring because of the advanced age of the mother, who has recently become pregnant. Amniocentesis and karyotype analysis of the fetus show it to be 47,XYY. Was the mother's age an important factor in producing the abnormality? What options would you suggest to the parents?
10. What is sociobiology? Why is it so difficult to be sure that a normal human trait is inherited? How would you set up a study to decide whether human male aggression is determined by our genes?

References and Further Reading

Ehrman, L., and P. A. Parsons. 1982. *The Genetics of Behavior*, 2nd ed. Sinauer Associates, Sunderland, Mass. A good general text covering all important aspects of the topic.

Fuller, J. L., and W. R. Thompson. 1978. *Foundations of Behavior Genetics*. C. V. Mosby, St. Louis. 533 pp. A well-rounded treatment of both animal and human studies.

Loehlin, J. C., G. Lindzey, and J. N. Spuhler. 1975. *Race Differences in Intelligence*. Freeman, San Francisco. 380 pp. A balanced discussion supported with data from many fields.

Lumsden, C. J., and E. O. Wilson. 1981. *Genes, Mind and Culture.* Harvard Univ. Press, Cambridge, Mass. An attempt to give sociobiology theoretically sound underpinnings.

McClearn, G. E., and J. C. DeFries. 1973. *Introduction to Behavioral Genetics.* Freeman, San Francisco. 349 pp. Another very sound general textbook.

Rosenthal, D. 1971. *Genetics of Psychopathology.* McGraw-Hill, New York. A primer that is somewhat out of date but sound in its approach.

Wilson, E. O. 1980. *Sociobiology,* Abridged Edition. Harvard Univ. Press, Cambridge, Mass. A shortened version of the original *Sociobiology,* but nothing important is omitted.

Since ancient times, humans have speculated on their relationships to each other, to the nonhuman animals, and to the gods. Such myths have often taken the form of humans having descended from some animal, either itself mythological or a godlike counterpart of a living form. It is doubtful though that any of the ancients ever believed themselves to have descended through a process of evolution from ancestors that were common to humans and to other existing animals. The ancestral animate forms were merely convenient guises for a deity whose kinship to humans gave the latter special status.

It was certainly apparent to the ancients—whether or not they thought much about it—that certain groups of animals were more closely related to each other than they were to other groups. Fish, birds, and snakes are natural classifications that would provoke little argument. The systematic classification of plants and animals based on their relationships to each other did not occur until the eighteenth century, however.

The Classification of Organisms

The Binomial System of Nomenclature

Current systems of classification trace their origins to the work of the great Swedish botanist and physician Carl von Linné (1707–1778), better known by the

Latinized version of his name, Carolus Linnaeus. Linnaeus introduced a hierarchical system of classification whose basic structure is still in use. He divided the living world into two kingdoms, plants and animals. A third kingdom, the protists, has been added to include bacteria and related groups that are not plant or animal. In fact, the dividing line between plants and animals is not sharp. Animals get their energy from consumption of organic substances, they tend to be motile, and they do not have cell walls made of cellulose, but these descriptions also apply to some plants.

The basic unit in Linnaean classification is the *species*. (The word species is both singular and plural. An animal *specie* is a coin with an animal on it, such as the nearly extinct buffalo nickel.) In the classical approach, a species is defined by representative individual specimens that are presumed to embody the important characteristics of the species. Since most such specimens were preserved, the comparisons were entirely anatomical. The existence of local *races* within a species was recognized, but all members of a species were assumed to be essentially identical and distinct from members of other species. This approach was *typological*, in that there was considered to be an ideal type for each species. The deviations of individuals from the ideal type were considered errors of nature.

Certain species are more closely related to each other than to other species. For example, lions and tigers are both cats and are more similar then either is to a fish. Species that are very similar are placed in the same *genus* (pl., *genera*). Many genera have numerous species, others have only one. The genus *Homo* has the single living species *Homo sapiens*. Genera that are similar are grouped into *families*, families into *orders*, orders into *classes*, and classes into *phyla*, the largest division within a kingdom (see Box 19.1).

19

Human Evolution

Box **19.1**

Linnaean Classification of the Animal Kingdom

Only selected taxons of living species are shown.

Kingdom Animalia
  Phylum Protozoa (amoebae, paramecia, malaria parasites)
  Phylum Porifera (sponges)
  Phylum Coelenterata (sea anemones, corals, jellyfish)
  Phylum Platyhelminthes (flatworms, flukes, tapeworms)
  Phylum Nematoda (roundworms, trichina)
  Phylum Mollusca (snails, clams, squids, nautilus)
  Phylum Annelida (earthworms, leeches)
  Phylum Arthropoda (joint-legged animals: lobsters, insects, spiders)
  Phylum Echinodermata (starfish, sea urchins)
  Phylum Chordata (during development, a notochord, a hollow nerve cord, and gill slits)
    Subphylum Vertebrata (vertebrates)
      Class Agnatha (jawless fish: lampreys, hagfish)
      Class Chondrichthyes (cartilage fishes: sharks, rays)
      Class Osteichthyes (bony fishes)
      Class Amphibia (salamanders, frogs, toads)
      Class Reptilia (turtles, snakes, lizards, alligators)
      Class Aves (birds)
      Class Mammalia (all have hair, mammary glands, and are homeotherms)
        Subclass Prototheria (duck-billed platypus)
        Subclass Metatheria (pouched animals: opossums, kangaroos)

All organisms are referred to in scientific works both by the name of the genus and the species. Thus, humans are *Homo sapiens*, abbreviated *H. sapiens* if the genus has already been identified. This use of two names is called the *binomial* system of nomenclature. Since each species is identified through a published description, the use of binomial names in scientific writing makes it possible to know exactly which species is meant when scientists in different parts of the world communicate. By convention, the binomial name is always italicized. The name of the genus is capitalized, but the name of the species is not. Higher orders of classification are not italicized.

Organisms in a particular classification are said to belong to that *taxon* (pl., *taxa*), whether reference is made to a single species or to an entire phylum. Taxonomy is thus the study of the classification of organisms.

Box **19.1**

(*continued*)

Subclass Eutheria (placental mammals)
    Order Insectivora (shrews, moles)
    Order Chiroptera (bats)
    Order Edentata (amadillos, sloths)
    Order Lagomorpha (rabbits, hares)
    Order Rodentia (mice, rats, squirrels)
    Order Cetacea (whales, porpoises)
    Order Carnivora (dogs, bears, cats, seals)
    Order Proboscoidea (elephants)
    Order Sirenia (manatees, dugongs)
    Order Perissodactyla (horses, tapirs, rhinoceroses)
    Order Artiodactyla (pigs, camels, giraffes, cattle, sheep)
    Order Primates
      Suborder Prosimii
        Infraorder Lemuriformes (lemurs, aye-ayes)
        Infraorder Lorisiformes (lorises, galagos)
        Infraorder Tarsiiformes (tarsiers)
      Suborder Anthropoidea
        Superfamily Ceboidea (New World monkeys)
        Superfamily Cercopithecoidea (Old World monkeys)
        Superfamily Hominoidea
          Family Pongidae
            Genus *Hylobates* (*Hylobates lar:* gibbons)
            Genus *Pongo* (*Pongo pygmaeus:* orangutans)
            Genus *Pan* (*Pan troglodytes:* chimpanzees)
            Genus *Gorilla* (*Gorilla gorilla:* gorilla)
          Family Hominidae
            Genus *Homo* (*Homo sapiens:* humans)

A hierarchical classification such as that introduced by Linnaeus is based on human perceptions of the relationships among the millions of known species, both living and extinct. These perceptions are based on the information available at a particular time, which may be incomplete or wrong. It is not surprising then that taxonomists specializing in a particular group of plants or animals may disagree on many aspects of classification. It is not uncommon to find that what one author considers an order is labeled a subclass or class by another, or that genera may be grouped differently into families. Taxonomists working with different groups of organisms have also tended to use different criteria to define the various levels of classification. For example, species may be grouped together into the same genus in one part of the classification, while equally similar species are separated into separate genera in other parts. Despite these shortcomings, the use of the

Linnaean system of classification has brought order into a field that could easily be chaotic.

Discrete Species and Indiscrete Evolution

Once the hierarchical classification was seen to work, it could not be long before the possibility of evolution of species was suggested. Just as family members, or dogs of a particular breed, resemble each other because of common descent, the resemblance of species should also be due to their having descended from a single ancestral species. Furthermore, such ancestral species were sometimes observed as fossil remains of extinct species. If this were true of closely related species, why not all species, both living and extinct, including *Homo sapiens*? As discussed in Chapter 17, there was much awareness of the concept of evolution prior to Darwin. His contribution was to provide the driving force, natural selection.

A key element of Darwin's view of evolution is its slow, incremental nature. Over tens of thousands of generations, a species may adapt to some ecological niche, undergoing sufficient changes so that it can legitimately be called a new species. But there is no one generation at which the old species suddenly crosses the threshold to become a new species. Evolution is continuous, and each offspring generation closely resembles the parent. This is not a large problem with living species. What is alive now provides the description of the species. It can be a problem in discussing the relationships of fossil species to living species or to each other.

Evolution does not occur at a constant rate in different species insofar as body form and adaptation are concerned. (The evolution of DNA may be another matter.) A species that is well adapted to a stable environmental niche may not tolerate mutations. Few would improve the adaptation. Many species seemingly have not changed for enormously long periods. For example, the horseshoe crab, *Limulus*, is sometimes referred to as a living fossil, this genus being virtually unchanged for at least 200 million years. Our own ancestors at the time had not yet acquired mammalian status. Other species that are less well adapted are subject to more selection and change. And the environment to which a species is adapted may change and create new selection pressures and opportunities for evolution.

Species and Speciation: Current Concepts

The Linnaean approach to classification, as noted earlier, was typological in concept. This approach to classification can be traced at least to Plato. Variations within a species were regarded as deviations from the ideal type, which was unchanging, with no relationship between species. Such a philosophy has no place for evolution, and Linnaeus was a creationist, not an evolutionist, in spite of the implications of the hierarchical groupings.

With the discovery of Mendelian heredity and the development of the genetics of populations, the idea of a typological species became even more untenable, and clearly no one individual from a species could serve as the type specimen for all members of the species. What one human being would be representative of all

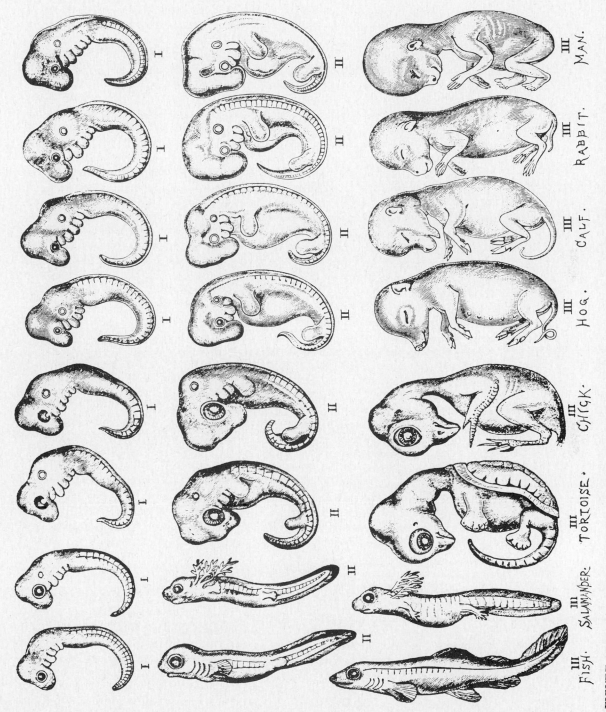

FIGURE 19.1 Diagram showing similarities among early embryos of various vertebrates and divergence in later embryonic stages. (Reproduced from G. J. Romanes, *Darwin, and After Darwin*, Open Court Publ. Co., Chicago, 1910. Based on a figure of E. Haeckel.)

III MAN.
III RABBIT.
III CALF.
III HOG.
III CHICK.
III TORTOISE.
III SALAMANDER.
III FISH.

members of the species *Homo sapiens?* The fact is, a species is a population of genetically diverse individuals. Even the horseshoe crab shows individual variability at the biochemical level.

Efforts to define a species have not been wholly successful. One definition was based on the ability of members of a species to interbreed readily. Only matings within the same species would produce fertile offspring. This definition is not consistent with all biological facts, however. There are many well-documented cases of fertile *interspecific* (between species) hybrids, cases in which no one argues that the parents should be placed in the same species. Such hybrids may be rare in nature; indeed, they would have to be for the two species to remain distinct. Conversely, in the case of certain extensive species, populations at the extremes of the geographic distribution cannot interbreed with each other, even though they can interbreed with the intervening populations.

The definition of a species must take into consideration the genetic diversity within a species as well as the lack of sharp separations between closely related species. Ernst Mayr has defined species as "groups of interbreeding natural populations that are reproductively isolated from other such groups." (E. Mayr, *Populations, Species, and Evolution*, Harvard Univ. Pr., Cambridge, Mass., 1970, p. 12.) A key part of the definition is "*reproductively isolated.*" Such isolation may be geographic, behavioral, or genetic. Under laboratory conditions, different species may be induced to interbreed. But in nature they do so rarely if ever. Different species have different gene pools that do not exchange, even though the gene pools may be very similar.

The term *race* is often used to describe subpopulations of a species that differ by some number of inherited characteristics from other members of the same species. In some instances, races may be incipient species, that is, subspecies that are evolving separately and that will likely achieve reproductive isolation from the remainder of the species. In other instances, races may represent adaptation to local environments without sufficient divergence from the parent species to merit designation as a subspecies. There are only arbitrary guidelines to determine when a population should be designated a race, and the term has been used in a variety of ways. Human races will be discussed further in the following chapter.

If one accepts the hypothesis that all living forms originated from a single primordial life form, then the several million living species that descended from that form must have diverged from smaller numbers of parental species. Indeed, this must have happened often since the beginning of life. This process of *speciation*, or species formation, happens when the gene pool of a derived population changes so that the derived population is distinct from the parent population and will not ordinarily interbreed with it.

In order for speciation to occur, there usually must first be isolation. That is, the parent species must be divided into noninterbreeding populations. Isolation typically is geographic. Mountain ranges and oceans have been especially important, but glaciers, rivers, deserts, and forests all have been important, depending on the species, its modes of transportation, and its environmental requirements.

Once two populations are separated, each may evolve along different pathways. This does not meant that they will. If the populations are small, they may change genetically purely by chance (*genetic drift*). If the separated populations are large, then selection must operate to change the populations in different directions. The two habitats may differ in the type of food available, climate, predators, and a competition for a particular ecological niche. Those individuals best able to function in an environment will contribute the larger share of genes to the gene pool of the next generation. A small evolutionary step will have occurred.

A new species must not only develop different adaptations compared to its parent or sibling species, it must also become reproductively isolated from them. This can occur through development of different behavioral patterns, incompatible body morphology, or genetic incompatibility. Examples of behavioral differences are mating calls. Animals may respond only to the highly specific mating call of their own species. Or certain insects will feed and mate on a single species of plant. In many insects, the size and shape of the genitalia are highly variable in closely related species, and only members of the same species can mate successfully. Mechanisms that operate at the gene level include chromosomal differences as well as gene differences. If the karyotype of two related species has diverged, the hybrid may be sterile because of the inability to segregate normal haploid chromosome complements at meiosis. At the gene level, there are many examples in nonhuman species of genes that are *coadapted*; that is, selection has favored alleles at several loci that function well together. When hybrids are formed, these coadapted genes are separated, and the hybrid is not as fit as either parent.

Although evolution is generally a very slow process, it may occur fairly rapidly on occasion. One such occasion was the introduction of finches into the Galapagos Islands from the South American mainland (Box 19.2). Darwin's study of the Galapagos during the voyage of the *Beagle* strongly influenced his thinking about evolution.

The Evidence for Evolution

In this section, we will not be concerned with the evidence in favor of evolution as opposed to that for special creation. Evolution is the only scientific explanation for the existence of the known living and extinct species. Rather, we shall be concerned with the kinds of evidence that help decide where a species should be placed in relation to other species—how we know that two species are "closely related" or that A and B are closer to each other than either is to C.

Morphology

No feature has been more important historically in assigning a species to its proper classification than shape: of the whole plant or animal and of various tissues and organs. Linnaeus certainly had little else to go on, and additional techniques became available only in the twentieth century, some in the last decade. Morphology is in fact an excellent means of studying evolution. The myriad variations are under genetic control for the most part, and differences between two closely related species may reflect differences at a number of loci. Morphology is the only approach that can be used on fossil remains.

Morphology is not just a description of the static state of the adult organism. It includes the stages of development of the embryo, the fetus, and the young animal (Fig. 19.1). It also includes the function of the organ or structure. For living species, soft tissues are of special interest. Even for fossils, one can deduce much about the soft tissues. The attachment of muscles to bones tells about the size of the muscle and its function, which may in turn reveal many things about the way the animal carried out its activities. Examples of the use of morphological characteristics in the study of human evolution will be given later.

Protein Structure

In the 1960s, it became technically possible to determine the primary amino acid sequences of proteins without an impossibly large effort. Since proteins are translations of information coded in DNA, the amino acid sequence of a protein provides an excellent record of primary genetic information. Homologous proteins are those that have the same evolutionary origin in different species. Usually they will also have the same function, although many examples are now known in which homologous proteins have diverged somewhat in their functions. Comparison of the primary sequences of homologous proteins can tell us which of several species are closely related and which are distant. For example, comparison of the hemoglobin α chains in several species (Table 19.1) shows similarities and differences that agree with the evolutionary distances among these species.

A single protein sequence is not very helpful for closely related species. The α-globin sequences are identical for humans and chimpanzees, and that of gorillas differs from these by a single amino acid. When the sequences for a number of

--- Box **19.2**

Darwin's Finches

The Galapagos Islands, located a thousand miles off the west coast of South America, are uniquely important in the history of science. The Galapagos provide a remarkable diversity of landscapes—barren rocky deserts, tropical forests, rocky shorelines, and mountains. It was there, during the voyage of the *Beagle*, that Darwin observed many examples of animal adaptation to local habitats. The most striking adaptations were found in the finches, and it is the finches that are generally given credit for suggesting to Darwin—after his return to England—the theory of natural selection.

The Galapagos finches arrived from mainland South America many centuries ago—no one knows when. A variety of foods and virtually no predators awaited them. The original finches were probably birds that lived mostly on the ground and ate seeds. Their beaks were short and heavy for crushing seeds. Several present-day species in the Galapagos still forage on the ground, but others have moved primarily to the trees, some eating insects and others cactus or other plant

Box 19.2

(continued)

materials. The most remarkable adaptation is one species that behaves like a woodpecker. Lacking a woodpecker's beak, it uses a cactus thorn or twig to probe the bark for insects. Since much of the adaptation is associated with feeding

habits, the beaks have evolved more strikingly than plumage or other features. (Drawings from D. Lack, *Darwin's Finches*, Cambridge University Press, Cambridge and London, 1947.)

Looks and acts like a warbler.

Certhidea olivacea
warbler finch

Acts like a chickadee.

Camarrhynchus parvulus
small insectivorous
tree finch

A cactus spine or twig is used to probe for insects in much the same manner as a woodpecker.

Cactospiza pallida
woodpecker finch

Platyspiza crassirostris
vegetarian tree finch

The heavy beak is adapted for crushing seeds.

Geospiza magnirostris
large ground finch

The long thin beak is used for probing into cactus flowers.

Geospiza scandens
cactus ground finch

TABLE 19.1   Sequences of the First Twenty Amino Acids of the α-Globin Chain for Several Species.

A vertical line indicates the same amino acid in that position as in the prior sequence.

Some of the positions (2, 3, 6, 7, 11, and 16) are invariate for these species. Positions 1, 9, and 20 are identical for all except the carp, which is the most distantly separated of the species. The next most distant from the mammals is the chicken. The closest species by this comparison are human and chimpanzee, which are identical not only for the first twenty amino acids but indeed for the entire 141 amino acids of the α-globin chain.

The comparison of amino acid sequences is relatively insensitive for comparing related genera but it is effective for comparing different Orders within the same class or different classes within the same phylum. These generalizations vary somewhat with the particular protein.

In the table below, a blank cell represents a vertical line in the original (i.e., the same amino acid in that position as in the prior sequence). Positions are numbered 1–20.

Classification	Organism	1	2	3	4	5	6	7	8	9	10	11	12	13	14	15	16	17	18	19	20
Class Mammalia																					
Order Primates	human	Val	Leu	Ser	Pro	Ala	Asp	Lys	Thr	Asn	Val	Lys	Ala	Ala	Trp	Gly	Lys	Val	Gly	Ala	His
Primates	chimpanzee																				
Artiodactyla	pig				Ala				Ala											Gly	
Artiodactyla	cow								Gly												
Carnivora	dog				Pro				Thr		Ile		Ser	Thr		Asp		Ile			
Rodentia	mouse				Gly	Glu			Ser				Ala	Ala		Gly					
Class Aves	chicken				Asn	Ala			Asn		Val		Gly	Ile	Phe	Thr				Ala	
Class Osteichthyes	carp	Ser			Asp	Lys			Ala	Ala			Ile	Ala	Trp	Ala			Ser	Pro	Lys

proteins are considered, the discrimination for closely related species should be very good. For less closely related species, fewer proteins are required to yield useful information.

Another aspect of proteins that may be useful is the duplication of genes. For example, the α-globin gene is duplicated in humans. Since there has been no divergence in structure, the duplication must be recent on the evolutionary time scale. One may ask therefore whether other primates have the same duplicated α gene. If neither gorilla nor chimpanzee has a duplicated α locus, then the duplication must have occurred after the human line separated from the great apes. On the other hand, if only the chimpanzee has a duplicated α locus, it would suggest that the duplication occurred after the gorilla line split off and before the chimpanzee and human lines split. The answer to this is not yet known, and the sequence of splitting of these lines has yet to be established with certainty.

DNA Structure

It is now possible to isolate segments of DNA, insert them into bacteria, and make many copies (see Chapter 14). With large quantities of pure DNA, one can determine the nucleotide sequences. Since the differences among all species is in the DNA, this should be the ultimate solution to determine how closely related various species are. As yet, the results are very limited and have not resolved any issues on evolutionary relationships. Of particular interest should be sequences of DNA that are not translated into protein. First, these sequences would provide information not available from analysis of proteins. Second, there is evidence that many of the sequences that do not code for proteins are less constrained insofar as evolutionary change is concerned. Base-pair substitutions may occur, but such mutations are less likely to be selected against.

Information on DNA sequences is also given by the sensitivity of DNA to restriction endonucleases (Chapter 14). Each of these enzymes can break a DNA double strand only at highly specific sequences. For example, the restriction endonuclease Eco R1 recognizes only the sequence

$$5'\text{—G A A T T C—}3'$$
$$\cdot \ \cdot \ \cdot \ \cdot \ \cdot \ \cdot$$
$$3'\text{—C T T A A G—}5'$$

When such a sequence occurs, the covalent bonds connecting G and A are broken in both DNA strands. Any mutation that leads to substitution of any of these base pairs makes the sequence no longer sensitive to Eco R1. Such a sequence of six nucleotide pairs will occur only a limited number of times in a piece of DNA that codes for several genes. If DNA of that size can be isolated or otherwise identified, the presence or absence of these sequences can be established. The distribution of such sequences in homologous DNA from different species can indicate how close the species are.

Karyotypes

The banding patterns of chromosomes reflect the nucleotide sequences. Therefore, one might expect karyotypes also to vary among species depending on the closeness of relationships. This has proved to be true and to be a powerful means of comparing species within the same order. Chromosomal rearrangements are one of the isolating mechanisms of speciation. As a consequence, very close species may show rearrangements. Furthermore, the rearrangements accumulate in subsequent speciations and are therefore useful in working out the order in which various species separated.

For best results, very high resolution banding techniques are used. These techniques are based on late prophase chromosomes that are not fully condensed and that show many more bands than the standard metaphase karyotype (Chapter 4). With these preparations, it has been possible to identify a number of rearrangements that have occurred during primate and anthropoid evolution.

The Evolution of *Homo*

Using the kinds of evidence discussed in the previous sections, scientists have been able to arrive at the main features of evolution, including primate evolution. Although many details have yet to be established, we nevertheless know a great deal about both the remote and immediate ancestors of *Homo sapiens*.

The Geological Calendar

Our concepts of the evolutionary time scale come from geologists. In areas of sediment, the oldest rocks and layers are at the bottom and the most recent on top. Inclusions in the sediment tell us what existed at the particular time the rocks were forming. Figure 19.2 lists the various geologic eras since early forms of life arose. Some of the major biological events associated with each period also are given.

We are most familiar with fossils of bones and other hard body parts. These are likely to persist long enough for firm impressions to form, which are then impregnated with or replaced with minerals. But soft tissues and cells can also form fossils, and ancient fossils of bacteria and other soft organisms have been found. Some date back 3 billion years, but these are very rare and are of very simple forms. The major fossils date from the beginning of the Cambrian period. By that time, there were many invertebrates and marine plants, although as yet there were

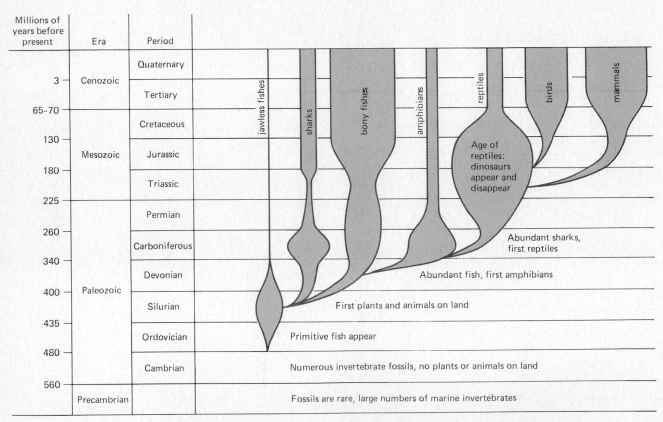

FIGURE 19.2

Geological ages and vertebrate evolution. The width of the bar associated with each group of vertebrates reflects the relative abundance of that group.

no forms on land. The modern plants that we know arrived much later, but the progenitors of the plant kingdom arrived very early, since only through photosynthesis can we account for the appearance of large amounts of oxygen in the atmosphere and the disappearance of ammonia, hydrogen, and methane. Without this change, the animal life that we know could not have arisen.

Vertebrate Evolution

Early fish.   The main features of vertebrate evolution have been established by comparison of the morphology of living and fossil forms. Where newer techniques have been applied, they confirm the morphological results. The geologic periods during which the various classes evolved are shown in Figure 19.2. The most primitive forms were the jawless fish (class Agnatha), which retained the rod-like structure known as a *notochord* throughout life. (All members of the phylum Chordata have a notochord sometime during early development.) Some were covered with heavy external armor, but these are all extinct. As the name suggests, the Agnatha lack a mouth, or rather a mouth with a hinged jaw. Existing species include the lampreys.

Sharks and rays (class Chondrichthyes) developed from the Agnatha by developing a cartilaginous skeleton and a mouth, along with numerous other adaptations. At about the same time, the bony fishes (class Osteichthyes) developed, probably from the Agnatha rather than the Chondrichthyes. A principal characteristic is a bony skeleton. This very successful group is still represented by very large numbers of fresh and salt water species.

Amphibia.   Among the early fish were the lobe-finned fish, of which the coelacanth still exists. The class Amphibia developed from lobe-finned fish by adapting the lobes and fins to legs and claws. They also had to develop the ability to breathe air, at least as adults. Present-day amphibians resemble fish in many ways. With rare exceptions, they still must lay their eggs in fresh water. The eggs hatch into tadpoles, which are entirely aquatic. Some amphibians, such as newts and salamanders, spend most or all of their adult life in water. Toads and frogs, on the other hand, are primarily terrestrial after their metamorphosis from tadpole to adult form.

Reptiles.   Reptiles were the first completely terrestrial vertebrates, developing from Amphibia. The typical early reptiles were probably very much like present-day lizards. Various adaptations gave rise to snakes, turtles, crocodiles, and dinosaurs. With the exception of snakes and a few lizards (which lost their legs), all reptiles are four-legged air-breathing animals with internal fertilization. The largest terrestrial animals ever to have lived were reptiles, the dinosaurs, that were present throughout the Mesozoic Era.

Birds.   Reptiles gave rise to two classes, Aves and Mammalia. The birds still have many reptilian features. Their forelimbs became adapted into wings and their scales into feathers. Many other secondary adaptations have also occurred, but the homology with reptiles is apparent. Because flight places many restrictions on weight and shape, birds deviate less in overall morphology than do other large classes.

Mammals. Mammals arose from a reptile early in the Mesozoic when the reptiles had increased and greatly diversified. Among the many traits that distinguish mammals from their reptilian ancestors are hair rather than scales and nourishment of the young by milk from mammary glands. The latter means also that all young must be cared for by their mothers, probably a key adaptation in development of social interactions in mammals. The suckling of young also removes some of the urgency for young animals to develop skills at finding food in order to survive. This permits slower development and a much longer period during which the young animal can learn. Behavioral traits need not all be transmitted as genetically coded instincts. The brain can afford to be more flexible and hence more adaptable to new situations. The beginning of what we would call true intelligence may well have been possible because of the enforced mother-child relationship.

Another important trait of mammals is warm bloodedness (homeothermy). There are arguments that this trait actually arose among extinct reptiles from which mammals descended. Whatever the truth, a regulated temperature permits an organism to be much more independent of its environment. And it permits the evolutionary selection of proteins that are required to function only over very narrow temperature ranges. The metabolism can be "fine tuned" in ways that are not possible in cold-blooded animals.

Mammalian Evolution

Of the three subclasses of mammals, *Homo* clearly belongs to the Eutheria, mammals with placentas. The development of a placenta removes the risk faced by the egg-laying Prototheria or the pouched Metatheria. In the latter, very immature young must crawl from the vagina to the pouch, where they attach to nipples and develop until they are able to function more independently. A placenta permits the fetus to develop slowly and in a well-protected environment. The temperature is just right, and there is a very efficient system of nutrition and waste removal. The period of gestation can be prolonged if it is adaptive to do so (Fig. 19.3).

All placental mammals share these and other advantages, but the sixteen orders of the subclass Eutheria have a remarkable variety of adaptations. One might characterize the primitive stock as being very flexible and adaptive. Mammals underwent a remarkable radiation at the beginning of the Tertiary period. By this is meant that many different species were formed and that these evolved in different directions to produce the variety of mammals that now exist. The increased independence from the environment due to warm bloodedness, the placenta, and mammaries, as well as a high mobility, all put the primitive mammals in a position to exploit an enormous variety of ecological niches, opportunities that were not ignored. The more generalized the animal, the more likely it is to evolve with a changing environment. The highly specialized organism cannot cope when the environment to which it is adapted changes. Since we are here and are very successful, at least numerically, we may presume that our ancestors were the more generalized members of the species at their level.

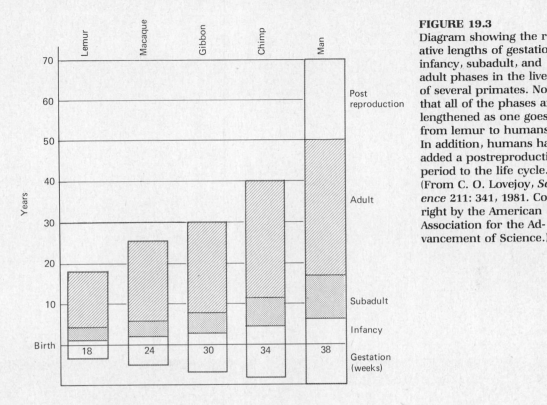

FIGURE 19.3
Diagram showing the relative lengths of gestation, infancy, subadult, and adult phases in the lives of several primates. Note that all of the phases are lengthened as one goes from lemur to humans. In addition, humans have added a postreproductive period to the life cycle. (From C. O. Lovejoy, *Science* 211: 341, 1981. Copyright by the American Association for the Advancement of Science.)

Primate Evolution

Prosimians. The earliest primates have been traced to the Cretaceous period (Fig. 19.4). The progenitor of the primates seems to have been a small insectivore about the size of a squirrel or smaller. Our ancestral insectivores may have been somewhat like their descendants, the modern tree shrews of southeast Asia. These are small, intensely active mammals with voracious appetites (Fig. 19.5A). When the ancestral tree shrews left their terrestrial home for an arboreal life, a premium was placed on certain adaptations that are characteristic of primates. The eyes moved toward the front of the face, providing stereoscopic vision and much better ability to judge distances to nearby tree branches. Opposable thumbs provided a much better grip of tree limbs. Greater visual acuity and the ability to pick up objects and examine them or eat them is a combination that sets primates apart from all other animals.

Some authorities place the Tupaioidea (tree shrews) as well as our common shrew-like ancestor among the Insectivora rather than the Primates. This honor of being the most primitive living primate would then pass to the Lemuriformes and Lorisiformes, represented today by a variety of species of lemurs and lorises (Fig. 19.5B, C). The lorises are found primarily in southern Africa, but some genera are

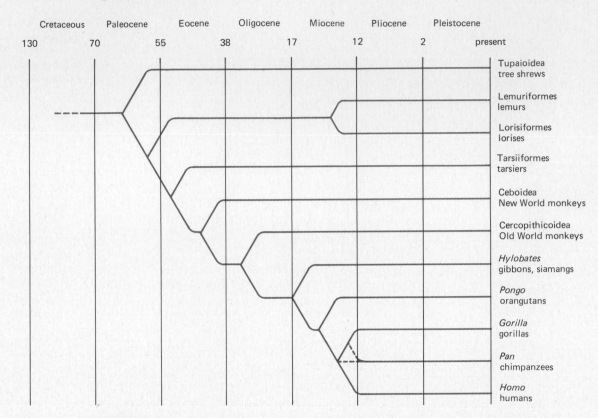

FIGURE 19.4
Diagram showing origin of the various taxons of Primates. It is not certain whether the *Gorilla/Pan* separation occurred at the same time as the hominid separation or slightly later.

also found in Asia. They are nocturnal arboreal animals that will eat almost anything, although their native diet is primarily insects. The lemurs, found largely on the island of Malagasy, are larger than the lorises. Some of the species of lemurs have quite striking coats. Some are nocturnal but most are diurnal. Lemurs primarily eat fruit and other plant materials.

Next on the evolutionary ladder are the tarsiers (Fig. 19.5D). This small nocturnal animal is found at present only in southern Asia and Indonesia. The eyes are very large, consistent with their nocturnal habits, and face definitely forward. The paws are like small hands and are clearly used to handle objects.

Monkeys.   There are two major groups of monkeys, the New World and the Old World, that are not at all closely related. The New World monkeys, known scientifically as Ceboidea or platyrrhine monkeys, evolved from the prosimians

FIGURE 19.5A
Living prosimians: a tree
shrew. (Figures 19.5A–D
copyright © Zoological
Society of San Diego.)

during the Eocene. They are relatively primitive compared to the Old World monkeys and some have developed one very characteristic feature—a prehensile tail (Fig. 19.6A). The pictures of monkeys hanging by their tails are all of New World monkeys. No others can do it, Hollywood notwithstanding. As their name suggests, the New World monkeys are found in the Western Hemisphere in Central and South America. The Old World monkeys, the Cercopithecoidea or catarrhine monkeys, are found in tropical and some temperate areas of Africa and Asia. There are many species, including the macaques, which are much used in medical research, and baboons. The baboons live primarily on the ground (Fig. 19.6B).

Hominoids.   The superfamily Hominoidea consists of the apes and humans, often referred to collectively as *hominoids* (man-like). The most primitive of the apes is the genus *Hylobates*, consisting of the gibbons and siamangs of Southeast Asia. These rather small apes have developed brachiation—swinging through the trees with the arms—as the primary means of locomotion. They have very long

FIGURE 19.5B
Living prosimians: *Galago senegalensis* or bush baby, from one of the two subfamilies of the infraorder Lorisiformes.

arms and can move rapidly and accurately (Fig. 19.7A, B). The ancestors of the present *Hylobates* were the first of the *pongid* (ape) stock to separate from the main evolutionary line of pongids, the split having occurred approximately at the beginning of the Miocene epoch (Fig. 19.4).

The remainder of the apes are referred to as great apes because of their large size. The orangutan, *Pongo pygmaeus*, is also a Southeast Asia species, occurring only on Borneo and Sumatra, although fossils indicate a more widespread distribution in earlier times (Fig. 19.7C). The orangutan is about the size of a chimpanzee. It is slow in its movements and generally stays in the trees. On the ground it may assume an upright position, but *bipedalism* is not a significant means of travel. Orangutans have become rare, and it is estimated that half those living are in zoos.

The gorillas, *Gorilla gorilla*, and chimpanzees, *Pan troglodytes*, are inhabitants of Africa. The chimpanzees (Fig. 19.7E) live almost entirely in trees but often are on the

FIGURE 19.5C
Living prosimians: black-
and-white ruffed lemur.

ground, where they may assume an upright position on occasion. The primary means of movement on the ground is by means of "knuckle-walking," that is, using the knuckles of the hands for support rather than the palms. A pygmy variety of chimpanzee, *Pan paniscus*, also occurs in an area south of the Congo.

Gorillas (Fig. 19.7D) are the largest of the great apes and live primarily on the ground. They are quadrupedal, with long arms and short legs. An upright position is occasionally assumed as a defensive posture but not for locomotion. Two subspecies are recognized, the predominant lowland gorilla and the rare mountain gorilla. The latter is somewhat more adapted to terrestrial life, with shorter upper limbs, than the lowland gorilla.

The relationship of modern humans to the great apes has been much debated. Which are our closest relatives among the apes? All have specialized to some extent since the divergence, and arguments based on comparative anatomy could be (and have been) made for any of the great apes. Biochemical studies of proteins and DNA sequences have settled the issue as diagrammed in Figure 19.4. For the proteins that have been analyzed, primarily hemoglobins, those from gorillas,

FIGURE 19.5D
Living prosimians: Min-
danao tarsier. Note that
the large eyes of the tarsier, adapted for noctur-
nal vision, face forward,
providing binocular vision
useful for judging dis-
tances. This is a valuable
adaptation in an arboreal
animal. Note also the use
of the front paws for
grasping objects, such as
lunch.

chimpanzees, and humans are identical or nearly identical, with small but significant differences between this group and the orangutans. Thus the orangutan line must have diverged earlier than the others.

A comparison of karyotypes has also been informative. Karyotypes appear to evolve slower than species but faster than proteins. The similarities among the karyotypes of humans and the great apes are striking (Fig. 19.8). All except humans have 48 chromosomes. The evolutionary line that led to *Homo sapiens* decreased the number to 46 by joining two acrocentric chromosomes together by fusion at the centromeres, forming human chromosome 2. A number of other less conspicuous changes, mostly inversions around the centromeres, have also occurred. By comparing the karyotypes of many species, it is possible to construct a *cladogram*

FIGURE 19.6A
Monkeys: a New World monkey, the howler monkey, showing the prehensile tail wrapped around the support. (Figures 19.6A–D copyright © Zoological Society of San Diego.)

(b)

(c)

(d)

FIGURE 19.6B–D
Monkeys: B. Pygmy mar-
mosets, another species
of New World monkey;
C. Lion-tailed macaque,
one of the many species
of Old World monkeys;
D. A Hamadryas male ba-
boon (Old World monkey)
comforting an infant.

FIGURE 19.7A
The anthropoid apes: a gibbon. The very long arms are adaptations to the usual mode of travel by brachiation (swinging by the arms through trees). (Figures 19.7A–E copyright © Zoological Society of San Diego.)

FIGURE 19.7B
The anthropoid apes: the siamangs, a family of which is shown, are closely related to the gibbons and also move primarily by brachiation.

(b)

FIGURE 19.7C
The anthropoid apes: the orangutans are members of the genus *Pongo* and are the most primitive of the great apes. This photograph illustrates the difference in appearance of the female (left) and male (right).

(c)

FIGURE 19.7D
The anthropoid apes: a
family of lowland gorillas.
The male is about twice
the size of the female. Although gorillas often
stand more or less erect
on two feet, they move
about in the position of
the male, using the
knuckles of the hands for
support.

(a diagram showing the relationships of existing species to each other and to common ancestors). Such a cladogram based entirely on chromosome analysis is shown in Figure 19.9. It confirms the ideas based on morphology and extends our knowledge of primate evolution considerably.

DNA sequences are a more sensitive way to make observations on the relationships of closely related species. The complete base-pair sequence of mitochondrial DNA has been established for humans. The presence of specific short sequences can be recognized by the ability of restriction endonucleases to break DNA at those sequences, as described earlier. Mutation may cause such a sequence to change and no longer be a cleavage site, or mutation may cause a new cleavage site to appear. Comparison of the restriction cleavage sites in mitochondria of humans and apes confirms the patterns of relationships in Figure 19.4. The patterns of the common chimpanzee and the pygmy chimpanzee were closest, with most of the cleavage sites (approximately 80%) being identical. The 20% difference does confirm though that these groups are quite distinct and should be in different species. The gibbon was the most distant from human, with the orangutan being next. As with protein comparisons, gorillas, chimpanzees, and humans were about equidistant, with a suggestion that the gorilla and chimpanzee split may have occurred after separation of the hominid line. Additional studies should eventually clarify this ambiguity.

The primates are somewhat unusual in that each of the major evolutionary steps seems to be represented by living species. It is quite improper to think of ourselves as having descended from a living species of chimpanzee, gorilla, or monkey. We descended from common ancestors, and each line has evolved away from that

FIGURE 19.7E
The anthropoid apes: a chimpanzee (*Pan troglodytes*). This photograph shows especially well the body form of chimpanzees, with their well-developed musculature. Chimpanzees spend a large amount of time on the ground, and their movements in trees are better described as climbing than brachiation. The position of the chimpanzee in the photograph therefore is not typical as a means of travel.

ancestral form. Nevertheless, the array of major ancestral forms, based on extensive fossil evidence, would be rather like an array of the living forms.

The explanation for this unusual situation seems to be that each major step in evolution was associated with movement to a new ecological niche. When the early shrews took to the trees, no one was there except insects, their food, and of course some birds. These very small shrews lived among the leaves on small branches. When lemurs evolved, they moved lower in the trees to branches that would hold them. The monkeys continued this trend, and the much larger apes were close to the ground on big limbs or were actually on the ground. Thus, in a tropical forest with several kinds of primates, the various species would be somewhat stratified vertically. A gorilla would never compete with a tree shrew for the same insect.

Hominid Evolution

We know who our living relatives are. What about our more closely related dead ancestors, the ones who were not quite *Homo sapiens* but who were progressing along the pathway toward this recent evolutionary experiment? (The term *hominid* refers to human forms, whether *H. sapiens* or an ancestral species.) As seems to have been true in all other areas of evolution and speciation, there is evidence of a number of "experiments," most of which were not successful in the long run, but a recent variant (*Homo sapiens*) may yet be. We are not entirely sure which of the extinct populations of early humans may have been a direct ancestral line and which were aunts and uncles who have no surviving heirs. But we know who some of the principals were at various prehistoric times, and we can speculate as to which were most likely to have passed on their genes (with occasional mutations) to us.

Interpretation of Fossils.   All reconstruction of hominid antecedents is based only on fossil bones and teeth. There are no proteins, DNA, or karyotypes still around for our scrutiny. Fossil bones can tell us a great deal about the lives of their owners, however. For example, one can recognize whether a particular fossil form is from an animal that walked upright or quadrupedally (Fig. 19.10). The form of the bones reflects muscle attachment and size. Modern gorillas have a small gluteus maximus. Humans have a large gluteus maximus, which holds them in an upright position. The area of the pelvis to which the gluteus maximus is attached increases as the size of the muscle increases, an indication of evolution to a more upright position. The shape of the pelvis also changes from a set of bones from which internal organs are suspended to a set that supports them from below.

Another change is in the position of the *foramen magnum*, the hole at the base of the skull through which the spinal cord passes (Fig. 19.11). The head has approximately the same orientation whether the body position is horizontal or vertical. If the body is horizontal, the head is attached to it at the back of the skull and is held in position with large neck muscles. If the body is upright, the head balances on top of it and requires less muscle. Thus both the position of the foramen magnum and the shape of the skull where muscles are attached indicate whether the head is in front of the body or on top of it—or in some intermediate position.

Fossil teeth can also be very informative. The teeth of the great apes are in a U-

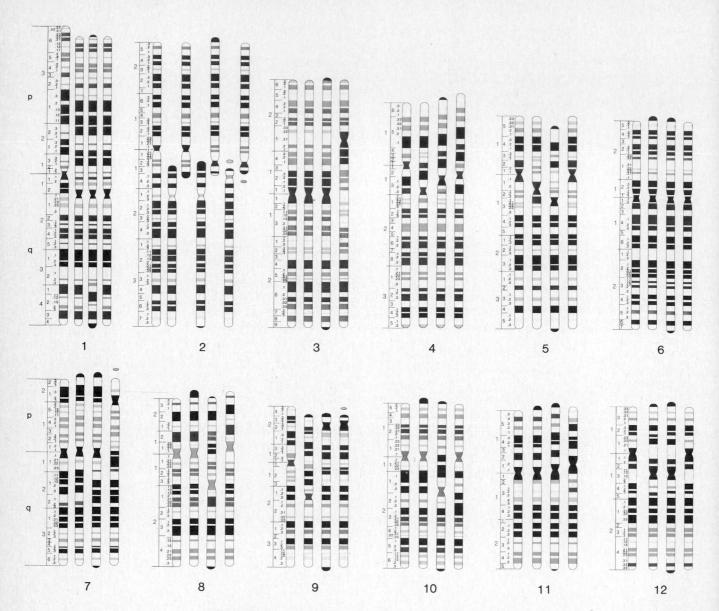

FIGURE 19.8

A comparison of human and ape karyotypes. The diagrams are based on late prophase chromosomes, which show many more bands than the conventional metaphase chromosomes. The chromosomes are arranged in the order human, chimpanzee, gorilla, orangutan. The chromosome numbers are based on the human karyotype. (From J. J. Yunis and O. Prakash, *Science* 215:1525, 1982. Copyright 1982 by the American Association for the Advancement of Science.)

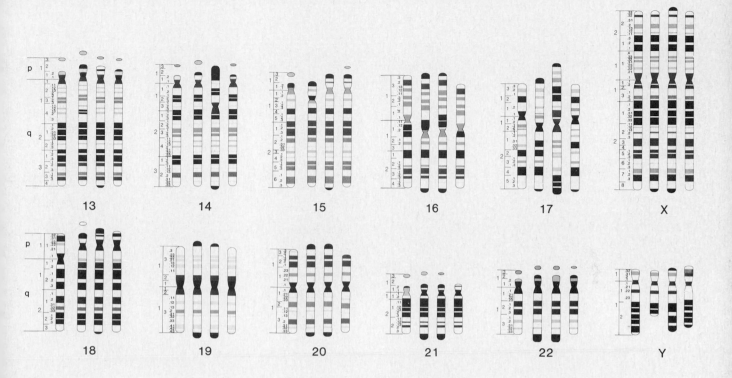

13   14   15   16   17   X

18   19   20   21   22   Y

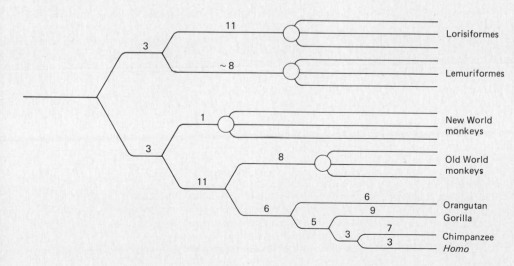

FIGURE 19.9
Evolutionary tree of primates based entirely on comparison of karyotypes. The numbers are the approximate number of structural changes between separation of lines. Circles are the beginnings of extensive speciation, and the lines emanating from circles are symbolic of many genera and species within that taxon. (Based on data of B. Dutrillaux and of J. J. Yunis and O. Prakash.)

FIGURE 19.10
Drawing of skeletons of gorilla and man. Many differences reflect adaptation to the different modes of walking. The pelvis of the gorilla is relatively long and narrow, while the human pelvis is wide. The head of the gorilla is carried in front of the body and requires strong neck muscles to hold it in place. The human head is balanced on top of the spine. These and other adaptations in fossil primates tell us whether the owners walked upright or on all fours. (From P. V. Tobias, *Human Genetics, Part A: The Unfolding Genome*, B. Bonné-Tamir, ed. Alan R. Liss, New York, 1982. pp. 195–214.)

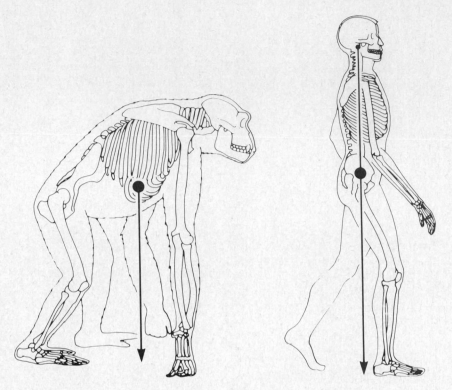

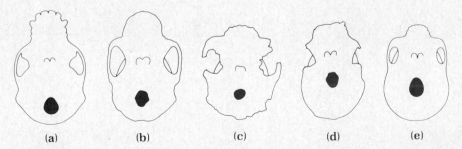

(a) (b) (c) (d) (e)

FIGURE 19.11
The base of the skulls of (a) a gorilla, (b) and (c) *Australopithecus*, (d) *Homo habilis*, and (e) modern *Homo sapiens*. The foramen magnum (in black) is the opening through which the spinal cord is attached to the brain. With more upright posture, the foramen magnum moves forward so that the head is balanced above the spine rather than suspended in front of it. (From P. V. Tobias, *Human Genetics, Part A: The Unfolding Genome*, B. Bonné-Tamir, ed. Alan R. Liss, New York, 1982, pp. 195–214.)

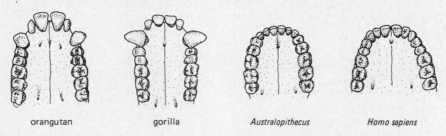

orangutan   gorilla   *Australopithecus*   *Homo sapiens*

FIGURE 19.12
Upper jaws of two great apes, an early hominid, and a modern person. The teeth of
Australopithecus resemble human teeth much more than they do pongid (ape) teeth.
The arch is curved rather than U-shaped and the canines are much less prominent.

shaped array, compared to human teeth, which are in a curved arch (Fig. 19.12).
Apes have large canine teeth, which are important in defense. The molars are very
large, as expected in animals whose principal diet is of vegetable origin, since plant
materials require a great deal more chewing than meat (at least when not cooked
into mush). Human canines are small, having been replaced as defensive weapons
by rocks, spears, and triggers manipulated by the hands. Human molars are also
much smaller than ape molars relative to the front teeth, since our ancestors were
hunters who ate meat.

Ramapithecus. Many other skeletal changes tell us about our hominid ancestors, but these will suffice to establish several points. At the time of divergence of
the pongids from the hominids (Fig. 19.13), the predominant hominoid was an ape-
like creature, genus *Ramapithecus*, whose teeth seem to have been very much like
those of modern humans rather than like apes. The arch was curved, the canines

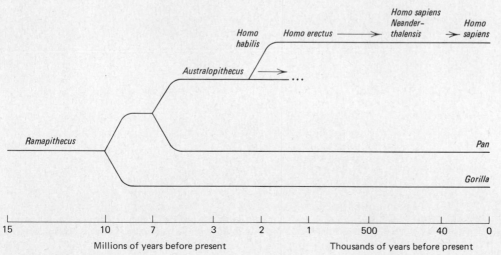

FIGURE 19.13
A diagram of hominid ev-
olution. There are still
many uncertainties about
the exact times and se-
quences of ancestral
forms of *Homo sapiens*.

were small, and the molars were somewhat reduced. *Ramapithecus* probably walked upright with the hands free for defense and hunting. Their brains were much smaller than modern human brains. Fossils have been found in Africa, Europe, and Asia, an indication of the widespread distribution of this species (or these species).

    Australopithecus.   The next recognized group is known by the generic name *Australopithecus;* the various members are Australopithecines. Fossils from this group are found only in Africa, but many diverse specimens have been recovered. They were clearly upright without large canine teeth. But the brain size was still small. The first such specimen was assigned to the species *Australopithecus africanus* and this designation is often used. It is difficult to decide to what extent the different fossils represent normal variation within a single large population compared to distinct populations and species. For our purposes, we shall only consider the hominids of that period at the level of the genus.

    Homo erectus.   In the late Pliocene and early Pleistocene, two separate populations of *Australopithecus* are found, one of which soon evolves into what was earlier called *Pithecanthropus erectus* but now is generally assigned to our own genus with the designation *Homo erectus* (Fig. 19.14). An earlier form, *Homo habilis*, is also recognized by many authorities. *Homo erectus* was widely dis-

FIGURE 19.14
A skull of *Homo erectus.* This particular skull was originally designated *Sinanthropus pekinensis.* Although the brain case is rather ape-like in form, the teeth are quite human. (Courtesy of the Library Services Department, American Museum of Natural History, New York. Photographed by Charles H. Coles.)

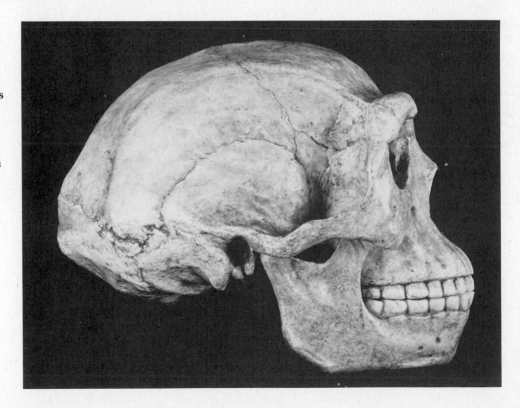

tributed in Africa, Asia, and Europe. The first discoveries of this group were in China (Pekin man, initially called *Sinanthropus pekinensis*) and Java (*Meganthropus palaeojavanicus*). A variety of names have been given to various African forms, but these all are generally grouped now into the single species *Homo erectus*. Some scientists prefer other classifications depending on their various interpretations of the fossil evidence. *Homo erectus* was beginning to have a larger brain, 1000 mL rather than the 500 mL of *Australopithecus* but not yet the 1450 mL of modern males. The pelvis also changed further to allow the birth of infants with larger brains. They were hunters and some at least used fire. There is evidence that primitive stone tools were deliberately fashioned and used.

   Homo sapiens.   The evolution to our own species seems to have occurred some half million years ago. At that time, there appeared the famous Neanderthal man, or as we shall call him, *Homo sapiens Neanderthalensis*, giving him subspecies status. Neanderthal man was very much like modern man except for the huge brow ridges (Fig. 19.15) and the dentition, which was still much like *H. erectus*. The brain size was very nearly that of modern man and indeed reached that size about 100,000 years ago. Some scientists believe that Neanderthal man was eventually displaced by modern *H. sapiens* and became extinct. Sufficient numbers of late Neanderthal fossils are now available to suggest that Neanderthal

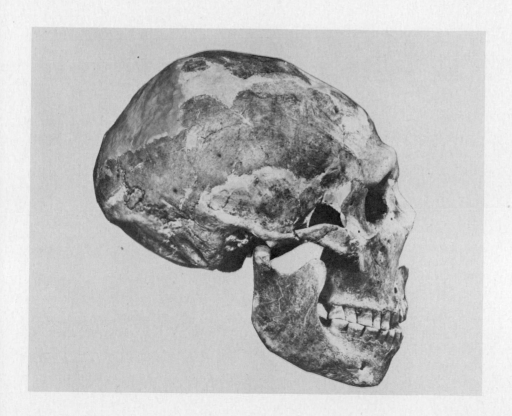

FIGURE 19.15
A skull of *Homo sapiens Neaderthalensis.* This skull was found in Iraq and is known as Shanidar I. (Photograph supplied by United States National Museum of Natural History, Smithsonian Institution, Washington, D.C.)

man evolved into modern man. Aside from the paucity of evidence for a separate line to modern man, it seems unlikely that Neanderthal man would easily be pushed aside. Major Neanderthal fossils have been found at many sites in Europe, Africa, and Asia. Were a different species of *H. sapiens* to appear, it is hard to imagine that they would remain genetically isolated from each other. And one still sees individuals whose heavy brows suggest a Neanderthal heritage.

The last clearly Neanderthal fossils are 50–60,000 years old. From approximately that time forward, *Homo sapiens* existed as we now know the species. The earliest modern finds are 40,000 years old. This is a very short time compared to the several hundred thousand years of Neanderthal and the million years or more of *Homo erectus*, so we are truly newcomers.

Whither?

Few would argue that humans are perfect. Most of us can think of a number of ways of improving our species, certainly the other members of it. Whether these "improvements" would confer Darwinian fitness or whether they would rapidly disappear through natural selection is an experiment we cannot undertake. But we can look at recent changes—those that we know about—that have occurred in hominid evolution and ask whether these changes are likely to continue in the same direction. This may leave a still cloudy crystal ball, but, like other forms of foretelling the future, it may provide some amusement.

Sexual Dimorphism. One well-documented change has been the decrease in sexual dimorphism, that is, differences in appearance, size, and so forth, between males and females. Gorillas show marked sexual dimorphism: Males are about twice as large as females. This is associated with their very different roles. Gorillas live in small families with the male responsible for the safety of the group. They must ward off attacks from intruders and project a fierce combative attitude. Females look after the kids and forage for food for them. Many of our fossil hominid ancestors appear also to have had marked sexual dimorphism in size. This has diminished, but human males are still taller *on the average* than females, are stronger, and hairier. Perhaps in another million years, these differences will have disappeared.

Operating against complete identity in appearance is the Darwinian idea of sexual selection. Noting the gaudy plumage of males in many species, he proposed that such males were more successful than their duller counterparts in attracting females. The plumage was advantageous in terms of natural selection, since the more colorful males left more offspring. Should humans choose their mates on the basis of differences in sexual appearance, such differences would be selectively advantageous. One trait suggested in this context is breast development. *Homo sapiens* is the only species in which nonlactating females have large breasts. They serve no obvious purpose except to draw the attention of males, and there is anecdotal evidence at least that the system works.

Information and Communication. No trait distinguishes humans from other primates, indeed all other organisms, more than speech. We do not know when the early hominids began to let abstract sounds stand for concrete objects or actions. It seems likely that it must have preceded *Homo erectus* and that the ability to communicate was an important selective force in the increased brain size of *Homo*. Indeed, the transition to *Homo* is associated with relative enlargement of the speech areas. With speech came a level of culture far beyond anything available without speech. Our heritage is no longer just in our genes. We can learn about things without experiencing them first hand. We can read books and learn about genes that most of us have never seen.

But is there a limit to how much information we can absorb, manipulate, and communicate, given strong selective pressure to increase these abilities? The question obviously cannot be answered, but the limits can at least be set to include the persons with unusual abilities who now exist—not just those with superior intelligence but those with strange abilities. Rare persons with *eidetic* memories can glance at a page, then read it mentally at a later time. Or they can recall details of something they have seen—details that were unnoticed at the time. There are many records also of *idiot-savants*, persons with less than normal intelligence who can do complicated calculations in their heads. These people may seem like freaks, but they set the minimum limits of biological possibility. But will selection favor them? Many have argued that selection does not favor high intelligence now. And it is only by having more children that persons with exceptional abilities (and genes) will influence evolution.

Some science fiction writers visualize future generations communicating by some form of extrasensory perception (ESP). The selective advantages of ESP would seem to be so great that, were it physically and physiologically possible, we should already practice it widely. Yet there is no widely accepted scientific evidence that ESP exists at all. It would not seem to be among the possible directions of evolution.

Neoteny. It has been suggested that humans are evolving to a more child-like form (*neoteny*) compared to our ancestors, both hominid and pongid. This suggestion is based to some extent on the more pronounced sexual dimorphism of early hominids and present-day gorillas. If that is eliminated, the principal change seems to be a reduction in facial growth coupled with an increase in cranial growth. Continuation of this trend would mean still smaller faces and larger heads (and larger pelves to accommodate them). Since our mouths are no longer used for physical combat, there would seem little pressure to keep them large. A reduction in size would not compromise our nutrition if we can maintain our present cultural standards.

A larger brain is more dubious. Neanderthal brains were slightly larger than present-day brains, so perhaps the trend is actually toward smaller brains. And the evidence suggests that quality of a brain is not related in the present human population to quantity. It is difficult to predict therefore how selection might alter brain size. Perhaps a better picture of our descendants would be with brains much like at present but with still smaller faces.

Etc. Whatever the future changes, humans have survived because they are very generalized animals, able to cope with an increasingly wide range of environments. We have been to the moon and can live for long periods underwater. We can see around the world, and we can talk to hundreds of millions of people at once. In a sense, we have evolved much faster than any other organism. Our special adaptation has been our adaptability. Can this be increased by selection?

Review Questions

1. What is the binomial system of nomenclature? Who developed it?
2. What is a taxon?
3. What is meant by typological thinking as opposed to population thinking?
4. What is one generally accepted definition of *species*?
5. A current theory about speciation is that it occurs when a population becomes reproductively isolated. What factors are then thought to operate to change the isolated population from the main population from which it was derived?
6. What is meant by *coadapted* groups of genes?
7. What is meant by the statement that humans are eutherian? What are some of the advantages of being eutherian? Are there disadvantages?
8. What sorts of evidence make it certain that humans and great apes are very closely related?
9. Our chromosome 2 appears to have arisen in an early ancestor by what process?
10. Which of the living primates appear to be our closest relatives?
11. The hominid line from which we came split off from the pongid line probably 5 to 10 million years ago. What is thought to be the earliest hominid as revealed in the fossil remains?
12. The first hominid assigned to the genus *Homo* exists as the fossil *Homo erectus*. How did *H. erectus* differ from earlier hominids? From the present *H. sapiens*?
13. What features in a skull are associated with bipedalism?
14. Various hypotheses have been suggested as to why our human ancestors began to walk upright. Can you suggest one or two adaptive changes that might have led to this important evolutionary step?
15. Part of the success of the human animal is that it is a generalized animal. What is meant by this?
16. What do you think our descendants living 200,000 years from now will be like? (Assume that their ancestors will not have been obliterated by atomic bombs or other such environmental catastrophes.)

References and Further Reading

Edey, M. A. 1977. *The Missing Link*, Revised Edition. Time-Life Books, New York. Many interesting photographs.

Grouchy, J. de, C. Turleau, and C. Finaz. 1978. Chromosomal phylogeny of the primates. *Ann. Rev. Genet.* 12:289–328. An excellent review of the principles by which evolutionary affinities are traced from chromosome structure.

Johanson, D., and M. Edey. 1981. *Lucy. The Beginnings of Humankind.* Simon and Schuster, New York. An entertaining account of the trials of being a hunter of early hominid remains. Centers around Lucy, the remains of a species of Australopithecus.

Leakey, R. E., and R. Lewin. 1977. *Origins.* Dutton, New York. A nontechnical account of human origins.

Mayr, E. 1963. *Animal Species and Evolution.* Harvard Univ. Press, Cambridge, Mass. 797 pp. A major summary of modern ideas of evolution and of how species are formed.

Stebbins, G. L. 1982. *Darwin to DNA, Molecules to Humanity.* Freeman, San Francisco. 491 pp. Genetics and evolution for the nonscience student told in an entertaining and personal way.

Tobias, P. V. 1981. The emergence of man in Africa and beyond. *Phil. Trans. Royal Society London* B 292:43–56. A condensed but informative discussion by an anthropologist who has been closely involved with many of the findings.

Yunis, J. J., and O. Prakash, 1982. The origin of man: A chromosomal pictorial legacy. *Science* 215:1525–1530. Using the highest resolution banding techniques for chromosomes, these scientists have provided the most compelling evidence on the primate family tree.

Human Variation and Race

Much of the preceding has dealt with individual variability. Each of us is genetically unique. Yet we share more genes in common with close relatives than with unrelated persons. On a larger scale, we also belong to populations with whose members we share more genes than with members of other populations. Whether we are discussing variation in terms of ethnic groups, races, or some other euphemism, we can each be identified at the very least as not being an African Pygmy, Tibetan, aboriginal Australian, or whatever it is that we are not. People who live in different parts of the world or whose ancestors did are different. Populations do differ genetically, and understanding the nature of these differeces will do much to clear up misconceptions about race and races.

The Nature of Races

Species and Races.   In the previous chapter, the point was made that all human beings belong to the same species, *Homo sapiens.* Members of all human populations have the same basic karyotype, and all apparently can interbreed without evidence of reduced fertility. Taxonomists often divide a species into *subspecies* if distinct populations exist that do not interbreed or that do so rarely in nature. Another term used to recognize subdivisions within a species is *race.* The definition of race is not particularly different from that of subspecies, although populations that are related as subspecies are generally thought of as more distinct from each other than are races. Th. Dobzhansky defines a subspecies as "a race that a taxonomist regards as sufficiently different from other races to bestow upon it a Latin name." *Local populations* are those that differ from their neighbors by only a few traits.

One may argue whether human populations differ sufficiently to merit the designation of races as used by taxonomists with other taxons. Some feel strongly that the different human groups are local populations at best. Such arguments are motivated, perhaps subconsciously in some instances, by distaste for *racism* rather than objection to the scientific concepts of race. Furthermore, taxonomists working in different taxons often use quite different standards for dividing populations into races, species, genera, and so on. *Race* has long standing use in the description of human populations. We prefer to join the majority of anthropologists and geneticists in its continued scientific use. Racism, on the other hand, is an attitude that can best be overcome by knowledge and education.

The Definition of Race. To begin the discussion of race, let us first try to define it. It will not be easy or entirely satisfactory, and the definition will consist of several parts. A race is foremost a population, not an individual. Just as the old typological idea of a species had to be discarded, so also must we avoid thinking of races typologically. By this is meant that no one person or small group embodies the ideal type for a particular race. All who are included in the race help define it.

Another part of the definition is that the populations that comprise a race differ

20

Genetic Variations in Human Populations

in their gene pool from other populations. This may mean that the frequencies of certain alleles are different. Or in some instances an allele may be found in one population and not in others, although such sharp differences are uncommon. The Diego blood group is found in substantial frequencies among Oriental and American Indian populations but is virtually nonexistent in others.

Finally, a race does not interbreed freely with other races. Prehistorically this was due primarily to geographic barriers. Even without sharp barriers, persons choose mates close to home, and prior to modern mobility, close to home was very close indeed.

One added factor is that the term race is limited by most scientists to fairly large populations. A small isolated group may be as genetically distinct from surrounding populations as one major race is from another. But the term would have little utility under those circumstances.

Whatever may have been true from time to time in the past, present-day races are not completely isolated from each other geographically. There is always gene flow between neighboring populations, and the geographically intermediate persons often are genetically intermediate. The typologist would consider them hybrid populations, but the truth very often is that they simply are intermediate in a continuously varying array.

If the above definitions sound somewhat insecure, that is because there is no simple set of rules that tells us exactly when a population should be designated a race. It depends on the purposes and proclivities of the scientist. One may find it useful to identify a large number of races. Others may view most of these as "local races" or populations, part of a larger racial grouping. The disagreements are not so much on the evidence or the interpretation but on the labels.

Evolution and Human Races

If all humans are members of a single species, when and how did the present races arise? Are they relatively recent or ancient? The earth's great land masses, with the exception of the Americas and Australia, have been continuously occupied by early man and hominid antecedents. Australopithicine fossils have been found in China and Africa, an indication of the very wide distribution of these particular ancestors. The Americas were occupied 20,000–30,000 or more years ago by Mongoloid populations. Did the present populations descend from a fully evolved *Homo sapiens* that expanded from some local population, pushing aside the existing populations to occupy the entire world, following which genetic divergence occurred? Or did the early hominids evolve in a parallel manner into modern humans, still differentiated geographically and exhibiting distinctive racial traits that antedate the most recent evolutionary steps?

The truth probably is somewhere between the two suggestions posed above. The idea that each evolutionary step involves displacement of all existing genotypes by a new advantageous genotype does not take into consideration adequately the features of sexual reproduction. While in one sense the individual who has a favorable mutation has a selective advantage, that individual ceases to exist after a limited number of offspring are produced. From the point of view of populations, the advantage can best be associated with the mutant allele, which can displace other alleles over time. Alleles at other loci would not be affected except for a temporary advantage conferred on closely linked alleles on the same chromosome as the mutant allele. Even though gene flow between populations is severely limited, so long as an advantageous mutation can get a toehold in a new population, it will spread in that new population without sweeping away alleles at other loci that may be distinctive for that population.

It may be, therefore, that the present varieties of human beings have very ancient antecedents. This would not prevent favorable genes from spreading throughout the entire world. Thus quite different populations could evolve along similar pathways. This is not the same as parallel evolution in which quite separate but related species evolve along similar pathways under the influence of similar ecological pressures. Rather, this is simply evolution of a large diversified population, evolution that does not destroy the diversity.

Selection and Adaptation in Human Races

In examining the differences between races, one is often struck with the triviality of the differences. In other words, the traits that characterize races seem to be generally unimportant for normal function. This led some anthropologists earlier to take the view that differences among races are largely nonadaptive; that is, the different traits are neutral with respect to natural selection. Many geneticists, on the other hand, have held the view that no two alleles are exactly equal in selective advantage; therefore, all differences should be the result of natural selection. Neither view appears to be acceptable any longer. Some of the genetic differences between races clearly are adaptive. Many other mutations involving substitution of an amino acid for a similar amino acid almost certainly are neutral. Neutrality is very difficult to establish, however, since one cannot know all the possible phenotypic effects of a mutation. One would not predict, for example, that sickle cell hemoglobin would provide protection against malaria in heterozygotes. Perhaps the same genes that caused the thick brow of Neanderthal man also caused other skeletal effects that were a disadvantage to later *Homo sapiens.*

It also is not possible to assess all environmental effects and interactions with genotype, particularly those that existed in the past. Adaptation occurs in specific environments, and a trait that has no selective advantage in a crowded urban environment may have been useful in an isolated habitat during the last ice age. And traits that we consider a disadvantage today may not have been during hominid evolution. Familial hypercholesterolemia often leads to heart attacks at an early age, say 40 years. It would not have been a disadvantage under conditions where few survived the trauma and infections to live to such an advanced age. Indeed, even today the disadvantage is more social than genetic, since most child bearing is completed before age 40.

Bergmann's Rule; Allen's Rule. Some traits for which races differ do seem to have adaptive value related to the environment. Bergmann's rule, developed for mammals and birds, states that races that live in cold areas are larger than their close relatives in hot areas. Heat loss occurs in proportion to the body surface area. Heat is generated as a function of body mass or volume. The ratio of volume to surface area for a sphere is $\frac{1}{3}\pi r^3/4\pi r^2 = r/3$. In other words, the greater the radius, the greater the ratio of mass to surface area. For an animal, the greater the size, the smaller the fraction of heat that would be lost due to surface area. This rule holds for many species, although it is only a rule and there are many exceptions. It appears to hold also for human populations if one places the major populations in their original locations. In general, the more northerly populations are larger than the tropical populations.

Allen's rule states that races of animals of equal mass tend to have a smaller surface area in cold regions compared to the surface area of related races in hot climates. This difference is achieved by the animals having a more spherical form with smaller appendages in the cold areas (Fig. 20.1). Body build is stockier and the face is more spherical with a flat nose. Tropical peoples tend to be more linear in body form, since their problem is to get rid of heat rather than to conserve it.

Skin Pigmentation. Many anthropologists and other scientists have observed that native groups in tropical regions are more heavily pigmented than groups that have lived in the north for long periods. Africans, Australian aborigines, Melanesians, and Micronesians all have darkly pigmented skins. Northern Europeans have very light skins. Several possible adaptations have been suggested. Pigment presumably serves as protection from the ultraviolet light of the sun, which is more intense near the equator. Persons with light skin are much more likely to get skin cancer from exposure to sunlight. On the other hand, vitamin D is formed in the skin from precursors that are converted by the action of ultraviolet light. Prior to the modern supplementation of foods with vitamin D, this was the major source of the vitamin. Persons with light skins presumably can make the conversion much more efficiently. Light skin may thus have been an advantage in regions of low sunlight.

Other Evidence of Adaptation. It is usually difficult to prove beyond reasonable doubt that a polymorphic trait is adaptive. In any event, the adaptations may have been to past environments rather than to the present. The difficulty is especially great if a particular allele is found at high frequencies only in one population. It could have attained high frequency by chance, or there could be environmental factors unique to this population. Finally, a gene of general adaptive value could have originated in this population and still be in the process of spreading to other populations.

Genetic Variation Among Human Populations

Genetic variants of blood groups, proteins, and other traits have proved to be helpful in understanding the differences that characterize human populations as well as in understanding the origins of present day populations. The informative gene systems are those that are polymorphic, with two or more alleles present in substantial frequencies. The polymorphism need not be present in all populations, so long as the frequencies differ between two populations being compared. Among the world populations a large number of loci differ in allele frequencies. Only a few can be cited as examples.

(a)

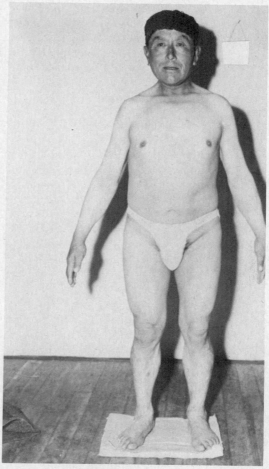

FIGURE 20.1
Some extremes of body form found in hot and cold climates. (a) A typical Nilotic male from the southern Sudan in Africa. The average height of these males is 71.5 inches, and all have the very thin body build shown. (b) An Aliute (American Indian) male from Alaska. This man, who is 62.2 inches tall, has the short limbs and stocky build found in many northern peoples. (The photograph in (a) was supplied by Dr. D. F. Roberts, The University of Newcastle Upon Tyne; (b) was supplied by Dr. W. S. Laughlin, The University of Connecticut.)

In addition to simple Mendelian traits as population markers, complex traits have also been used. Indeed, historically, complex traits such as body size and morphology, skin pigmentation, and hair color and form have played the major role in differentiating both individuals and groups of people from each other. Many such traits are subject to environmental influence, but with care one can select traits that are entirely or almost entirely hereditary. Such "anthropometric" traits may sometimes lack the intellectual satisfaction of single Mendelian loci, but they may be more informative because of the large number of loci involved in their expression.

The following are several examples of traits used to characterize different populations.

Blood Groups

ABO Groups.   The inheritance of red cell antigens has already been discussed in Chapter 15, with several examples of gene frequency differences among racial groups. The ABO groups are the most thoroughly studied in this regard. Figure 20.2 shows maps of the distributions of the *A* and *B* alleles in the aboriginal populations before Columbus messed things up. The *O* allele is the most prevalent in all populations and accounts for what is not represented by *A* and *B* alleles.

The *A* allele is found in all populations except many of the Amerinds (American Indians). It has substantial frequencies among North American Indians but is extremely rare in Central and South American Indians, virtually all of whom are blood type O. The *B* allele is so rare in Amerinds that it probably exists only by introduction from other populations that came after Columbus. It has its highest frequencies in Asia, especially India, which is interesting because the Amerinds are descendants of Asian populations. Apparently the *B* allele didn't make it across the Bering land bridge, or, if it did, it was eliminated through chance or selection. The total number of persons who were progenitors of the Amerinds may have been very small, which would allow the opportunity for genetic drift to operate strongly.

Rh Groups.   The Rh blood groups also show variation among different populations. The three principal alleles or haplotypes (Chapter 15) are R^1, R^2, and r. The r is especially high among the Basques. This frequency is thought to be a result of the distinctive origins of the Basques compared to surrounding European populations—the Basques being of ancient European stock, displaced and compressed by later arriving populations from the East. In the Rh and other genetic markers, they are distinct from the adjacent French and Spanish populations in the frequencies of different alleles (Table 20.1). One Rh haplotype R^o, is found almost entirely in persons of African descent including American blacks. For example, the R^o frequency in the Bantu (Leopoldville) is 0.60 and in American blacks it is 0.44. In American whites, the R^o frequency is 0.03.

Other Blood and Plasma Groups.   All polymorphic systems would be expected to show some variation among populations due to chance alone. Many systems have now been investigated and have borne out this prediction, although ordinarily one cannot distinguish between genetic drift and selection. One interesting red cell antigen is the Duffy blood group, determined by three alleles, Fy^a, Fy^b, and *Fy*. The *Fy* allele is found almost entirely in persons of African descent with the original black African populations being virtually 100% *FyFy*. This antigen is the receptor site for vivax malaria with homozygous *FyFy* persons being resistant to this form of malaria. Thus, selection seems to have been the force responsible for current distributions of this allele.

Various plasma proteins also have alleles that occur in polymorphic frequencies. The iron-binding protein transferrin has an allele, Tf^{D1}, that is found primarily in persons of African descent and in Australian aborigines, two populations that are

The values shown are the frequencies of the particular allele.

TABLE 20.1
**Distribution of the Rh
Blood Groups in Various
Populations**

Population	R^1	R^2	R^o	r	R^z	r'	r''
Norwegians	.42	.14	.01	.41	.00	.01	.01
U.S. whites	.42	.14	.03	.38	.00	.01	.01
French	.41	.12	.07	.39	.00	.00	.01
Spanish	.42	.12	.04	.38	.01	.03	.00
Basques	.38	.07	.01	.53	.00	.01	.00
Russians (Moscow)	.43	.16	.03	.36	.00	.02	.00
Africans (Congo)	.05	.08	.68	.17	.00	.02	.00
U.S. blacks	.16	.07	.49	.25	.00	.00	.03
Indians (Brahmans)	.59	.10	.03	.24	.01	.02	.00
Chinese	.73	.19	.03	.02	.00	.02	.00
Amerindians (Pima)	.59	.37	.03	.00	.01	.00	.00
Indonesians	.87	.09	.05	.00	.00	.00	.00
Polynesians (Tonga)	.69	.24	.06	.00	.00	.00	.00
Melanesians (New Britain)	.83	.14	.03	.00	.00	.00	.00
Micronesians (Marshall Islands)	.95	.04	.01	.00	.00	.00	.00
Australian aborigines	.55	.35	.04	.00	.06	.00	.00

Compiled from A. E. Mourant, A. C. Kopeć, and K. Domaniewska-Sobczak, *The Distribution of the Human Blood Groups*, 2nd ed. Oxford Univ. Press, London, 1976.

known not to be at all closely related. An allele, $Tf^{D\,Chi}$, is widely found in Asians and Amerinds. One variant of albumin, albumin Mexico, is due to an allele, Alb^{Mex}, at the albumin locus, that is widely distributed in Indians of the Southwest United States and Mexico. It is also found in high frequency in Mexican-American populations, which indicates their strong Indian heritage.

Probably no protein variants are better known than the hemoglobin variants. The only polymorphisms involve alleles that offer protection in the tropical areas against falciparum malaria, Hb's S and C in Africa and Hb's E and Constant Spring in Southeast Asia. These are important genetic characteristics of the populations in which they are found, including the descendant populations, such as American blacks, who are not exposed to falciparum malaria.

Persistent Lactase

One especially interesting trait that varies markedly among populations and that has practical nutritional importance is the persistence of an enzyme, intestinal lactase, beyond childhood. The principal sugar in milk is lactose, which is a disaccharide consisting of one molecule of glucose and one molecule of galactose bound covalently to each other. All except a few rare infants with inherited defects in lactase can break the covalent bond to produce glucose and galactose.

$$lactose \rightarrow glucose + galactose$$

The smaller sugar products are then further metabolized for energy.

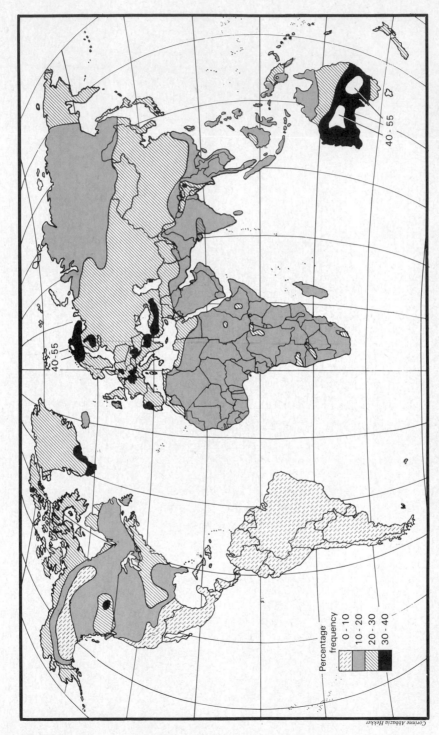

FIGURE 20.2A Distribution of the ABO blood group genes among the aboriginal populations. The *A* allele. (From A. E. Mourant, A. C. Kopeć, and K. Domaniewska-Sobczak, *The Distribution of the Human Blood Groups*, 2nd ed., Oxford Univ. Press, London, 1976.)

Percentage
frequency

0 - 10
10 - 20
20 - 30
30 - 40

40 - 55

40 - 55

Corinne Abbazia Hekker

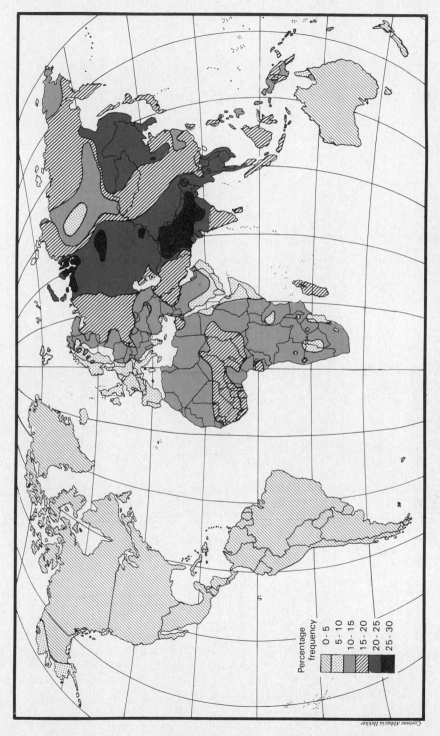

FIGURE 20.2B Distribution of the ABO blood group genes among the aboriginal populations. The *B* allele. (From A. E. Mourant, A. C. Kopeć, and K. Domaniewska-Sobczak, *The Distribution of the Human Blood Groups*, 2nd ed., Oxford Univ. Press, London, 1976.)

Percentage
frequency

0 - 5
5 - 10
10 - 15
15 - 20
20 - 25
25 - 30

Corinne Abbazia Hekker

TABLE 20.2
Lactase Persistence in
Various Populations

Population	Percent Lactase-positive	Frequency of Lactase-positive Allele
Thai	3	0.01
Arabs (Lebanon)	10	0.05
Chinese	12	0.06
Bantu (Uganda)	13	0.06
Amerinds (Oklahoma)	19	0.10
U.S. blacks	22	0.12
Mexico (rural)	26	0.14
Greece	53	0.31
India	71	0.47
U.S. whites	77	0.52
Spain	78	0.53
Finland	82	0.58
Australian aborigines	96	0.80
Sweden	97	0.83

Data are from G. Flatz and H. W. Rotthauwe, *Progress in Medical Genetics* n.s. 2:205–249, 1977.

At about the time of weaning, intestinal lactase drops to very low levels in a majority of children. After that time, large quantities of lactose, as in a glass of fresh milk, cause diarrhea and other intestinal upsets. Some persons, though, retain high levels of intestinal lactase throughout life. This is inherited as a simple Mendelian dominant trait. If we let L be the dominant allele, persons with persistent lactase are LL or Ll; adults without lactase are ll. The frequency of L varies from a low of 0.01 in Thai to a high of 0.83 in Sweden (Table 20.2).

This particular variation is important in efforts to provide nutrients by means of fresh milk or in foods that contain large amounts of fresh milk. Twenty-three percent of U.S. whites and 78% of U.S. blacks cannot tolerate such foods, and the percentage is higher in some other groups. Such efforts would be a very mixed blessing to these persons. Heating and fermentation of milk rapidly destroys the lactose, however, so that milk products such as yogurt, buttermilk, and cheese are tolerated.

Climatic Adaptation

Human beings live under quite diverse conditions of heat, cold, and humidity. Earlier in the chapter, it was suggested that the short extremities and round bodies of Eskimos and Mongolians are adaptations to cold. Various physiological measures of adaptation to climate support the idea that natural selection has made certain populations better able to survive in their usual or aboriginal habitats.

One must distinguish between physiological adaptation (nongenetic) and natural selection (genetic). For example, persons who move to the very high altitudes of Ecuador, Bolivia, or other Andean areas have difficulty at first with the low level of oxygen. The body responds over time by maintaining a higher level of red blood cells and thus compensates in part for the altitude. This kind of adaptation is nongenetic (athough the ability to adapt may well be influenced by genetic factors) and will not be of interest to us here. Rather, we shall consider inherited differences that appear with various climatic stresses. It is not always easy to prove that a particular adaptation is genetic or nongenetic. Ideally, persons from different groups should be reared together. Any variation that persists should then be genetic. Unless such comparison groups can be located, the suspicion remains that differences in experiences may account for the differences that characterize the populations.

Adaptation to cold has been especially intriguing because of the extreme examples known. Perhaps none seems so extreme to most of us as the Patagonians who live at the tip of South America in Tierra del Fuego. Magellan, among others, noted that these Indians are quite comfortable essentially nude moving about in canoes in freezing rain. The Australian aborigines are able to sleep nude in low desert temperatures that keep Europeans awake all night.

Both groups, as well as other populations that live in cold climates, have been studied to determine to what extent the adaptation to cold is a stable biological trait rather than a temporary response. The basal metabolism of the Tierra del Fuegans is high, and much larger amounts of heat are generated from their metabolism than is the case for other populations. The Australian aborigines seem to conserve their heat by reducing the peripheral blood circulation to the limbs. The internal body temperature remains normal during the cold desert night, although the arms and legs may become very cold. This adaptation persists during hot weather. Such studies do not prove a genetic basis for the adaptations, but the stability of the adaptations is easier to explain on the hypothesis of a genetic basis. There is no evidence that Europeans can adapt to such extremes of cold.

Adaptation to heat has been easier to study, since persons of diverse origins have been reared under similar conditions in hot and temperate climates. Studies comparing work performances of U.S. whites and blacks under various conditions of heat and humidity were carried out during World War II. Summarized briefly, the blacks were able to perform better in heat with high humidity, even when matched with whites for amount of subcutaneous fat, which serves as insulation. It is tempting to relate this difference to the selective advantage of certain genes in the ancestral tribes who lived in hot humid areas of Africa. Better information would be required before such a conclusion could be firmly drawn.

Human Races

How Many Races Exist?

From the foregoing discussions, no one should be surprised if there is disagreement on the number of races and their identification. If we avoid typological thinking and adhere instead to a population approach, we cannot draw distinct lines between races, local races, ethnic groups, or however else we may wish to distinguish populations. But we can recognize the existence of major populations

TABLE 20.3
The Major Human Races

1. African
2. European
3. Indian
4. Asiatic
5. Amerindian
6. Polynesian
7. Melanesian
8. Micronesian
9. Australian

From S. M. Garn, *Human Races*, 3rd ed., Charles C Thomas, Springfield, Ill., 1971, 196 pp.

that differ from other major populations in a variety of ways. For many purposes, it is useful to make such classifications, even though intermediate populations exist that underscore the arbitrariness of some of our efforts.

There are no exact number of human races. The list depends on the proclivities of the enumerator. Some prefer a short list with many subpopulations or local races. Others prefer to give each distinctive population recognition as a separate race. Thus various lists have anywhere from five to forty races. We shall follow the "short list" approach and identify nine races (Table 20.3). The original locations of these groups are shown in Figure 20.3.

There are clearly enormous differences in the numbers of persons in each racial group. The Asiatic group includes Chinese, Japanese, Filipino, Thai, and so forth, quite distinct local races. The African group also includes such distinct local races as the Bantu, Bushmen, Pygmies, and Sudanese. But these diverse populations share more in common than they share with other populations. This is because they arose from a common ancestral population in some instances, and there has been more opportunity for gene flow among them.

It may be argued that some of the major races do not merit separate status. Thus all evidence points strongly to the Amerindians' having originated from Asiatic populations some 20,000–30,000 years ago. The two groups share many genetic traits not found in other races. The Amerindians perhaps should be a local race belonging to the Asiatic race. Similar arguments can be advanced for the Polynesians, for whom some evidence exists that favors their origin from the Northwest Amerindians. Whether or not this is the case, they have close affinities with the Asiatics. The argument in favor of separate racial status is that both the Amerindians and Polynesians have been separated by geographic barriers from the main body of Asiatics for many thousands of years and have evolved quite distinct identities. To give them separate racial status emphasizes the lack of a biological yardstick with which to determine such status.

The Origin of Existing Races

One of the uses of genetic *markers* (variant alleles) is to help recognize the origins of human populations. This is done primarily by methods already discussed. For

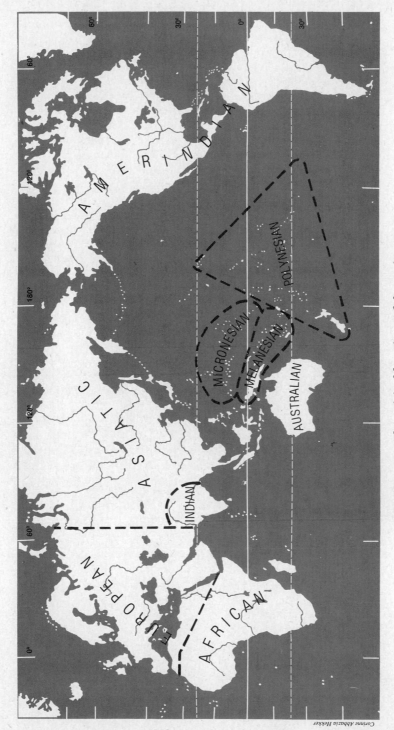

Corinne Abbazia Hekker

FIGURE 20.3 The original locations of the major races.

407

example, if two populations both have an allele not found in other populations, or found rarely, a reasonable explanation is that the populations descended from a single ancestral population that had the allele. This is presumed to be the case with the Diego[a] blood group found in Asiatics and in American Indians. The findings in this case are consistent with other genetic and geographic evidence.

Most of the examples of genetic variation are not all-or-none, as in the Diego blood group. Rather, they involve different frequencies of alleles. The ABO blood groups are an example. The hypothesis is that populations with similar allele frequencies are more closely related than are those with different allele frequencies. Such comparisons must be made with a number of different loci, however, since any one locus may give misleading results. Quite diverse populations may happen to have the same ABO frequencies, but they are not likely also to have the same Rh, MN, and HLA frequencies. By comparing a large number of such systems, a "best fit" can be calculated that shows the relative degrees of relatedness among a series of populations. Such a system is generally based on the idea that all populations arise by splitting, which is not always true. A computer-generated *phylogenetic tree* for some human populations is shown in Figure 20.4.

Hybrid Populations

Occasionally, quite distinct populations merge, or parts of them merge, to form a single new *hybrid* population. Historical examples are the U.S. black population, the Cape coloured of South Africa, and the Mexican-American population. The allele frequencies in the hybrid population will be the average of those in the parental populations, weighted according to the relative contributions of the parental populations. The allele frequencies in the hybrid population may also be used to reconstruct the relative parental contributions if their allele frequencies also are known.

This application can be illustrated with the U.S. black population. The frequency of the *Fy* allele in Africa is near zero, and in U.S. whites it is 0.429. In a large study of *Fy* in Oakland, California, the frequency was found to be 0.094 in blacks. One may set up an equation

$$a(0.000) + b(0.429) = 0.094$$

where a is the fraction of African alleles in U.S. blacks and b is the fraction of Europen alleles. The calculation is very simple in this case.

$$b = 0.094/0.429 = 0.219$$

Thus, 22% of the black gene pool is of European origin and 78% is African. Some isolated black populations in the Atlantic coastal regions of the South have higher proportions of African genes.

Hybrid populations must have formed many times in the past. Indeed, virtually all existing populations must be hybrid to some extent. It has already been noted that Neanderthal man was probably not swept away but rather contributed his genes to whatever invaders may have come his way. We return then to the early

(a)

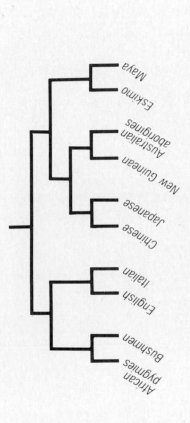

(b)

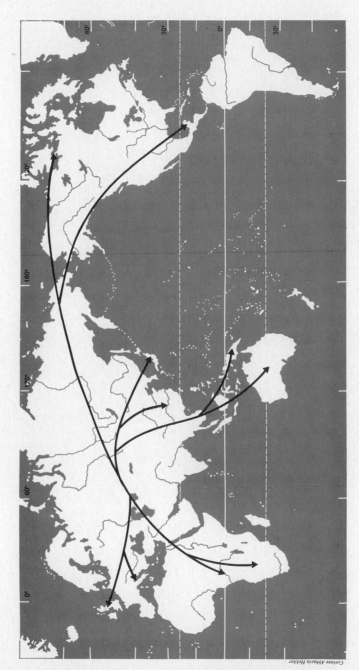

Corinne Abbazia Hekker

FIGURE 20.4 (a) A human phylogenetic tree. This "tree" is generated by comparing the gene frequencies at a number of genetic loci. Those whose frequencies are most similar are considered to be closest related. A number of slightly different trees can be generated depending on the mathematical model used. However, most are consistent with historical information. (b) Projection of the tree on a world map. The connecting lines should be interpreted as affinities and not as migration routes. (Data from L. L. Cavalli-Sforza and A. Piazza, *Theoretical Population Biology* 8: 127, 1975.)

409

theme of this chapter. A population is a particular group of people at a particular time and place. Classifications and labels have many uses, but one should not forget that they are creations of humans and, like many human inventions, have shortcomings that limit their use.

Review Questions

1. Distinguish among species, subspecies, and races.
2. Some anthropologists believe that the differences between the human races are nonadaptive. What is meant by this? How could these differences have come about? (Use the material in Chapters 17–20 to help answer this.)
3. What is meant by "selectively neutral"?
4. Alleles of some genes may have been "neutral" under the primitive conditions of life 50,000 years ago, but now they no longer are. Explain why and give some possible examples.
5. Some obviously deleterious mutant alleles of genes are present in an extremely high frequency. Cystic fibrosis is one example. Those who inherit this condition rarely have children, so that their Darwinian fitness is essentially zero. Can you give reasons why the frequency may be so high despite the strong selection against persons with the disease?
6. What happens to the volume of a body relative to its surface area as the body increases in size? What kind of adaptation to the environment would you expect to result from this?
7. Explain the statement "There are many inherited polymorphisms in the human population." What relation, if any, do these polymorphisms have with racial differences?
8. What selective pressure has been brought to bear on the human population by malaria parasites?
9. What does the intestinal enzyme *lactase* do? The lactase gene L determines its presence, and the recessive allele l its absence. In Sweden, the frequency of L is 0.83. Estimate the number of persons in the Swedish population who probably get sick from drinking fresh milk.
10. Twenty-three percent of U.S. whites cannot tolerate milk because of lactase deficiency. What is the frequency of l in this population?
11. The frequency of the B allele of the ABO blood groups is very low in Western Europe but increases in frequency as one proceeds east into central Asia. What explanations for this distribution can you advance?

References and Further Reading

Buettner-Janusch, J. 1973. *Physical Anthropology: A Perspective*. Wiley, New York. 572 pp. An introductory textbook that incorporates many of the ideas of genetics.

Coon, C. 1963. *The Origin of Races*. Knopf, New York. A sometimes controversial account, largely nontechnical, by one of the major figures in physical anthropology.

Garn, S. M. 1971. *Human Races*, 3rd ed. Charles C Thomas, Springfield, Ill. 196 pp. A very readable, balanced account of what races are.

Goldsby, R. A. 1977. *Race and Races*, 2nd ed. Macmillan, New York. 158 pp. An excellent introduction to the concepts and evidence, written for the student with little scientific background.

Mourant, A. E., A. C. Kopeć, and K. Domaniewska-Sobczak. 1976. *The Distribution of the Human Blood Groups*, 2nd ed. Oxford Univ. Press, London. 1055 pp. Hundreds of pages of raw data. The major resource for people who like to know about gene frequencies in obscure populations.

Osborne, R. H. (ed.). 1971. *The Biological and Social Meaning of Race*. Freeman, San Francisco, 182 pp. Papers by leading authorities on all aspects of race.

Our environment here on earth *is* and *has been* created to a great extent by the microorganisms, plants, and animals that share the planet with us. If they were to disappear so would we. Some are conspicuously more important than others, however, because they are our immediate sources of food. These are the domesticated and cultivated animals and plants originally derived from the wild species by our ancestors starting 10,000 or more years ago.

The first animal to be domesticated was probably the wolf. It has been bred selectively by several varieties of humans to produce many varieties of dogs over a period of many years. There is good evidence that this selection process may have been initiated 500,000 thousand years ago in China by an old hominid relative of ours called Peking Man or *Homo erectus.* The oldest dog remains associated with our species have been found in Iraq and are believed to be about 12,000 years old. Actually dogs may have been our companions since we became *Homo sapiens.* In the intervening period perhaps thousands of varieties have been bred, and a hundred or so still exist. Figure 21.1 illustrates some of the better known breeds. The surprising and important fact to be recognized is that this vast array of different types of dogs has been derived from a relatively limited pool of genes, the wild wolf pool.

Other wild species of animals began to be transformed about 10,000–20,000 years ago into the present-day domestic sheep, goats, pigs, chicken, cattle, and turkeys. Most of this domestication process occurred in central and southern Asia, but also the American Indian may have begun the domestication of the turkey as an important food source more than 5000 years ago.

The process by which a wild species is domesticated, or converted into a tractable, useful and edible form, is a kind of *artificial selection* called *selective breeding*. Now we practice what is called *scientific breeding* based on the knowledge of genetics gained over the past 80 years, but programs of empirical selective breeding have been carried on subconsciously and consciously for thousands of years. It wasn't necessary to know anything about genes, chromosomes, and cells to learn that if one wanted an animal with certain characteristics one should see to it that it had parents with these characteristics or close to them. The same applies to plants. At least 8000 or 10,000 years ago the grasses we now call wheat, barley, and millet began to be cultivated in plots of land and, as a result of selective breeding, started to undergo the changes that resulted in the present-day varieties of these important food plants. The improved versions have led to fantastic changes in the human condition.

The first humans lived as hunters and gatherers—hunters of wild animals, and gatherers of wild plants such as grasses, edible roots, seeds, and fruits. The beginnings of the domestication of sheep and goats probably led to the first real social organizations, and the start of herds led to a stabilization of a source of wool for clothing and meat and milk for food. Humans probably remained essentially as nomadic as they had been as hunters and gatherers, however, for the herds had to be moved to new fields as they gobbled up the vegetation on the old.

The beginnings of what we call modern civilization, including the growth of cities and kingdoms, started with the establishment of farming. People had to stop being wanderers to tend their crops, since their primary food source was now

21

The Breeding of Plants, Animals, and Microorganisms

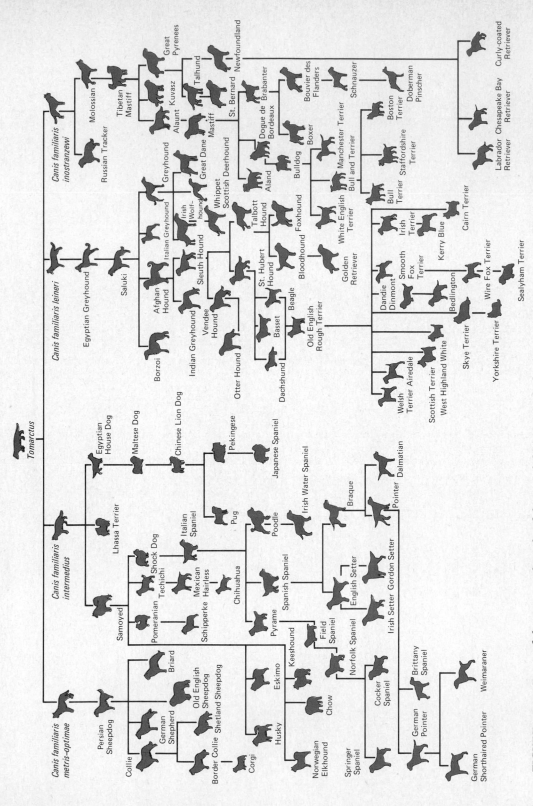

FIGURE 21.1 Some of the present breeds of dogs and their presumed ancestors. (From R. A. Goldsby, *Race and Races*, 2nd ed. Macmillan Publishing Co., New York, 1977.)

stationary. Crowding together for protection and unity of effort, they began the social interactions that became law and government. Cities were founded, kings reigned, and war became an important activity.

It is no exaggeration to say that the development of domestic food plants and animals was the most important step to be taken by the human animal after the invention of tools and the taming of fire. Large populations could not exist without agriculture. By selective breeding the yield of crops was increased enormously and consequently so did the numbers of humans dependent on these sources of food. The human population began its inexorable march toward overpopulation about 10,000 years ago because it had discovered a form of practical genetics.

Practical Breeding

Prior to 1900 and the rediscovery of the Mendelian techniques of analysis, many breeders of both plants and animals were highly successful in producing new forms, strains, and varieties to fit the various purposes of food, ornamentation, and useful work. Indeed, by 1900 nearly all of the present-day useful varieties of cattle, sheep, goats, swine, dogs, cats, fowl, and horses had been developed by empirical breeding. By empirical we mean using experience and observation handed down from previous generations without resorting to what we call science and theory. Granted, no sharp line can be drawn between empiricism and scientific theory, but it should be obvious that once the basic facts of inheritance through genes and chromosomes were recognized a power was gained that enable selective breeding to become much more efficient. We will illustrate this in the following sections by describing some of the methods now used to develop varieties of certain plants and animals and also some new techniques that show promise for the future.

Modern Plant Breeding

Here we will use examples from the three most important staple plants: wheat, corn (or maize), and rice. These three plants produce the basic source of food for the world's population. Other plants are important, too, particularly for certain populations. Beans, for example, are as important as corn in parts of Latin America. Table 21.1 lists some familiar plants. Some are used for clothing and other things besides food. But here we want to emphasize that all plants in the list, without exception, have been improved by the application of selective breeding, both before and since 1900.

Wheat

Wheat, like maize (we call it maize rather than corn, since to the British corn refers to wheat) and rice, is a grass. As you can see from Table 21.1, the grasses are our

TABLE 21.1   Plants Important to Humans that have been "Improved" by Selective Breeding Practices.

The starred plants originated in North or South America and were not known to Europeans and western Asiatics until after the "discovery" of America by Colombus.

Grasses	Staples	Other Vegetables		Fruits		Clothing, etc.
wheat	soybean	cabbage	artichoke	apple	citrus	cotton
rice	alfalfa	eggplant	okra	pear	plum	hemp
*maize	olive	onion	parsley	apricot	melons	flax
sorghum	beet	asparagus	parsnip	peach	strawberry	*rubber
barley	*potato	carrot	rhubarb	date	*avocado	*tobacco
oats	*kidney bean	celery	*cacao	fig	*pineapple	
rye	peanut	lettuce	*maté	grape	*tomato	
sugar cane		pea	*cassava	mango		
millet		radish	*squash	cherry		

most valuable food source. Roughly, wheat is the most important staple food for Europe, Africa, and North and South America, while rice is the staple for East Asia. The distinction is breaking down, but up to a few years ago it was quite sharp.

Wheat is probably the oldest of the cultivated grasses. Some believe its cultivation and domestication started 10,000 years ago, probably in that area encompassed by present-day Turkey, southern Siberia, Egypt, Israel, Jordan, Iraq, Iran, Pakistan, and western India. This should not be taken to mean that wild wheats do not grow in other areas. Actually, the wheat genus, *Triticum*, consists of 27 known species distributed over most of the Mediterranean basin and central Asia.

Four different species of wheat are cultivated for food. The most important economically is *Triticum aestivum* (Fig. 21.2), which is cultivated worldwide except in very low rainfall areas of the temperate zones where the pasta or macaroni wheat *Triticum turgidum durum* is grown. In this discussion, we shall confine our remarks mainly to the many cultivated varieties of *T. aestivum*, since they provide well over 90% of the world's wheat supply. It is estimated that at least 20,000 varieties of this wheat exist in cultivation in various parts of the world. These varities are the result of selective breeding programs designed to fashion strains adapted to the many different kinds of environments in which wheat is grown.

The 27 known species of wheat are grouped into three major categories based on chromosome numbers: diploid ($2n$), tetraploid ($4n$), and hexaploid ($6n$). The diploid strains have 14 chromosomes; therefore, the haploid number is 7. Thus the tetraploid wheat have $4 \times 7 = 28$ and the hexaploids $6 \times 7 = 42$ chromosomes, respectively. The major cultivated wheat, *T. aestivum*, is a hexaploid, which we are quite certain arose from a cross between two other species: the tetraploid cultivated *T. turgidum* and a wild species, *T. taushii*, that is a diploid. This event may have taken place in an Iranian farmer's field about 8000 years ago. How can we possibly know this? Consult Figure 21.3 and we'll try to explain. First of all, archeological evidence provides an approximate date. Second, genetic and cytological evidence provides a knowledge of *aestivum*'s ancestry. All wheat species

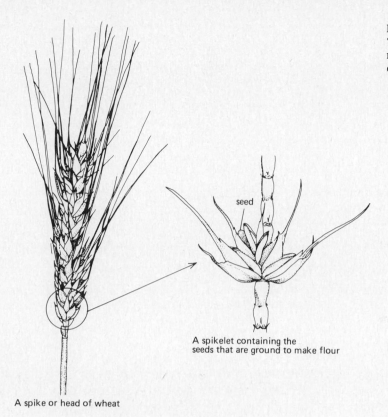

seed

A spikelet containing the
seeds that are ground to make flour

A spike or head of wheat

FIGURE 21.2
Triticum aestivum, the
most widely grown of the
cultivated staple wheats.

have identifiable wild forms except *aestivum,* which makes it highly probable that
this is a human-made species that we can, in fact, resynthesize today.

We start with a diploid wild species *T. monococcum* and cross it to another
diploid wild species *T. searsii* (named after the noted American wheat geneticist
Earnest Sears). Both of these species are found in the wild in southwest Asia, so
such a cross could also have occurred in the wild. Viable plants do result from the
cross, but the seeds of these offspring are sterile. An examination of the karyotypes
of the nuclei shows that at meiosis the fourteen chromosomes do not pair to form
seven sets as they should under normal circumstances. The set from *monococcum*
is designated the A genome and that from *searsii* we will designate the B genome.
Obviously, when *monococcum* (AA) is crossed to *searsii* (BB), the hybrid will be AB
(Fig. 21.3). But if the hybrid's A chromosomes do not pair with its B chromosomes, it
is a functional haploid that cannot undergo normal meiosis and produce functional gametes.

Occasionally in nature (and more frequently than that under the control of the
plant breeder), the chromosomes of a plant will double in number. In the case of
our *monococcum* × *searsii* hybrid, a plant with the genomic constitution AABB

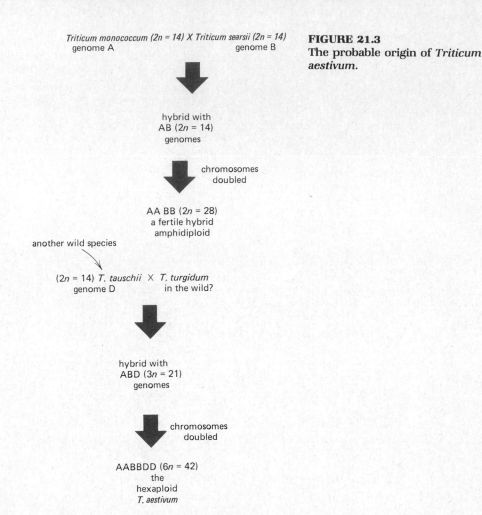

FIGURE 21.3
The probable origin of *Triticum aestivum.*

will result (Fig. 21.3). This plant is now a tetraploid since it has 4 × 7 = 28 chromosomes. However, since its A chromosomes will not pair with its B chromosomes, it is functionally a diploid whose gametes are AB in constitution. These gametes, pollen and eggs, will produce viable zygotes, AABB, on fertilization, and subsequently, good seed, so that the plants can be propagated by seeds. This is essential for raising large numbers of wheat plants.

A tetraploid that functions at meiosis as a diploid is called an *amphidiploid*. *T. turgidum* probably arose this way, since ½ of its chromosomes are A and ½ B. Most of our cultivated plants are amphidiploids, many of them created by plant breeders. Why create them? The answer is that it makes it possible to cross two wild species, each with its own desirable set of characteristics, and combine both sets in a single plant that now is far superior in its food or other commercially valuable qualities than either alone. Sometimes the hybrid will produce viable seeds if the chromosomes of the parents pair, but more often they do not, as in the case of different wheat species, and the chromosomes much be doubled in number to form the amphidiploids.

This is what apparently happened with the creation of *T. aestivum* about 8000 years ago. A cultivated plot of amphidiploid *T. turgidum* was invaded by a wild species *T. tauschii* with a genomic constitution of DD. D chromosomes will not pair with either A or B chromosomes so the hybrids had the constitution ABD and were sterile. The chromosomes doubled in some and the AABBDD genomic constitution characteristic of the present *T. aestivum* was formed. These plants are hexaploids (6 × 7 = 42) but still function as diploids and therefore are still classified as amphidiploids.

T. aestivum continues to exist after these many thousands of years because it produces good grain. It has been preserved by successive generations of wheat farmers. Left to itself it might well have disappeared long ago because it probably could not survive competition with other wild grasses. This is an important point to bear in mind. Cultivated plants are bred because they have characteristics important to humans as food, and so on. These characteristics are not necessarily of value to the plants themselves. The same is true, of course, for domesticated animals. Bessie the milk cow would not last long alone on the lone prairie without human assistance. Still in all, if *aestivum* is inbred for many generations there is the possibility of producing plants that are all of almost the same genotype. This can be dangerous, because a change in environment, such as a fungus becoming extremely virulent for the widely grown almost uniform genotypes, could cause a disaster to the crop and thus the human food it should supply. Even with the present 20,000 different genetic strains of *aestivum* in cultivation, this eventuality is still possible and must be guarded against. One way to guard against it and also make *aestivum* an even better wheat for humans is to bring in genes from wild wheats and other grasses that will cross with *aestivum*.

A few of the wild relatives of *aestivum* have chromosomes that pair with *aestivum* chromosomes making it easy to make hybrids with genes from both. But most wild species do not have chromosomes sufficiently similar to *aestivum* chromosomes to pair with them, and it is necessary to resort to special kinds of genetic engineering to effect the transfers of the desirable wild characteristics into the *aestivum* genome. Figure 21.4 diagrams crosses that made it possible to introduce into *aestivum* a chromosome from an "alien" species that could be any wild-type wheat species whatever its chromosome number. In the case illustrated, the alien species is diploid with fourteen chromosomes. The initial cross results in a sterile hybrid with 28 chromosomes (21 *aestivum* + 7 wild alien). The chromosome number is doubled by the application of colchicine, an alkaloid drug that arrests cells in metaphase (and is also used in the treatment of gout), and amphidiploid plants with 56 chromosomes are obtained. These plants are octaploid (8 × 7) and produce fertile seeds. However, only one of the chromosomes from the alien parent has the gene or genes that are desired in the *aestivum* line, the others are excess baggage and may in fact bear unwanted deleterious genes. The hybrid is now crossed to the *aestivum* parent strain and plants are obtained with diploid *aestivum* but haploid alien genomes as shown in the figure. These plants are again crossed to *aestivum* and a whole range of progeny are obtained with zero to seven alien chromosomes. The plant or plants with the desired alien chromosome(s) is

FIGURE 21.4
Transferring a chromosome from an alien species into *T. aestivum*.

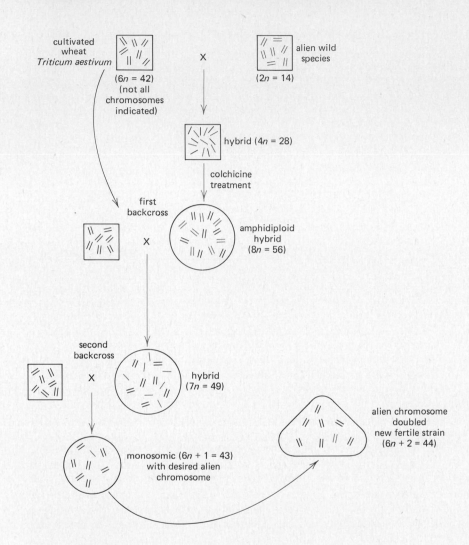

now manipulated to get a fully amphidiploid line with 21 pairs of *aestivum* and 1 pair of the desirable alien chromosomes.

These and other related techniques are making it possible to tap the gene pools of the many wild relatives of *aestivum* and by manipulating the chromosome constitutions create strains that are adapted to hot-dry, hot-humid, cool-dry, or cool-humid conditions, or strains that will mature in early summer versus those that will mature in late fall, or strains that are also resistant to certain fungi, bacteria, or insects. The possibilities are unlimited. The better the wheat plant is adapted to the local climate the more grain is provided to feed a hungry world.

These crosses need not involve only species of *Triticum*. For example, *T. turgidum* will also cross to rye grass (*genus Secale*; Fig. 21.5). The hybrid has certain agronomic characteristics that make it superior in some ways to either of the parent species.

FIGURE 21.5
The cross *T. turgidum* × rye grass.

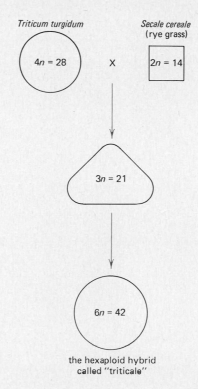

the hexaploid hybrid
called "triticale"

Rice

Together, rice and wheat provide about 90% of the staple food of humans—and in about equal proportions. Thus, rice is as important a source of human food as wheat. Like wheat, it exists in the wild with about 26 different species, only two of which are cultivated: *Oryza sativa* (Fig. 21.6) and *O. glaberrinia*. *O. sativa* is the major cultivated form in Asia, where most rice is grown, and in the United States. *O. glaberrinia* is cultivated mainly in tropical West Africa and a few other restricted areas in North and South America.

O. sativa, or Asian rice, as it is commonly called, probably arose in southern Asia in the general area of present-day India and China. Its origins are lost in the past, principally because it grows in moist humid areas in which the preservation of plant and animal remains as fossils is almost nonexistent. *O. sativa*, like the cultivated wheat *T. aestivum*, has many varieties, and it crosses with most of the many other wild rice species. All species of natural occurrence are either diploid with twelve pairs of chromosomes or tetraploid ($4n = 48$). The cultivated forms are diploid but may be made tetraploid with colchicine. This change does not seem to confer any superior agronomic properties upon them, however.

The present cultivated varieties of rice have all been crossed to wild species and

FIGURE 21.6
The domestic rice plant. *Oryza
sativa.* (From A. S. Hitchcock,
*Manual of the Grasses of the
United States*, 2nd ed. revised
by A. Chase, United States Government Printing Office, Washington, D.C., 1950.)

varieties to produce hybrids with enhanced yields. The breeding procedures are
much like the ones described for wheat.

Maize

Maize or Indian corn (*Zea mays*) is the world's third most important food crop.
Compared to wheat and rice it is relatively unimportant as human food, but when
fed to animals and transformed into milk, eggs, and meat it indirectly serves as a
principle source of human food. Of the maize grown in the United States, 90% is
used as animal feed. Most of the rest is used to produce syrup, oil, starch, meal, and
flour and in various fermentation processes designed to produce beer, distilled
liquors, antibiotics, vitamins, assorted chemicals (such as industrial alcohol), and
enzymes. Even corncobs are processed to produce important industrial chemicals. In fact, nearly all of the corn plant can be processed to produce things useful to

humans. It is said that all of a pig is used except the squeal. The same is true for corn except the squeal is not wasted. Most of the useful part of the corn plant is in the cob and its kernels, but many uses are now being found for the rest of the plant as well. The average American supermarket carries over 1000 food items in which corn derivatives are ingredients. Although it does not directly provide the total calories for humans that rice and wheat does, it certainly surpasses them in the number of useful products that can be derived from it.

Maize exists only as a cultivated species. Its closest relative is a grass called teosinte (*Zea mexicana*) found in the wild in Mexico and Central America (Fig. 21.7). It is probable that teosinte is the progenitor of corn. Corn crosses readily with teosinte, and because of this, certain morphological characteristics, and a great deal of archeological data, it is believed by many that corn was bred from teosinte or a related similar form some 8000–15,000 years ago by the native Americans living in the general area of Central America and southern Mexico (Fig. 21.8). Since these primitive beginnings, selective breeding has developed it into today's highly useful plant. George Beadle, the noted geneticist and corn aficionado, has called modern corn a "biological monstrosity." It is a human creation from a relatively useless grass and, without human intervention, it could never survive alone in nature.

Corn was unknown in Europe, Asia, and Africa before Columbus. It was introduced into Europe early in the sixteenth century and is now cultivated in all parts of the world where the climate is suitable. Before Columbus it was spread throughout North and South America by the native people and was truly a

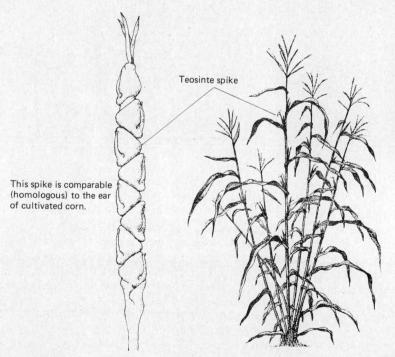

Teosinte spike

This spike is comparable (homologous) to the ear of cultivated corn.

FIGURE 21.7
Teosinte, a species of grass found in Mexico that is closely related to cultivated maize. (From G. W. Beadle, *Scientific American*, January 1980, p. 112.)

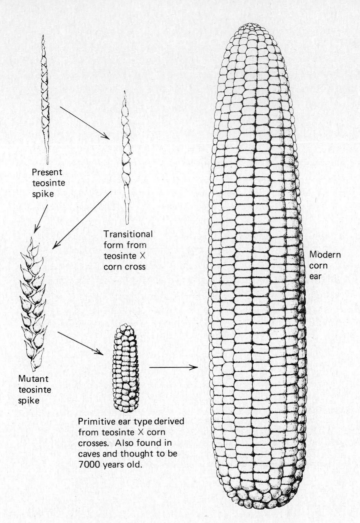

FIGURE 21.8
The possible derivation of maize from teosinte. The spikes and ears are shown in the proper relative sizes. (From G. W. Beadle, *Scientific American*, January 1980, p. 112.)

Present teosinte spike

Transitional form from teosinte × corn cross

Mutant teosinte spike

Primitive ear type derived from teosinte × corn crosses. Also found in caves and thought to be 7000 years old.

Modern corn ear

principal source of staple food. It is probable that the civilizations of the Mayans, Aztecs, and Incas were made possible by it, just as wheat and related grasses made the empires of antiquity such as the Roman Empire possible.

Modern corn, as cultivated in the principal corn belts such as the one in midwestern United States, is a product of genetic wizardry. Early in this century soon after the rediscovery of Mendelism, the American geneticist A. H. Shull began the experiments with corn that have led to the development of *hybrid corn*, one of the major triumphs of modern plant breeding. Shull first developed by self-fertilization several "pure bred" lines of corn. These he called homozygous, although in fact they were only partly homozygous. Their chief phenotypic characteristic was that their seed produced rather runty undistinguished plants. However, when these inbred lines were crossed to one another, the hybrid offspring are somewhat more vigorous than the parents. But when these, in turn, were crossed to hybrid offspring of a cross between two different inbred lines, the results were quite spectacular. Figure 21.9 outlines this breeding procedure, which is called the "double cross." The mating (A × B) × (C × D) results in a high yield of big plants with large cobs and kernels with highly desirable characteristics. This increase in vigor by crossing highly heterozygous plants is called *hybrid vigor* or *heterosis*. The theory is that the inbred lines A, B, C, and D each became homozygous for somewhat different sets of alleles for their many genes. On crossing them, the heterozygous offspring exhibit *heterosis* or *hybrid vigor*, the direct result of allelic gene interaction in heterozygotes. This vigor or heterosis is known to occur in many species of plants and animals, but it has been exploited to perhaps the greatest degree in corn.

The one great drawback is that the highly heterozygous, vigorous offspring

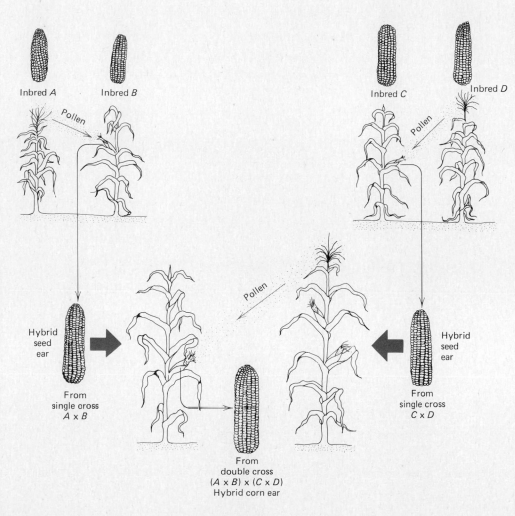

FIGURE 21.9
The procedure for breeding hybrid corn. (From R. P. Wagner, B. H. Judd, B. G. Sanders, and R. H. Richardson, *Introduction to Modern Genetics*, John Wiley & Sons, Inc., 1980.)

Inbred *A* Inbred *B*

Inbred *C* Inbred *D*

Pollen

Pollen

Hybrid seed ear

From single cross *A* x *B*

Pollen

From double cross (*A* x *B*) x (*C* x *D*) Hybrid corn ear

Hybrid seed ear

From single cross *C* x *D*

cannot be inbred profitably, for they soon start to go back toward a homozygous condition and lose their desirable qualities. Thus inbred lines must be maintained for carrying out the single cross leading to the double cross hybrid ear harvested by the farmer, as outlined in Figure 21.9. Farmers do not bother ordinarily to produce their own hybrid corn seed. Rather they purchase their seed from commercial seed growers. Hybrid corn seed production by these growers has become a highly profitable business.

Summary of Breeding Techniques

The production of new strains and varieties of food crop plants has become a major occupation involving the worldwide expenditure of billions of dollars per annum. In the United States the business of improving crops is shared by the Federal and State governments, private seed companies, and, to a lesser extent, nonprofit foundations. All of these organizations support breeders who resort to several approaches or combinations of approaches in order to obtain the results they desire. These approaches in summary are

1. Inbreeding plants with a desirable agronomic trait to enhance that trait.
2. Hybridization of inbred lines to obtain hybrid vigor or heterosis as in the case of hybrid corn.
3. Producing polyploids that may have more desirable qualities than the ordinary diploids.
4. Crossing related species to form amphidiploids that may have the most desirable characteristics of both parent species.
5. Introducing a single pair of chromosomes bearing the desired genes from a related species.

Benefits and Problems of Modern Plant Breeding

The sorts of breeding procedures described in the previous section plus various modifications on these themes have resulted in the enhancement of yields of practically every plant grown for human food. Table 21.2 compares the average yields per acre for most of the major plant crops in the United States for the years 1930 and 1975. It is obvious that every crop has increased significantly in yield in this 45-year period. Corn in particular has increased more than fourfold in yield, a truly stupendous achievement on the part of agriculturalists.

Not all of this increase is the result of modern selective breeding practices, however. A good part of it is the result of the application of modern agronomic practices by farmers. These include the use of fertilizers, improvements in the use of irrigation, the use of pesticides and herbicides, and improvements in farm management. It is almost impossible to determine exactly to what degree genetic manipulation has contributed to these increases, but various guess-estimates average about 50%. If true, this can only be reckoned as a most remarkable

Crop	Yield per Acre		Unit	Ratio 1975/1930
	1930	1975		
Wheat	14.2	30.6	bushels	2.2
Rice	46.5	101.0	bushels	2.2
Corn	20.5	86.2	bushels	4.2
Rye	12.4	22.0	bushels	1.8
Oats	32.0	48.1	bushels	1.5
Barley	23.8	44.0	bushels	1.9
Grain sorghum	10.7	49.0	bushels	4.6
Sugar cane	15.5	37.4	tons	2.4
Peanuts	649.9	2565.0	pounds	4.0
Soybeans	13.4	28.4	bushels	2.1
Cotton	157.1	453.0	pounds	2.9

**TABLE 21.2
Average Yield Per Acre
of Major Crops in 1930
and 1975.**

From: U.S. Department of Agriculture, *Plant Genetic Resources, Conservation and Use,* Washington, D.C., 1979.

achievement. In any case genetic application has contributed significantly to the so-called "Green Revolution."

In addition to increasing the overall yield of crops by introducing genes that work toward that end, significant advances have been made toward increasing the nutritional value of a number of different important crops. One example is the discovery that a new mutation in corn called opaque-2 increases the lysine content of corn. This amino acid is essential for humans as well as for the animals we depend upon for food. Corn protein is low in lysine, and an animal that has corn as its sole source of food can develop a serious nutritional deficiency. With the introduction of the opaque-2 mutant allele into the hybrid corn stocks, however, this deficiency is now being overcome.

Wheat grain is also deficient in lysine as well as several other amino acids. Attempts to increase the amino acid content along with the protein content by selective breeding have met with only partial success. Lines have been developed with very high protein content but not high lysine. Most of these lines, while having some highly desirable nutritional qualities do not however give high yields, and farmers avoid planting them. This is an example of one of several problems confronting breeders of both plants and animals—a selective breeding program to develop a particular desirable characteristic such as high lysine may also select concomitantly for an undesirable characteristic. One can't always have one's cake and eat it too.

The achievements of plant breeders, considerable as they have been in the struggle to get the highest yields with the least cost, have also contributed to a negative facet of the general problem of feeding the human population. This negative factor is perhaps best illustrated by an example drawn recently from an episode in the United States. In 1970 a catastrophic epidemic of southern corn blight, a disease caused by a fungus, wiped out a considerable portion of the United States corn crop east of the Mississippi River. Ordinarily about 5% of the corn crop is lost to viral, fungal, bacterial, and insect damage, but in this year the loss was closer to 50%. The outbreak was traced to a new mutant form of the pathogenic fungus (*Helinthosporum maydis*) that causes the blight. This new form or race was particularly virulent on corn with Texas male sterile cytoplasm (cmo-T). In 1970 about 90% of the corn grown in the United States carried this gene, which caused plants to produce sterile pollen. The gene was introduced into the hybrid corn lines by the hybrid corn seed producers to make it easier to develop hybrid strains. Its presence makes it possible to raise plants that are unable to self-fertilize or cross fertilize with others like themselves because they produce no viable pollen. Hence, inbreeding is eliminated. Without this gene present, it is necessary to go down the rows of the inbreds and remove the pollen-producing tassels by hand—an expensive and laborious process.

This incident makes it clear that although the development of strains with desirable agronomic characteristics by genetic techniques is in the best interest of increased food production, it can also lead to disastrous results without adequate monitoring. The wide use of strains with the same or almost the same genotype is definitely to be avoided. This is generally recognized by breeders and steps are being taken to avoid it.

A major problem related to inbreeding, not only of corn but all other food crops, is the continued use of the same strains in cultivation without preserving the gene pool or germ plasm of other seemingly less useful strains. These strains could die away by neglect and be lost. The preservation of these strains (and also of the many wild relatives of wheat and rice) is now looked upon as an important undertaking. The United States Department of Agriculture, recognizing the importance of preserving the germ plasm of all possible strains of food crops, has started to do this in collaboration with breeders through the National Germplasm System. Major accessions are at the National Seed Storage facility in Fort Collins, Colorado, and the Plant Genetics and Germplasm Institute at Beltsville, Maryland. Here thousands of different kinds of seeds are stored and continually tested to insure fertility so that when needed they can be used to start new stock.

The Future of Plant Breeding

Although the general methods of breeding described in the previous sections will probably never be abandoned, it is now possible to foresee more revolutionary advances in the development of new crop plants. The techniques are now at hand, and what remains are the developmental and engineering phases putting the techniques to work.

For the most part the new genetic technologies for plant breeding have been made possible by developing methods for culturing isolated plant cells in the laboratory much as animal somatic cells are cultured as described in Chapter 10. From a single cell or group of cells from a leaf, for example, a complete plant bearing flowers and seeds may be propagated, as shown in Figure 21.10. Completely homozygous plants can be obtained by including haploid eggs or pollen to start cell division and produce haploid plants. These plants can be induced to set seed by doubling the chromosome number with colchicine. The resultant homozygous diploids may have more desirable characteristics than the heterozy-

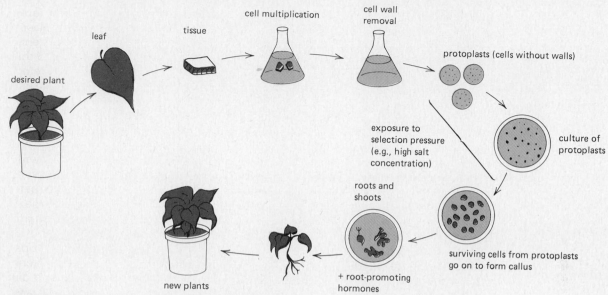

FIGURE 21.10
Growing plants from single isolated cells.

gotes from which they were derived. Asparagus has been improved and a new strain of high-lysine rice has been developed using this technique.

The fusion of cells from two different strains, much as animal cells are fused (Chapter 10), is now being regularly done in laboratories. First the plant cellulose cell walls are removed with enzymes to form protoplasts as shown in Figure 21.11. The protoplasts are fused and a plant is grown from the fusion cell. This plant may be an amphidiploid if the chromosomes from the two strains do not pair or synapse during meiosis (Fig. 21.11). Like amphidiploids formed by crossing the same two species, as illustrated in Figure 21.3 by conventional means, the amphidiploid or *allotetraploid* may possess the desirable qualities of both species. The advantage of protoplast fusion over the conventional cross is that the chromosome sets of two different strains or species may be combined by cell fusion but not by fertilization of an egg by pollen because of a sterility or incompatibility barrier. This phenomenon is common among plants in general and, of course, is not unknown in animals; a donkey crossed to a horse produces a sterile mule.

The isolation and propagation of single cells makes it possible to grow clones from a single cell (Fig. 21.12). The clones may be induced to form new plants complete with roots, stems, leaves, and flowers, or they may be maintained as a group of freely growing cells grown in large batches either in a liquid medium like bacteria or on a gel surface. Some species of plants can be propagated more economically and efficiently from clones than seeds. Asparagus and carrots are but two examples. Cells grown only as a disorganized mass in a callus are now being propagated to produce drugs such as anticancer agents, antiviral agents, steroid hormone precursors, flavorings, and insecticides, to mention but a few. The possibilities are almost limitless, and all major pharmaceutical companies are now

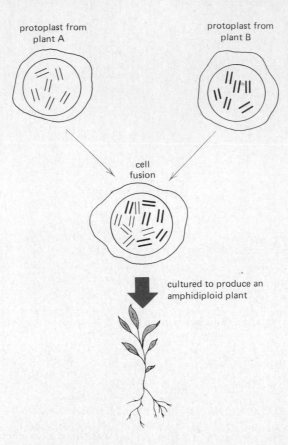

protoplast from
plant A

protoplast from
plant B

FIGURE 21.11
Fusion of plant cells from different strains or species. New varieties can be generated from the fused cells.

cell
fusion

cultured to produce an
amphidiploid plant

engaged in pursuing those many possibilities. Food for space-age future passengers on their way to distant planets will probably be grown on board ship using light engery from our sun or other suns if they leave our solar system.

The technique of introducing DNA from an animal cell into a plasmid or virus as described in Chapter 14 can also be applied to plant cells. Currently research is being carried out to introduce DNA into crop plants via plasmids or viruses in attempts to increase the nutritional value of these plants, or to introduce into them other desirable genes to increase their productivity or endow them with other agronomically valuable attributes. Figure 21.13 shows one procedure, now being exploited in various laboratories, that uses the plasmids found in the bacterium *Agribacterium tumefaciens*. This bacterium can infect crop plants in the plant class Dicotyledonae. This class includes nearly all crop plants except for the grasses and such things as bananas. When *Agribacterium* infects a plant, it induces the formation of a tumor called a crown gall. A plasmid carried by the bacterium infects the plant cells, and, remarkably, a segment of its DNA called T-DNA is inserted into one of the plant chromosomes. As a result, the plant gall cells now are able to produce nitrogen-rich compounds required by the bacterium. Experiments are now under way to introduce, by the gene-splicing techniques described

FIGURE 21.12
Growing clones from
plant cells.

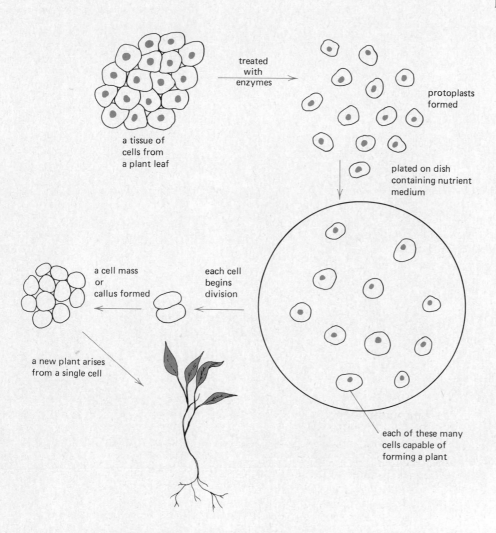

in Chapter 14, desirable genes into the plasmid T-DNA (Fig. 21.13) and then into a
plant chromosome. By means of the tissue culture techniques described in Figures
21.10 and 21.11 plants can then be produced that bear the desired gene. The theory
is good. We must now wait for the engineering to be accomplished.

   Box 21.1 gives one of the possibilities that may become reality by using techniques of this sort. Developments in these new plant breeding technologies may
prove to be as important to the human species as the rise of agriculture 8000–
10,000 years ago. It is quite possible that we are entering a new era comparable to
the one that saw the end of the hunting and gathering period in our history.

FIGURE 21.13
Infecting plant cells with *Agribacterium tumefaciens* and introducing its plasmid T-DNA into the plant's DNA.

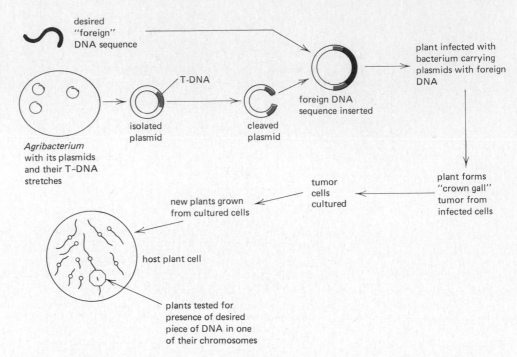

Box **21.1**

Genetic Engineering in Plants: Nitrogen Fixation

By using the energy from sunlight, plants are able to carry out just about all the chemical reactions and processes to make the food that they need and in turn what we animals need. Essentially, all they need is CO_2, H_2O, and the inorganic elements listed in Table 3.2. The one thing the higher plants cannot do on their own is to *fix atmospheric nitrogen*. They cannot convert the major source of nitrogen into usable form. Usable nitrogen, or fixed nitrogen, is in the form of NH_3 (ammonia) or its salts, such as am-

monium sulfate, $(NH_4)_2SO_4$, or ammonium nitrate, NH_4NO_3. Without these forms of nitrogen, plants could not grow because they could not synthesize amino acids and other nitrogen-containing essentials, such as the purines and pyrimidines of the nucleic acids.

The fixation of nitrogen is accomplished by certain bacteria, by blue-green algae and fungi, and by humans in chemical plants. The eukaryotic plants, including the fungi, get their fixed nitrogen primarily from the nitrogen-fixing bac-

Box **21.1**

(*continued*)

teria in the soil, which are able to carry out the transformations

$$\text{atmospheric } N_2 + H_2 \rightarrow NH_3$$
$$\overset{O_2}{\rightleftharpoons} NO_3^- + H_2O$$

Other bacteria carry out the process of *denitrification* in which the fixed nitrogen is converted back to free nitrogen, N_2. This prevents all the atmospheric nitrogen being converted by organisms.

An extremely important group of plants from the standpoint of nitrogen fixation is a family containing important crop plants, such as soybeans, beans, peas, peanuts, and alfalfa. These and their relatives in the family Leguminoseae all have bacteria growing in the root cells that are able to fix nitrogen and supply the host plant with a direct source of NH_3. These bacteria belong to the genus *Rhizobium* and the important role they play in agriculture cannot be overemphasized. They produce more fixed nitrogen than the host plant can use; as a result, a field in which a legume has grown is rich in fixed nitrogen at the end of the growing season. This surplus can then be used the next year by plants that do not have *Rhizobium* in their roots. Among these are our three most important crops: wheat, rice, and corn.

Currently, attempts are being made to convert nitrogen-fixing bacteria that do not attach to plant roots to forms that do for plants such as corn. The conversion of these bacteria to the state of intimate association with root systems will probably be accomplished by introducing genes into them through plasmids or viruses that change their growth habits. Also, the species of *Rhizobium* vary in their efficiency to fix nitrogen. The most efficient contain the gene *hnp* carried by a plasmid. Attempts are being made to introduce this gene into the strains of *Rhizobium* that do not have it so that the plants they inhabit will be faster growing.

Nitrogen fixation is the principal process in agriculture after photosynthesis. Making it more efficient for plants would be another major step forward in fostering the Green Revolution. The additional advantage in providing fixed nitrogen directly by microorganisms is that it would reduce or eliminate the industrial production of ammonia. The industrial process uses great amounts of energy, and most of the ammonia produced is lost to the atmosphere and converted to nitric oxides. These oxides, in turn, act on the ozone in our upper atmosphere and reduce its concentration. The effect is to permit more ultraviolet light to reach the earth's atmosphere.

Modern Animal Breeding

Although humans may have at one time in their past evolutionary history been essentially herbivorous, we have been omnivorous for some 1 million years and will probably continue to be. The meat of animals is highly nutritious, even though the production of that meat is highly inefficient compared to the production of food of comparable nutritional value derived directly from plants. The reason for this inefficiency, of course, is that animals must eat plants to make meat, and a great deal of energy derived initially from the sun is lost in the process. Still we seem content to pay for that inefficiency, and the production of meat and milk continues to be an important agricultural enterprise second only to growing food crops.

As we have stated earlier in this chapter, animal breeding by humans has gone on for thousands of years, and all the basic stocks of domesticated animals, with few exceptions, were developed by the beginning of this century. However, in the last 30 years major advances have been made in the production of animal food because of newly introduced breeding techniques. For example, the average milk yield of cows in the United States has more than doubled as the number of dairy cows has decreased by 50%. Probably about 25% of this increase is the result of breeding programs designed to raise milk yield per cow.

Many of the advances in meat production made in the past 80 years have been the result of activities of professional breeders who provide the breeding stock to the farmer or rancher. A prime example is the development of the Santa Gertrudis breed of cattle on the King Ranch in Texas. The selective breeding program was started about 50 years ago with a herd of purebred Shorthorn cattle. By rigorous selection procedures the new Santa Gertrudis evolved from the Shorthorn over a period of about 25 years. The selection program was designed to produce a strain that could withstand the rigors of southern and southwestern humid climates and still produce high grade beef. In this it proved to be quite successful. The result was the first new breed of cattle in recent history.

Breeding animals has become big business just as breeding plants has. About fifteen breeders provide the breeding stock that produces 3.7 billion chickens each year in the United States. A good part of the breeding operations consist in the production and distribution of semen. It is no longer necessary to keep a bull, ram, or billy goat on a ranch. For best results the farmer or rancher orders the semen of a prize male specimen who, over a period of time, may father 100,000 offspring. Artificial insemination programs are now major activities in most developed countries. In some cases the animals can only be bred by artificial insemination, because they have been fashioned by selective breeding into biological monstrosities like corn. The breast of the male commercial turkey is so large that he cannot mount a female. As a result, among commercial turkeys only artificial insemination of the females will insure fertile eggs.

The Future of Animal Breeding

Despite the possibilities of artificial insemination to develop better cattle, swine, and sheep, the technique has not been used to full advantage for a variety of reasons. New techniques are being developed, however, that will probably result in greater use and increased productivity. Since beef is the primary animal food for humans, at least in the United States, an enhanced use of artificial insemination over the 5% now generally used in range cattle could be a great boost to production, especially when combined with *superovulation* and embryo transfer as shown in Figure 21.14. The scenario, yet to be fully developed, is as follows.

A cow (as well as a sheep, pig, goat, or mare) can be superovulated by hormonal stimulation. This results in the release of six to eight more eggs than the one she would normally deliver to her uterus. These eggs are fertilized by sperm from a donor bull with superior qualities of beef. The early embryos cannot all develop to term in that one mother because of her inability to support all of them. These can be removed as shown in the figure, and surrogate mothers who are receptive (in estrus or heat) can be brought in to use. Each receives two embryos, which she can easily bring to term. Since twinning is uncommon in cattle, accounting for only 1–2% at most of live births, the calf yield can in this way be doubled. Also, the calves will have superior qualities, because both parents have been selected for those qualities. This can also reduce the number of spontaneous abortions, which is high in cattle just as in humans. Just as cultured mammalian cells and semen can be frozen and stored for periods of years, so can very early embryos be frozen and stored. Although not widely used yet, it may be that frozen embryos will someday be distributed much the same way that semen is now.

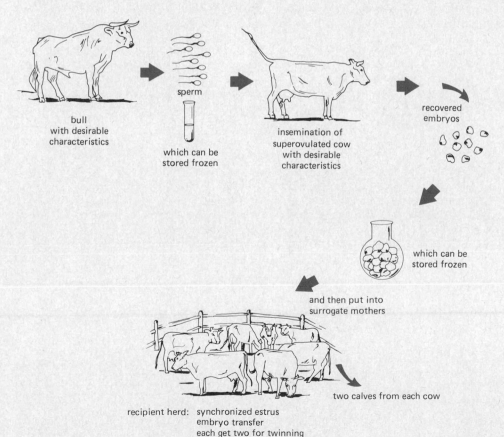

FIGURE 21.14
Superovulation and embryo transfer in cattle.

sperm

bull
with desirable
characteristics

which can be
stored frozen

insemination of
superovulated cow
with desirable
characteristics

recovered
embryos

which can be
stored frozen

and then put into
surrogate mothers

two calves from each cow

recipient herd: synchronized estrus
embryo transfer
each get two for twinning

Genetically identical individuals can be produced by cloning. This occurs naturally through twinning to produce identical twins, quadruplets, and so on, but these are rare. It is, however, possible to manufacture clones by manually separating the cells of very early embryos (two- and four-cell stages), as shown in Figure 21.15, and then reinserting the cells into surrogate mothers for subsequent development into complete adults. This has been done successfully in the laboratory for sheep and mice.

Manipulation of embryos has even been used to manufacture sheep-goat chimeras, that is, animals, part of whose cells are of sheep origin ($2n = 54$) and part of goat origin ($2n = 60$). This was accomplished by dissociating 4–8 cell early embryos obtained from the two species, mixing the cells together and allowing them to reaggregate, and then implanting them into a foster goat or sheep mother. The resulting chimera is a mosaic of goat and sheep tissue, a fact that is readily discerned by the variegated color and appearance of the wool. A chimera is quite different from a hybrid, in which the parental contributions are together in the same nuclei.

Theoretically, desirable genes can be inserted into the cells of an animal by the techniques already described for plants. This use of recombinant DNA and gene transfer techniques is yet to be worked out for animals, but there is no theoretical reason why such techniques will not be successful. The advantages will be considerable because single desirable genes can then be introduced instead of whole chromosomes that may (and frequently do) carry undesirable as well as desirable genes. As in the case of plants, the possibilities for genetic manipulation in animals are almost infinite in number.

The Uses and Breeding of Microorganisms

Beer, wine, bread, and cheese are products of microbial activity that have been among the main staffs of life for humans for thousands of years. By microorganisms we mean those organisms that are ordinarily classified as bacteria, fungi, or molds. Mostly these kinds of organisms live in the soil where they are engaged in the decomposition of plant and animal remains and excretions. Were it not for them, no dead trees or other plants would rot, and no animal excrement or remains would be converted to other useful ends. Without them we would soon suffocate in our own excrement or be buried beneath the corpses of dead animals.

In the last hundred or so years the considerable synthetic and degradative powers of microorganisms have been harnessed, and now a large number of chemicals and pharmaceuticals are produced on industrial scales. Box 21.2 lists some of these products and their uses.

Since the discovery of the life cycle of yeasts (Fig. 12.2) it has become possible to breed yeasts better adapted to the production of beer, wine, bread, and various kinds of industrial chemicals such as alcohol. The development of the recombinant DNA techniques described in Chapter 14 makes it possible to design bacteria and yeasts to synthesize compounds that they would never be capable of doing

FIGURE 21.15
The cloning of animals.

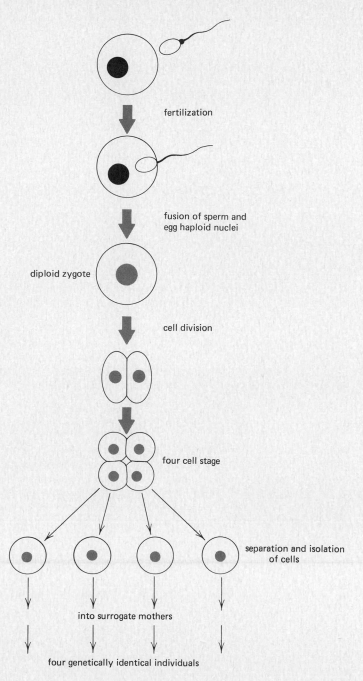

fertilization

fusion of sperm and
egg haploid nuclei

diploid zygote

cell division

four cell stage

separation and isolation
of cells

into surrogate mothers

four genetically identical individuals

Box **21.2**

Some Important Products of Microorganisms

Organism Type	Products
Yeasts of various kinds	Wine, beer, soy sauce, bread, industrial ethanol (alcohol), protein from certain waste materials such as paper pulp, vitamins.
Other molds and fungi	Cheese, soybean curd, citric acid, various enzymes used industrially, penicillins and many other antibiotics.
Bacteria	Cheese, yogurt, vinegar, acetone, butanol, microbial protein, vitamins, amino acids, many different kinds of antibiotics, and some insecticides.

otherwise. Theoretically, practically any gene from any organisms can be introduced into a bacterium or yeast, thus programming it to produce the desired result on command. Again, the possibilities are almost infinite.

What Has *Homo sapiens* Wrought?

We are now as a human species capable of destroying all life including ourselves by a number of different means. We can bomb ourselves out of existence with nuclear energy from either fission or fusion bombs, we can destroy our habitats and thus ourselves with polluting chemicals, or we can so alter the life on our planet by genetic means that an unforeseen environmental change could have a devastating effect on our well being. Our position grows more precarious with each passing year. However, since it is rumored that we are intelligent, we may possibly put this intelligence to use to save us and the life on our planet. The intelligent use of genetic techniques is one important facet to be considered in our attempt to save ourselves.

Review Questions

1. What are the basic principles of selective breeding of plants and animals that have been used for thousands of years?
2. In the Book of Genesis of the Bible, Chapter 30, versus 37–39, we find, "Then Jacob took fresh rods of poplar and almond and plane, and peeled white streaks in them, exposing the white of the rods. He set the rods which he had peeled in front of the flocks in the runnels, that is, the watering troughs, where the flocks came to drink. And since they bred when they came to drink, the flocks bred in front of the rods and so the flocks brought forth striped, speckled, and spotted." This is not selective breeding. What is it?
3. What is so remarkable about the development of the many dog breeds from the wolf? What genetic principle is proved?
4. Describe the probable origin of the present major wheat, *Triticum aestivum*.
5. What is an amphidiploid?
6. The cabbage ($2N = 18$) can be crossed to the radish ($2N = 18$). The hybrid is sterile, but when induced to double its chromosome number, fertile plants called *Raphanobrassica* are produced. (Unfortunately this hybrid has the roots of a cabbage and the leaves of a radish.) What is the chromosome number of *Raphanobrassica*, and how many pairs of homologs will synapse at meiosis?
7. Why is it important that as many wild species and cultivars (varieties) of cultivated food plants as possible be kept alive?
8. What is the apparent significant major difference between the origin of corn and that of wheat?
9. How is hybrid corn seed produced?
10. Hybrid corn demonstrates heterosis. What does that mean?
11. What kind of problem does the breeder encounter when inbreeding plants and animals with desirable traits?
12. What danger exists in the too wide use of a single strain or variety of a food plant?
13. Future plant breeding is going to include the application of genetic engineering as well as traditional breeding practices. What can be done with genetic engineering techniques that can't always be done by crosses?
14. Present day animal breeding practices include superovulation and embryo transplant. What are these processes? Why are they useful?
15. Develop a scenario for cloning mice. Can one use this procedure for humans too?

References and Further Reading

Beadle, G. W. 1980. The ancestry of corn. *Scientific American* 242(January):112–119, 162. An eminent geneticist demonstrates how genetics can be used to trace the origin of corn from wild grasses.

Clutton-Brock, J. 1981. *Domesticated Animals from Early Times*. University of Texas Press, Austin. A history of the development of animal breeds by humans.

Feldman, M., and E. R. Sears. 1981. The wild gene resources of wheat. *Scientific American* 244(January):102–112. A description of the probable origin of present day domestic wheats by two of the outstanding authorites.

Hopwood, A. 1981. The genetic programming of industrial microorganisms. *Scientific American* 245(September):90–125. How to make organisms on order to carry out specific tasks.

Nayar, N. M. 1973. Origin and cytogenetics of rice. *Advances in Genetics* 17:153–292. A rather technical treatment of the subject.

Office of Technology Assessment, Congress of the United States. 1981. *Impacts of Applied Genetics. Microorganisms, Plants, Animals.* U.S. Government Printing Office, Washington, D.C. A nontechnical description of current work and future prospects in the area of genetic engineering.

Rhoad, A. O. (ed.). 1955. *Breeding Beef Cattle for Unfavorable Environments.* The University of Texas Press, Austin. Some of the techniques used to develop breeds like the Santa Gertrudis.

Rutger, J. N., and D. M. Brandon. 1981. California rice culture. *Scientific American* 242(February:42–51. A description of how crop yields can be increased enormously by proper management.

Scientific American 245 (September), 1981. This issue is devoted to various aspects of genetic engineering with microorganisms.

Shepard, J. F. 1982. The regeneration of potato plants from leaf-cell protoplasts. *Scientific American* 246(May):154–166. A good guide to the general techniques of plant cell culture.

Sprague, G. F. (ed.). 1977. *Corn and Corn Improvement.* No, 18 in a series in agronomy, American Society of Agronomy, Madison, Wisconsin. All you want to know about corn is in here.

Medical genetics includes all aspects of genetics that impinge directly on human health. Many of the preceding chapters have dealt with aspects of medical genetics. In this chapter, we will be concerned with the way some of these genetic topics are integrated into the practice of medicine.

The usual steps in management of a disease are (1) diagnosis and (2) treatment. Genetics provides added dimensions to diagnosis, not only in the identification of the basic defects in affected persons but also in predicting risks for persons as yet unborn or unconceived. Understanding the nature of the primary genetic defect has led to new ideas for treatment. The new technologies of gene manipulation suggest possibilities for treatments that may become as important to medicine as antibiotics have been.

The Burden of Genetic Disease

The measure of the health burden attributable to any particular factor—infection, environment, or heredity, for example—is difficult. There are no yardsticks to measure such burdens. One may calculate the cost of providing treatment, but one cannot realistically estimate the cost to the economy of the loss of a productive tax-paying person. And there is no way to assess the cost of emotional stress from a genetic or other disease. Therefore the following discussion addresses only the more limited costs that can be calculated, even though crudely. Readers must add to those costs their own assessment of the emotional and other burdens.

The Frequency of Genetic Disease

Each year about 3.2 million babies are born in the United States. Approximately 30,000 of these will be born with a disorder inherited as a simple Mendelian trait, either dominant, recessive, or X-linked. An additional several percent will have congenital defects of various types, many of which have a strong genetic predisposition. Many of these children will die as infants; others will require special medical attention for the rest of their lives. In some cases, special institutional care may be required all their lives.

Table 22.1 gives the incidence of major genetic disorders in the United States. Except for chromosomal abnormalities, the data are not very reliable. They especially do not give adequate weight to inherited degenerative diseases that have late onset. As has been noted, ultimately we all die of some genetic defect, especially if we remember that aging itself is built into our genetic makeup.

22

Human Genetics Applied: Counseling, Screening, Engineering

TABLE 22.1
The Frequency of Genetic Disorders Detectable in Newborns

There have been relatively few reliable surveys of genetic disorders except for the frequency of chromosome abnormalities. Therefore these data, which are compiled from various sources, are only crude estimates.

Type of Defect	Frequency per 100 Liveborn
Autosomal dominant	0.2
Autosomal recessive	0.2
X-linked recessive	0.04
Complex inheritance	1.5
Congenital malformations	2.5
Chromosomal abnormalities	0.5
Total	4.9

The Cost of Detrimental Genes

The Biological Cost. By biological cost, we mean the cost in ill health to the individual and especially the effect on the genetic well-being of the affected person (or zygote) and of the family or population that share his genes. Some of the points to be made have already been discussed in Chapter 17.

A major point to remember is that the severity of the genetic defect in terms of viability does not determine the impact on the family. Undoubtedly, many mutations are capable of causing the zygote not to implant in the uterus properly or the embryo not to develop beyond the earliest stages. These unrecognized pregnancies can hardly be called a health burden at all. A pregnancy delayed by one or even a few months is not viewed as a problem. Only if the zygote can develop to a recognized pregnancy resulting in miscarriage is there a perceived problem. In most instances, embryos or fetuses lost by spontaneous abortion are ultimately replaced by successful pregnancies.

If a defective child is born, the situation is quite different. There are much more emotional and financial costs which lead to a reduction in the number of subsequent pregnancies in many families. If the defect typically leads to early death, especially in the newborn period, the child more likely will be replaced with a subsequent pregnancy than if lifelong care is required. This leads us to the interesting conclusion that the impact of genetic defect on the family and population is inversely proportional to the severity of the defect for the individual. This, of course, does not apply if the defect is so mild as to be consistent with "normal" development.

One may calculate the effect of various combinations of degree of genetic defect and child-bearing patterns on the frequency of detrimental genes. We will illustrate this for the two simple cases involving an autosomal recessive lethal defect in which (1) homozygotes are not replaced by additional conceptions and (2) homozygotes are replaced. For a mating *Aa* × *Aa*, where *a* is the recessive lethal allele, the usual three types of zygotes are expected in the ratio 1 *AA* : 2 *Aa* : 1 *aa*. The relative frequency of *a* among alleles at the *A* locus is 0.5, as it is in the parents. If *aa* produces a normal phenotype, the frequency of alleles in the parents and their offspring should be the same. But since *aa* is lethal, the offspring consist of 1 *AA* : 2 *Aa*, with only 3/4 as many offspring. The frequency of *a* among them is 0.33, less than in the parents. The gene frequency in the population will drop until the number of alleles lost through natural selection is equal to the number generated by new mutations. If the *aa* zygotes are replaced by additional conceptions, maintaining the total number of offspring but in this case either *AA* or *Aa*, the frequency of *a* in the offspring, weighted by the increased number of offspring, will be $(4/3)(0.33) = 0.44$. The frequency of *a* will continue to decrease under these circumstances but more slowly than if there is no replacement of *aa* offspring.

The Economic Burden of Detrimental Alleles. There is no way to measure accurately the cost of genetic disease. But it is a substantial portion of the total cost of health care. Many of the diseases that we do not ordinarily think of as genetic, such as cancer and heart disease, appear to have strong genetic predispositions. We cannot say what portion of the costs of these and other major health problems is properly attributed to genetic defects. However, we can assign all the costs of certain simply inherited conditions to the genotype of the affected persons. Mendelian traits such as PKU and chromosomal defects such as trisomy 21 are examples of defects whose costs for care are entirely attributable to the genotype.

It has been estimated that the cost of maintaining children with Down syndrome is over 1 billion dollars per year in the United States. The annual cost of special care for all children afflicted by serious and severe genetic disease may be more than 10 billion dollars in the United States alone.

We may be willing as compassionate persons to assume this economic burden. However, it is only part of the burden. The other part cannot be measured quantitatively. It is the burden born by those who are the parents and others who must sacrifice a good part of their lives to care for these children.

Genetic Defect: Solutions to the Problems

Two general approaches are currently available for coping with genetic defect. The first is therapy; the second is prevention.

Therapy. We do not know how to alter human genes yet. Therefore, all therapy is directed toward changing the phenotype. It can be changed in a number of ways, including manipulation of diet, supplying needed hormones such as insulin or growth hormone, or prescribing glasses for those with inherited eye defects. Means of therapy will be discussed more fully later in the chapter.

Prevention. By prevention we refer here to prevention of the genotype rather than the phenotype. This can be translated into elimination of zygotes with undesirable genotypes though avoidance of matings that would produce them or through termination of pregnancies in which they have occurred. Both methods raise many social and legal questions, some of which will be addressed in Chapter 23. Here we limit our considerations primarily to the biological questions.

The avoidance of offspring from matings that are high risk for a genetic defect can only occur if the potential parents are recognized as being at high risk for producing such a child. Unfortunately, recognition is most likely to occur because a defective child has already been born. The question may also arise because one or both of the parents are related to persons who have specific genetic defects. In these instances, the prospective parents may seek *genetic counseling* to help them evaluate the risk of a defective child. If they are found to be at high risk for a genetic defect, they may choose not to have children or, if the defect can be diagnosed prenatally, they may abort all affected fetuses.

Every mating in the population has a risk of producing children with genetic defects. The risk for any specific defect may be very low. But, in a large population, a few matings will occur that have a high risk of a defective offspring. If the defect is a rare autosomal recessive trait, such as PKU, a few of the heterozygous persons will have affected relatives to serve as a warning. Virtually all $Aa \times Aa$ matings will involve persons who have no reason to suspect their genetic risk until they produce an affected *aa* child. In a few instances, they can be detected by mass *genetic screening* of the population and advised of their risk. Where recognition of high-risk parents is not technically possible, screening of newborns may identify those with defect, and therapy can be instituted to prevent the defects that would otherwise develop. Some of the prospects and problems of genetic screening are explored further in the following sections.

Genetic Counseling and Medical Genetics

The Purpose of Counseling

The primary purpose of counseling is to help people understand the nature and risk of genetic disease. Counseling includes diagnosis, estimation of the occurrence or recurrence of a particular trait or of any genetic defect, and assistance in coping with a particular problem. Many aspects are like any other medical practice. But genetic counseling is different in one important aspect—often the medical problem has not occurred. The "patient" is a perfectly normal person or family with the concern that they may transmit genes for defective development to persons yet to be born or conceived. Genetic counseling is thus much more family oriented than traditional medical practice.

A variety of problems may be brought to the genetic counselor. Many involve simple Mendelian traits. A child has been born with a defect; will it recur in subsequent children? A parent has a defect; will the children inherit it? A sib of one of the parents has a defective child; will it recur in children of these parents? The parents are first cousins; will their children be normal? The parents are members of an ethnic group with a high frequency of a genetic disorder; will their children be affected?

To answer these questions, the counselor must learn as much as possible about the nature of any medical or other problems in the family. This means that accurate diagnoses must be obtained, often by additional studies. An extensive pedigree must be compiled not only of living persons but also of deceased relatives and miscarriages. Biological relationships will sometimes need to be verified by discreet and diplomatic inquiry and by genetic tests. And the conclusions must be transmitted to persons who often are under great stress and who have little appreciation for statements about probability.

In the following sections, we will discuss some of the common problems faced by genetic counselors and the solutions that are available for some.

Estimating the Risk of Inherited Disease

Mendelian Traits. The simplest risk estimates are those associated with Mendelian traits. Once a condition is established as an autosomal dominant trait, and assuming the condition is sufficiently rare so that all matings under consideration are *Dd* × *dd*, the likelihood that any particular offspring from that mating will carry *D* is ½. For rare recessive disorders, the typical mating that has already produced one affected child is *Aa* × *Aa*, and the likelihood that any child will be homozygous recessive *aa* is ¼.

In some instances, analysis of a pedigree will assist with the diagnosis. Muscular dystrophy occurs in several forms, most often as an X-linked recessive but also as an autosomal recessive and an autosomal dominant. Each of these has a quite different pattern of transmission and quite different risks (Fig. 22.1).

Even simple Mendelian traits may present special problems to the genetic counselor. Some traits, of which Huntington disease is an example, are inherited as simple dominant traits. However, the disease does not appear usually until the person is an adult (Fig. 22.2). The person seeking counseling may be an adult with no evidence yet of difficulty but whose parent had Huntington disease, or it might be the child of such a person. Estimates of the risk are more complicated in such circumstances, and the counselor must take into consideration the fact that the person whose risk at birth was 50% now has a smaller risk.

Empiric Risk Estimates. Often it is necessary to counsel people about diseases or conditions that are not transmitted as simple Mendelian traits. These traits could be genetically complex (Chapter 11) or influenced by environmental as well as genetic factors. In such cases, there may be no *theoretical* basis for predicting the risk of a child's being affected, but one can base the estimate on prior experience with similar cases. Such *empiric risks* must take into consideration the presence of other family members with similar defects, their relationship to the child in question, the number of sibs who are not affected, sex, and so forth.

An example of a table of empiric risks for cleft lip (with or without cleft palate) is given in Table 22.2. The occurrence of cleft lip is strongly influenced by heredity, but it is not a Mendelian trait. Boys are more frequently affected than girls, the overall risk in England being 0.13% for boys and 0.07% for girls. If a child is born to a family with no history of cleft lip among close relatives, the average likelihood that the child will be affected is 0.10%, ignoring the differences between boys and girls. However, if the father was affected, the risk rises to 3.0%, and if both parents were affected it is 35.4%. If neither parent was affected but one affected female has already been born, the risk is 3.1%.

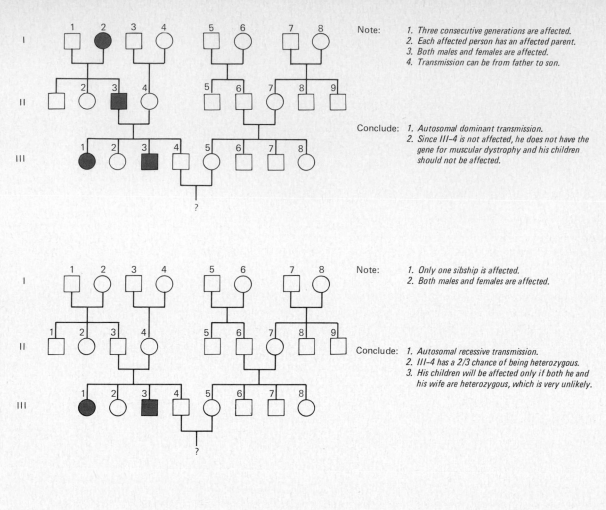

Note:
1. Three consecutive generations are affected.
2. Each affected person has an affected parent.
3. Both males and females are affected.
4. Transmission can be from father to son.

Conclude:
1. Autosomal dominant transmission.
2. Since III-4 is not affected, he does not have the gene for muscular dystrophy and his children should not be affected.

Note:
1. Only one sibship is affected.
2. Both males and females are affected.

Conclude:
1. Autosomal recessive transmission.
2. III-4 has a 2/3 chance of being heterozygous.
3. His children will be affected only if both he and his wife are heterozygous, which is very unlikely.

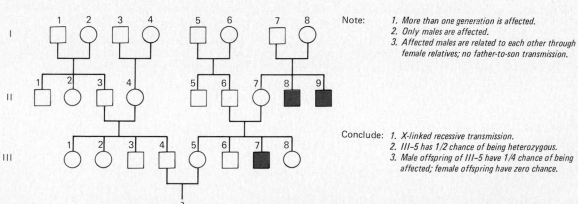

Note:
1. More than one generation is affected.
2. Only males are affected.
3. Affected males are related to each other through female relatives; no father-to-son transmission.

Conclude:
1. X-linked recessive transmission.
2. III-5 has 1/2 chance of being heterozygous.
3. Male offspring of III-5 have 1/4 chance of being affected; female offspring have zero chance.

FIGURE 22.1
Three pedigrees of muscular dystrophy that might be encountered in genetic counseling. In each case, the information obtained by the geneticist leads to different conclusions as to the mode of transmission and the risk of an affected child of the couple III-4 and III-5.

448

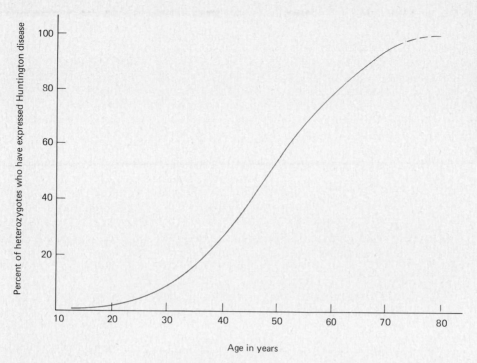

FIGURE 22.2
Graph showing the cumulative probability that a person heterozygous for Huntington disease will have expressed the disorder at each age. (Data from R. G. Newcombe, *Annals of Human Genetics* 45: 375–385, 1981.)

Values are given in percent risk of a child of either sex being affected with cleft lip. The figures are based on studies in England.

TABLE 22.2
Empiric Risks for Cleft Lip (± Cleft Palate)

Index Family[a]	Neither Parent Affected	Father Affected	Mother Affected	Both Parents Affected
No sibs	0.1	3.0	3.5	35.4
1 sib U	0.1	2.8	3.2	32.3
1 M sib A	2.6	10.0	10.8	40.8
1 F sib A	3.1	10.9	11.8	41.6
1 F sib A + 1 sib U	2.9	9.8	10.6	38.9
2 M sibs A	8.1	18.1	18.9	44.4
1 M + 1 F sib A	8.8	19.1	19.9	45.1
2 F sibs A	9.5	20.1	20.9	45.7
2 F sibs A + 1 sib U	8.8	18.2	18.9	43.1

[a]U = unaffected, A = affected, M = male, F = female.

Information taken from C. Bonaiti-Pellié and C. Smith. 1974. Risk tables for genetic counselling in some common congenital malformations. *J. Med. Genet.* 11: 374–377; based on data of C. O. Carter. 1969. Genetics of common disorders. *Brit. Med. Bull.* 25: 52–57.

Prenatal Diagnosis

Many genetic conditions can now be diagnosed prenatally; that is, they can be recognized before birth. In some instances, the diagnosis is done by means of ultrasound. In this technique, very high frequency sound waves are passed through the uterus by means of a sound source placed against the mother's abdomen. The reflected sound waves are detected as in "sonar" used in underwater detection. They can be fed into a computer that reconstructs a three-dimensional image of the abdomen, giving quite good details of the anatomy of the developing fetus (Fig. 22.3). Various congenital malformations, genetic and nongenetic, are readily detected by this technique. Some can also be detected by X rays, but radiation is now known to cause some congenital defects as well as mutations. Its use on embryos and fetuses is to be avoided whenever possible.

A second method of prenatal diagnosis is provided by amniocentesis (Box 22.1). The amnion contains fluid in which the embryo or fetus is suspended, as are cells that detach from the fetal tissues. Both the fluid and the cells can be tested for a variety of genetic defects, some of which are described below. Since amniocentesis is a surgical procedure and is expensive, it is only carried out when there is a high risk of defect.

In most instances, prenatal diagnosis shows the fetus to be normal, at least with respect to the conditions tested. If an abnormality is detected, several options are open.

1. Many minor developmental defects can now be corrected by surgery during the fetal period.
2. The pregnancy is allowed to come to term with treatment started soon after birth.

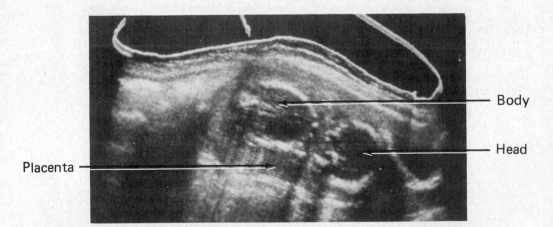

FIGURE 22.3
Ultrasound image of a human fetus. The mother's head is toward the left and her abdominal surface near the upper layer of the image. This technique produces an image that is comparable to a cross section of the fetus, placenta, and maternal organs. (Provided by Dr. R. Santos-Ramos, The University of Texas Health Science Center in Dallas.)

3. The pregnancy can be terminated.

The last is often the choice when no treatment is possible.

The ability to make diagnoses by ultrasound and amniocentesis during the fetal period and, in some instances, to institute surgical or other treatment has led to a new field of medicine, *fetal medicine*. At present the possibilities for treatment are still very limited. As examination procedures become more sophisticated and widespread and as new methods of treatment are developed, fetal medicine will become an important part of medical practice.

Prenatal Diagnosis of Chromosomal Defects. The most frequent use of prenatal diagnosis is to detect chromosomal abnormalities. Cells from amniotic fluid can be cultured for 2–3 weeks, then treated with colchicine and stained as with white blood cells or cells from other tissues (Chapter 4). Any change in chromosome structure or number that is visible in the microscope can be detected.

The procedure is applied to high-risk pregnancies only. These are (1) pregnancies of older mothers, usually defined for this purpose as 36 years of age or older, (2) pregnancies where one of the parents is known to have an abnormal chromosome complement, such as a balanced translocation, and (3) pregnancies in families that have already produced a chromosomally abnormal child. Approximately 1 per 200 live births involves an abnormal complement of chromosomes. However, among the children of mothers over 40 years of age, the risk of trisomy 21 is approximately 1 in 20, and other nondisjunctions are also increased. Even so, most such pregnancies will be karyotypically normal. If one of the parents has a balanced translocation, the frequency of abnormal offspring is approximately 10%, although the risk varies with the specific translocation. Offspring would be at high risk regardless of the age of the parents. Most fetuses will be normal, even though the risk of defect is much higher than in the general population.

Prenatal Diagnosis of Metabolic Defects. The cells isolated from amniotic fluid—sometimes called *amniocytes*—are rather generalized cells. They express many of the genes that are active in a variety of body tissues, although they do not express genes that characterize highly differentiated tissues, genes such as those that code for hemoglobin, normally synthesized only in red cells, or plasma proteins, normally synthesized only in liver. If a gene is active in amniocytes, defects in that gene can often be detected in cultured amniocytes and used to diagnose the presence of the defect in the developing fetus. Typically these involve loss of activity of a specific enzyme, associated in liveborn children with a defect in metabolism (Chapter 9). A list of the more common defects that can be diagnosed prenatally by analysis of enzymes in amniocytes is given in Table 22.3.

A complete list would include more than 100.

Fabry disease	I-cell disease
Galactosemia	Lesch-Nyhan syndrome
Gaucher disease	Maple syrup urine disease
Glucose-6-phosphate dehydrogenase deficiency	Niemann-Pick disease
	Tay-Sachs disease
Glycogen storage diseases	Xeroderma pigmentosum
Homocystinuria	

TABLE 22.3
A Partial List of Inherited Enzyme Defects That Can Be Detected Prenatally by Amniocentesis

Box 22.1

Amniocentesis

The principal methods of prenatal diagnosis involve a process called *amniocentesis* (from *amnion*, the membrane surrounding the fetus, and *centesis*, puncture). This surgical procedure consists of the insertion of a needle into the amniotic cavity surrounding the developing fetus and the removal of some of the amniotic fluid, as illustrated.

Amniocentesis can be done as soon as the position of the fetus and placenta can be ascertained and the amnion is large enough for the surgeon to place the tip of the needle into the cavity. Puncturing the embryo or placenta could have serious consequences. Ultrasound is used to locate the placenta and fetus so that the surgeon can carry out the procedure with minimal risk. With this precaution,

the risk of miscarriage appears to be no greater than in pregnancies without amniocentesis. There is no evidence of damage to the fetus.

Generally amniocentesis is performed in the twelfth to fourteenth week of pregnancy, which corresponds roughly to the beginning of the second trimester and the change in status from an embryo to a fetus.

The amniotic fluid contains fetal cells, called *amniocytes*, and various dissolved products of fetal metabolism. The cells can be cultured and tested for activity of various genes or for the complement of chromosomes, as described in the text. The fluid can often be assayed directly for metabolic abnormalities. The variety of genetic disorders detectable prenat-

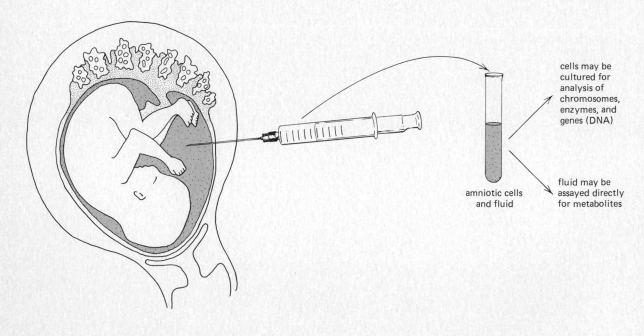

cells may be cultured for analysis of chromosomes, enzymes, and genes (DNA)

amniotic cells and fluid

fluid may be assayed directly for metabolites

Box **22.1**

(*continued*)

ally is very large. Hence, only those disorders for which the fetus is at high risk are tested.

For most abnormalities detected prenatally, no means of therapy are currently available. Therefore the only choices of the parents are to continue the pregnancy or terminate it, should the fetus be found to have such a disorder. Under the circumstances, amniocentesis is not ordinarily performed unless the parents have expressed their wish to abort the fetus if it is abnormal. Fortunately, the majority of fetuses at high risk are found to be normal.

In some instances, the enzyme defect in the fetus causes the chemical composition of the amniotic fluid to be altered. In such cases, culture of amniocytes is not necessary. One can assay for abnormal levels of metabolites directly. This provides a rapid answer rather than the several weeks delay associated with cultured cells.

As with other prenatal diagnoses, only high-risk pregnancies would be tested. In practice this means that the parents have been identified as carriers of the detrimental genes. Too often this information comes from their having already produced an affected child. For any recessive trait, the birth of an affected child means that both parents must be heterozygous, and any subsequent child will have a 25% chance of being affected. Such a high risk clearly warrants prenatal diagnosis. Many parents who already have produced one defective child will not consider letting another pregnancy come to term unless they can be reassured that the fetus is not affected. Such assurance is provided three-fourths of the time.

DNA Analysis and Prenatal Diagnosis. The most promising new tool for prenatal diagnosis is direct analysis of DNA. The technique depends on the use of restriction endonucleases, enzymes that break DNA at very specific sequences (Chapter 14). The principal application to date has been with hemoglobin disorders, primarily thalassemia but also sickle cell anemia. The procedure for thalassemia is illustrated in Figure 22.4. Amniotic cells are cultured to increase the amount of DNA available. The DNA is then extracted and treated with a restriction endonuclease, which cuts the DNA into thousands of fragments. These fragments are spread out in an electrophoresis system that separates according to molecular size, but because of the enormous number of fragments, it is impossible to see individual fragments without a "probe." A probe is a small piece of DNA, usually prepared by copying a specific messenger RNA with reverse transcriptase (page 261). This *cDNA* (complementary DNA) is prepared from highly radioactive nucleotides. When added to DNA fragments that have been separated by electrophoresis, the cDNA forms a complex with those fragments that have nucleotide sequences complementary to it. These fragments have the genes corresponding to

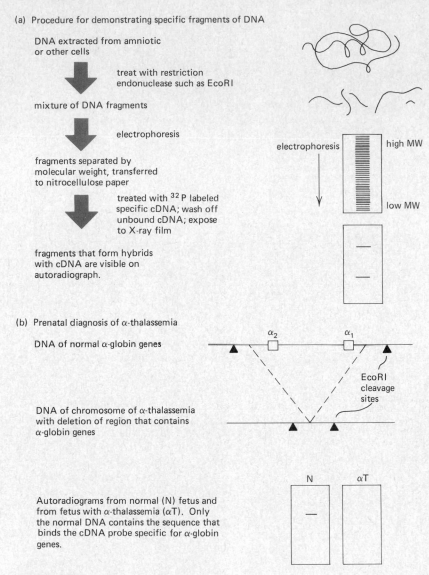

(a) Procedure for demonstrating specific fragments of DNA

DNA extracted from amniotic or other cells

    treat with restriction endonuclease such as EcoRI

mixture of DNA fragments

    electrophoresis

fragments separated by molecular weight, transferred to nitrocellulose paper

    treated with ^{32}P labeled specific cDNA; wash off unbound cDNA; expose to X-ray film

fragments that form hybrids with cDNA are visible on autoradiograph.

electrophoresis   high MW   low MW

(b) Prenatal diagnosis of α-thalassemia

DNA of normal α-globin genes

$α_2$ $α_1$ EcoRI cleavage sites

DNA of chromosome of α-thalassemia with deletion of region that contains α-globin genes

N   αT

Autoradiograms from normal (N) fetus and from fetus with α-thalassemia (αT). Only the normal DNA contains the sequence that binds the cDNA probe specific for α-globin genes.

FIGURE 22.4

Prenatal diagnosis of α-thalassemia based on analysis of DNA from amniotic cells. α-Thalassemia is caused by deficiency of α-globin chain synthesis. In its most severe form, both α-globin loci have been deleted, and the fetus homozygous for this mutation has no α-globin genes from which to manufacture fetal or adult hemoglobin. The condition known as hydrops fetalis results, involving generalized edema of the fetus and placenta with spontaneous abortion or death soon after birth.

(a) DNA can be extracted from the amniotic cells from amniocentesis and treated with a restriction enzyme selected to cleave the DNA at suitable sites near the gene. The DNA fragments are then separated by electrophoresis and transferred by "blotting" to paper made of nitrocellulose. A radioactive cDNA "probe" is added that will hy-

the mRNA used to prepare cDNA. One can identify the fragments by the fact that they bind the highly radioactive cDNA (Fig. 22.4).

Certain thalassemia mutations involve the deletion of a globin gene. If the α-globin genes are lost, and if the person is homozygous for this deletion, there is no way to make hemoglobin and severe anemia results. Fetuses with this condition are usually aborted spontaneously. Figure 22.4 shows how the presence or absence of the α-globin gene is detected in this form of thalassemia. The important point to remember is that the gene can be detected *whether or not it is active in the cells studied.* Therefore one may use white blood cells, cultured skin cells, or, in the case of prenatal diagnosis, cultured amniocytes or chorionic villi as the source of DNA.

This method has also been extended to prenatal detection of the gene for sickle cell hemoglobin. In adults, the presence of the gene is readily and inexpensively detected by electrophoresis of hemoglobin from red blood cells. However, the early fetus makes fetal hemoglobin primarily (Chapter 13), and substantial quantities of adult hemoglobin are not made until near the time of birth. In addition, taking a blood sample from the early fetus (actually from the placenta) is difficult and carries a substantial risk of damage to the fetus. But the Hb S mutation can be detected in amniocytes as shown in Figure 22.5.

The use of DNA analysis for prenatal diagnosis or other genetic counseling is still very limited. Few laboratories are trained in this rather sophisticated technique. But the great versatility of the procedure and the ability to analyze genes directly rather than the phenotype make it likely that the procedure will become much more widely available. The only limitation is the availability of probes, but with time these will become much more numerous.

Linkage and Prenatal Diagnosis. Often a particular mutant gene cannot be recognized as an enzyme defect, either because the gene is not expressed in the tissues available or because the primary gene defect has not been identified. An improvement in risk estimates can sometimes be achieved if the mutant gene is closely linked on the same chromosome to a polymorphic locus. As yet there are few examples where close linkage is used in prenatal diagnosis, but as the chromosomal locations of more and more gene loci are established, linkage analysis should become a very useful technique. An example of its use is given in Figure 22.6.

In order for linkage to be useful, the closely linked markers must be segregating in the family under study. This can be a major limitation at present. Any marker is useful, however, and the many variations in restriction endonuclease sites that are being discovered may make linkage studies more often feasible.

---

bridize only with DNA fragments whose sequence is complementary to the cDNA. Such fragments are detected by exposure of the radioactive hybrid bands to photographic film.

(b)The α-globin genes are in a single DNA fragment if the restriction endonuclease Eco RI is used. DNA from normal persons shows a band that contains these genes. DNA from a fetus homozygous for this α-thalassemia deletion shows no such bands. This technique will work as described only for deletions of DNA, but not all forms of α-thalassemia are due to deletions.

FIGURE 22.5
Diagnosis of sickle cell anemia by restriction enzymes. The enzyme DdeI cleaves DNA at the sequence -CTNAG-, where N is any nucleotide. In the β chain gene of normal Hb A, this sequence occurs in the fifth and sixth codons (underlined). The mutation of β^A to β^S changes the sixth codon so that this site is no longer cleaved by DdeI. The DNA fragments are separated by electrophoresis and detected with radioactive cDNA probes as in Figure 22.4. The three genotypes are clearly distinguishable. This procedure can be done on any source of DNA, including amniotic cells.

(a) Cleavage sites of β^A and β^S globin genes (5' end only):

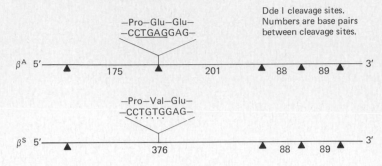

(b) Restriction endonuclease patterns of the 5' end of β-globin DNA from homozygous normal (AA), heterozygous sickle cell (AS), and homozygous sickle cell (SS) persons.

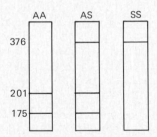

Genetic Screening

As the term has come to be used, *genetic screening* means mass testing of persons for a condition for which they are not known to be at high risk individually. Thus it is population oriented rather than family oriented. The population as a whole may well be at high risk for the condition being tested. For example, the United States black population is at high risk for sickle cell anemia, the Ashkenazi Jews are at high risk for Tay–Sachs disease, and the populations of western European descent have a high frequency of cystic fibrosis. Testing for these conditions outside the high-risk population would yield very few affected persons.

Screening of Newborns for Inherited Disease

An inherited disease need not be common to justify screening. The first disease to be routinely screened was phenylketonuria (PKU). Most states now require

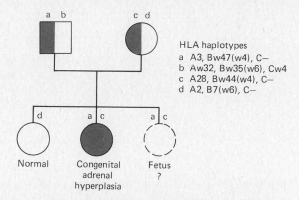

HLA haplotypes
a A3, Bw47(w4), C—
b Aw32, Bw35(w6), Cw4
c A28, Bw44(w4), C—
d A2, B7(w6), C—

Normal

Congenital
adrenal
hyperplasia

Fetus
?

FIGURE 22.6

Prenatal diagnosis of congenital adrenal hyperplasia (21-hydroxylasedeficiency) based on its close linkage to the major histocompatibility complex (MHC) on chromosome 6. This autosomal recessive disorder is located within the MHC region. In the pedigree shown, one child had already been born with the disorder, indicating that both parents must be heterozygous. Both parents also are heterozygous for HLA haplotypes, symbolized in the pedigree by *a, b, c,* and *d.* The affected child received the deficiency allele coupled to *a* from the father and coupled to *c* from the mother. Amniotic cells from the fetus were found also to express the HLA types represented by *a* and *c,* which indicated that it should also be homozygous for congenital adrenal hyperplasia. The prediction was confirmed after birth of the child (From M. S. Pollack et al., *Lancet i:* 1107, 1979.)

screening of newborns for PKU so that effective therapy can be instituted as soon as possible when it is most critical in avoiding mental retardation. Only about 1 infant in 15,000 will prove to have PKU. But the cost of 15,000 tests and treatment for one patient is small compared to the cost of a lifetime of institutionalization for that person, not to mention the emotional and the humane costs in allowing a person to be mentally retarded when it is avoidable.

A list of inherited diseases commonly screened in newborns is given in Table 22.4. The most common, hypothyroidism, can be treated with thyroid hormones to produce a completely normal person. Without early detection and treatment, the condition known as cretinism develops, involving mental retardation as well as other physiological changes. With conditions such as hypothyroidism and PKU, there is very little argument that the cost of screening is fully justified. With diseases that are extremely rare, say fewer than one affected per million persons, the decision is less clear-cut. If the real cost of testing is one dollar per test (which is rather a low cost), over 1 million dollars would be spent to detect each affected child. One must balance this cost against the other uses of a million dollars, considering also the humane and ethical issues. These decisions cannot be made solely on the basis of genetic considerations; rather, they must be made by the general public, balancing the various societal values against each other.

TABLE 22.4
A Partial List of
Inherited Diseases That
Can Be Screened in
Newborns and for
Which Therapy is
Available

Disease	Frequency in U.S. Population
Hypothyroidism	1:3,800
Iminoglycinuria	1:12,000
Phenylketonuria	1:15,000
Vitamin D resistant rickets	1:18,000
Galactosemia	1:65,000
Branched-chain aminoaciduria	1:400,000
Hereditary fructosuria	1:600,000

Screening of Normal Persons for Carrier Status

Another application of population screening is to detect persons who carry recessive genes, either autosomal or X-linked, in heterozygous combination. If one could identify matings in which both partners are heterozygous for the same detrimental autosomal recessive gene, they could be advised of the 25% risk of an affected offspring. At present such counseling is virtually limited to persons whose carrier status is revealed by having produced an affected child.

An excellent example of the possibilities and problems of screening is provided by sickle cell anemia. Recognition of heterozygotes is very simple and inexpensive. A small blood sample subjected to electrophoresis reveals whether a person is heterozygous (or homozygous) for any of a number of variant hemoglobins. In U.S. blacks, the frequency of Hb S heterozygotes is 8%. Thus the cost per heterozygote identified is trivial. Marriages between heterozygotes would be expected to occur $(0.08)^2 = 0.0064$ or once in every 156 marriages. If either or both parents are homozygous for normal Hb A, the likelihood of a child with sickle cell anemia is zero.

Many screening programs for sickle cell heterozygotes have been carried out, producing some results that were not anticipated. Considered purely from the genetic aspects, the programs have been successful. From other aspects, they have been less so. Perhaps most important has been the failure on the part of many to discriminate between heterozygous carriers, who are normal, and homozygotes with sickle cell anemia. As a result, heterozygotes have been discriminated against by employers and insurance companies. Most of these occurrences are in the past, but they illustrate the great importance of providing information and education along with the classification.

A more successful program of heterozygote detection is illustrated by Tay-Sachs disease, a degenerative disease of the central nervous system that invariably leads to death at 2–4 years of age. This autosomal recessive trait is found predominantly among Ashkenazi Jews, the group who migrated primarily into eastern Europe after the dispersal of the Jews from Palestine. Most American Jews are Ashkenazic in origin. The frequency of the Tay-Sachs gene in this population is approximately 0.017. Homozygotes occur in 1 per 3600 births. Heterozygotes can be identified by a fairly simple blood test for hexosaminidase A, the enzyme that is defective in Tay-Sachs disease. The tragedy of the disease is such that persons potentially at risk to become parents of a child with Tay-Sachs disease have welcomed the opportunity to avoid such risk. Tay-Sachs disease is one that can be diagnosed prenatally also, so that affected pregnancies can be terminated.

It has been estimated that each of us on the average carries four to eight recessive genes that are severely detrimental or lethal when homozygous. Screening for heterozygosis for detrimental genes could not be used effectively to prevent such persons from reproducing. Few would be left to carry on the species. Screening can help potential parents in family planning and can alert them to specific risks, some of which can be tested by prenatal diagnosis.

Genetic Therapy: Some Prospects

The popular press has almost daily reports on genetic engineering. Articles covering progress in plant and bacterial projects designed to make more commercially useful strains for producing the many different kinds of products described in Chapter 14 appear most often along with reports on the financial status of the companies involved. The most spectacular reports, however, are those that describe the possibilities of changing the genotypes, at least in part, of humans with the sorts of inherited metabolic disorders that we have been discussing in this and most of the preceeding chapters. Less spectacular, but equally provocative, are reports that it may be possible to introduce active enzymes to carry out metabolic reactions in a patient who has an inherited block in one of them because the normal enzyme is deficient. Actually something akin to this has been done with diabetics for at least 50 years by supplying them with the protein insulin extracted from pancreases of sheep and other animals.

Enzyme Therapy

The introduction of an active enzyme into patients deficient in that enzyme is theoretically quite feasible at present, especially since the advent of recombinant DNA techniques. Previously it had been thought that purified enzymes from plant or animal sources would induce the formation of antibodies to the foreign proteins when they were injected into humans. This would not only cause an immune reaction that could inactivate the enzymes, it also could endanger the patient's life.

Now, however, it is possible to isolate human structural genes for human enzymes and introduce these genes into plasmids, as described in Chapter 14. The plasmids, after infection of a host bacterium, will multiply along with the host, and, provided conditions are right, the enzyme can be synthesized and isolated in relatively large quantities. It can then be supplied to those who need it at a reasonable price. For example, Lesch-Nyhan babies are deficient in the enzyme hypoxanthine phosphoribosyl transferase (HPRT). Could these babies be saved by providing the enzyme encapsulated in some form in the body to be released slowly and prevent the buildup of metabolites that cause such a problem?

The use of enzymes that are human in origin might well overcome the problem of immune response to the "foreign" protein. Actually most mutations of structural genes that result in deficient enzyme activities do result in the formation of proteins that are closely related to the active enzymes. In fact, they generally will have only one amino acid difference from the active enzyme formed by the "wild type" allele. Therefore their presence may well be recognized as "self" and the introduction of a closely related active form of the protein or enzyme will not elicit an immune response. But it may also be true that the immune systems of many persons who produce none of a particular enzyme may regard even normal human enzyme as foreign. Only experience will tell us how useful such enzyme therapy will be.

This is all speculative, of course, but it should be evident that enzyme therapy is a possibility for the future. Exploratory experiments with experimental animals are even now being done along these lines in several research laboratories. Some enzymes may work properly only if they are located at specific sites within the cells of a particular tissue. This requirement may limit some applications of enzyme therapy, but other enzymes appear to work effectively when injected into the bloodstream. This general approach is not far removed from use of an external source of insulin to treat diabetics.

Genetic Engineering: Changing the Genotype

This procedure might be given the name *gene therapy* as distinct from the enzyme therapy we have discussed. It consists of replacing a defective gene with a functional "good" gene in the cells of an individual. Technically, it is a transformation or transduction rather than a mutation.

One form of genotypic transformation is now being practiced with cells in culture. Animal cells can be transformed by purified DNA. The cells will take up the DNA from the surrounding medium, or the DNA can be injected directly into the nucleus of a cell. In either case, the DNA can become incorporated into the recipient's chromosome and be expressed phenotypically.

An even more spectacular result has been obtained with adult rabbit β-globin DNA. The rabbit β-globin gene was incorporated first into a λ (lambda) phage chromosome. The phage in this case was used like a plasmid. The phage DNA containing the rabbit gene was then injected into fertilized mouse eggs, which were returned to the oviducts of female mice. Some of the offspring from the injected mice showed the presence of rabbit β-globin. When these were mated to other mice, the rabbit β-globin turned up in *their* offspring, which showed that the rabbit β-globin genes got into the germ cells of the mice. Strains of mice that produce rabbit β-globin are now on hand.

Just how these procedures can be used to change the genotypes of humans in a controlled way is not immeditely evident. However, it is evident that tools are coming into our hands to make it possible to do amazing things with the animal genotype, including that of humans. These sorts of experiments will raise very serious ethical questions with which we will have to grapple.

Review Questions

1. A man and his wife are heterozygous for the same recessive lethal gene. When homozygous, this lethal allele causes death of the fetus in the first trimester. This couple now has six living, healthy children. About how many conceptions were probably aborted due to homozygosity for the recessive gene?
2. What is meant by the economic burden of detrimental alleles?
3. What steps can be taken to cope with the problems of genetic defects?
4. What purpose does genetic counseling serve?
5. A woman whose pigmentation is normal for her race marries an albino man not known to be related to her but of the same racial group. The frequency of the albino gene in the population from which they came is about 0.01. What is the probability that their child will be an albino?
6. In the white population, the frequency of babies born with galactosemia is 1 in 65,000 births. A couple planning to have a child finds that the wife has a case of galactosemia in her family background. If you were a genetic counselor, how would you advise the couple on the risk to their child?
7. Prenatal diagnosis is rather an expensive medical procedure. When is it not indicated?
8. What distinction can be made between prenatal screening and genetic screening?
9. What kinds of genetic therapy are now in use? What kinds may be in the offing?
10. What dangers may exist in the extensive use of genetic therapy?
11. What is amniocentesis?

References and Further Reading

Committee for the Study of Inborn Errors of Metabolism. 1975. *Genetic Screening: Programs, Principles and Research.* National Academy of Science, Washington, D.C. A discussion of the scientific and political issues raised by screening.

Kelly, P. T. 1977. *Dealing with Dilemma.* Springer-Verlag, New York. 143 pp. A guide to the counselor, focused on social and psychological issues rather than genetic.

Porter, I. H., and R. G. Skalko (eds.). 1972. *Heredity and Society.* Academic Press, New York. 324 pp. A series of papers dealing with issues in genetic counseling and the impact of counseling strategies on human populations.

The following are primarily technical resources to help genetic counselors.

Fraser, F. C. 1974. Genetic counseling. *Amer. J. Human Genetics* 26:636–659.

Fuhrmann, W., and F. Vogel, 1976. *Genetic Counseling,* 2nd ed. Springer-Verlag, New York/Heidelberg/Berlin. 138 pp.

Lubs, H. A. and F. De La Cruz (eds.). 1977. *Genetic Counseling.* Raven Press, New York. 598 pp.

Stevenson, A. C., and B. C. C. Davison. 1976. *Genetic Counseling,* 2nd ed. Lippincott, Philadelphia. 357 pp.

A major social consequence of Darwin's theory of natural selection was the beginnings of the eugenics movement in the latter part of the nineteenth century. Eugenics may be defined as the use of public policy to guide the genetic future of the human population. Momentum was added in the first part of this century when it was realized that the principles of Mendelism could be applied to humans as well as to animals and plants of economic importance.

Eugenics

Eugenic principles have been advocated and applied in the human population for thousands of years. Plato in his *Republic* strongly urged selective breeding of the "best" to provide rulers for his state. He advised that offspring of the "inferior," or of the "better" when they chanced to be deformed, "be put away in some mysterious, unknown place, as they should be." He thus advocated both selective breeding of people and elimination of those considered inferior, presumably by infanticide. The Spartans of Plato's time were also much concerned with the breeding of their people. They wanted to produce superior warriors. Therefore, their program was aimed at selecting for offspring with the physical attributes to make good soldiers. They commonly practiced infanticide of "weaklings."

Royal families of most countries have limited marriages (although certainly not procreation) to individuals within the royal or aristocratic line. This was not just for political reasons alone. Implicit was the belief that they were the best by inheritance or "blood." The incestuous practices of the Ptolemaic line of Pharoahs of Egypt are a good example of this. They were descended from Ptolemy, a Macedonian Greek who was one of Alexander the Great's most trusted generals. He founded a dynasty that ruled Egypt from 323 to 30 B.C.E. The last of the line as a ruler was the famous Cleopatra. It was common practice among these Hellenic Egyptians for the Pharoah to marry his sister or other close collateral relative (Fig. 23.1). It is not recorded how many aberrant offspring resulted from this extreme pattern of inbreeding, but at least Cleopatra had enough physical endowments and intellectual ability to keep Julius Caesar and Mark Antony on the string. She outbred with them and had a son by Julius and three children by Mark before taking her own life and ending the Ptolemaic line.

The importance of "good birth" or being "well born" was debated in the early days of the Republic of the United States of America. At the time of the founding of the Republic, the framers of the Constitution and other founding fathers were divided over how much trust could be put in the wisdom and judgment of the masses of the people as opposed to that of the cognescenti or the intellectual aristocracy. The Federalists, as represented by John Adams, the second president under the new Constitution, believed for the most part in an aristocracy derived from "good families." On the other hand, the Republicans, as represented by Thomas Jefferson, the third president, tended to put great faith in the "common people" and the "aristocrats" among them. The masses, Jefferson said, "will elect the wise and the good."

23

Genetics, Eugenics, Public Policy, and Law

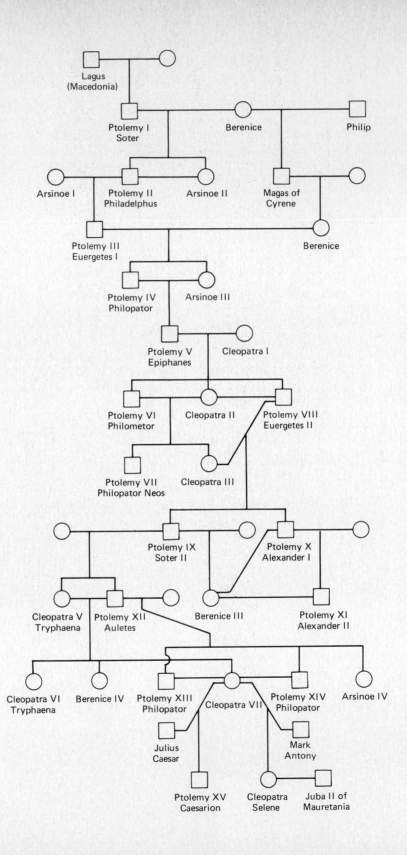

FIGURE 23.1
A pedigree of the Ptole-
maic Pharoahs of Egypt.
In spite of the close in-
breeding, many of the
offspring appear to have
been normal.

In 1813, Jefferson and Adams had extensive correspondence in which this dichotomy of views occupied a central position. In one of the most significant of Jefferson's letters to Adams at the end of this period in 1814, he wrote

I agree with you that there is a natural aristocracy among men. The grounds of this are virtue and talents. . . . There is also an artificial aristocracy founded on wealth and birth, without either virtue or talents; . . . The natural aristocracy I consider as the most precious gift of nature for the instruction, the trusts and government of society. . . . May we not even say that that government is best which provides most effectually for a pure selection of these natural aristocrats into the offices of government? The artificial aristocracy is a mischievous ingredient in government, and provisions should be made to prevent its ascendancy.

Thus we see that a fundamental premise in the political philosophy underlying the American system of government was a genetic one. This debate between Adams and Jefferson was never resolved completely. Although it can generally be said that birth is not of utmost importance in the United States, it can still be important if you are not a descendant of white Europeans.

The Eugenics Movement in England

The first modern major attempts to codify and organize the thinking in this matter in order to apply it to society in general were made by a cousin of Charles Darwin, Francis Galton, who was mentioned in Chapter 11 as the founder of modern statistics. Galton coined the term *eugenics* in 1883 in his highly influential book, *Inquiries into Human Faculty and its Development.* His analyses in this and an earlier treatise, *Hereditary Genius, an Inquiry into its Laws and Consequences,* led him to the conclusion that systematic efforts should be made to improve the breed of humans by limiting the reproduction of the less fit and encouraging reproduction of the more fit. He, of course, drew on the theories of his cousin, Darwin, to support his thesis as well as on his own data gathered to show that ability, as defined by him, runs in families.

Galton's ideas received much support from a contemporary, Herbert Spencer (1820–1903), who applied Darwin's phrase "survival of the fittest" to human society. This was a misuse of Darwin's theory, which defined fitness in no uncertain terms as reproductive ability (Chapter 17). Other kinds of what might be called fitness, such as intellectual ability, physical strength or agility, and the like, are not fitness in the Darwinian sense. Spencer's and Galton's ideas led to the movement called *Darwinian Socialism.* This movement became quite popular, especially among the upper classes of English society, which was highly class conscious. These ideas lost most of their popularity in England in the early 1900s, but they retained an influential following among a group of intellectuals who were the forerunners of the modern facism that flowered in Germany under the name of Nazism. The Aryan "master race" conjured up by Hitler and his followers was to be promoted and maintained by the practice of eugenics. The destruction of "inferior" peoples such as the Jews and Slavs by the Nazi's definition was part of this grand plan of eugenical practice to better humanity.

Eugenics in the United States

Soon after the rediscovery of Mendelism in 1900, eugenics became a popular subject in the United States, and various states began to pass laws designed to prevent the breeding of the "unfit" by sterilizing them. One of the results of this has been an abrogation of the civil rights of thousands of individual Americans and a revulsion against human genetics on the part of many Americans.

Compulsory sterilization laws began to be passed by state legislatures early in this century. Indiana was the first to enact a sterilization law in 1906, to be followed by fourteen other states in the next five years. The general feeling on the part of the legislators in these states was perhaps best expressed by Madison Grant in his book, *The Passing of the Great Race*, published in 1916. Referring to defective individuals, he wrote

The individual himself can be nourished, educated, and protected by the community during his life time, but the state through sterilization must see to it that his line stops with him, or else future generations will be cursed with an ever increasing load of victims of misguided sentimentalism. This is a practical, merciful, and inevitable solution of the whole problem, and can be applied to an ever widening circle of social discards, beginning always with the criminal, the diseased, and the insane, and extending gradually to types which may be called weaklings rather than defectives and perhaps ultimately worthless race types.

The sterilization laws were repeatedly challenged in the courts on constitutional grounds. Some were struck down, but not all. The high courts in Michigan and Virginia upheld the laws concerning sterilization in their states. The Virginia decision was appealed to the United States Supreme Court, which upheld the state court's decision. The opinion was written by none other than Justice Oliver Wendell Holmes, Jr. The essence of his opinion was that the state had the right to prevent its citizenry from being "swamped with incompetence" just as it had the right to require all to be vaccinated against smallpox.

This decision stimulated further state legislation in favor of sterilization, and thousands of institutionalized persons were sterilized annually. Revulsion against these practices mounted constantly during the 1930s, but it was not until the 1950s that the number of sterilizations began to decline. Meanwhile, many additional cases were taken to the courts, which, in general, responded positively to the pleas for protecting civil rights but did not strike down the rights of the states to sterilize under certain conditions.

Sterilization still continues in the United States, but generally only with permission of the courts. The effectiveness of the sterilization programs in preserving the "genetic purity" of the population is dubious at best. It must now be admitted that a good part of the support for the eugenic programs early in this century was from "native" Americans who believed that many of the immigrants entering the country at the time were genetically inferior to those already here. This reaction is not unusual, even though it may be completely unsound and unjustified scientifically.

A good part of the concern of those who supported sterilization practices was based on the fear that mentally retarded persons would overwhelm the general population with their retarded offspring. Actually, we now know that only a small percentage of the mentally retarded inherit this condition from mentally retarded parents. Furthermore, as was discussed in Chapter 17, retarded persons have low Darwinian fitness. Hence, their contribution of genes to future generations is extremely low. There is no reason to believe that the average intelligence of the general population will decrease because present retardates are not sterilized.

Positive and Negative Eugenics

The American geneticist H. J. Muller (1890–1967) was also greatly concerned about the genetic future of the human race. He never recommended sterilization programs, such as recommended by Madison Grant in the quotation given earlier, but he did recommend that sperm banks be created to store the sperm of highly intelligent and healthy men, just as the sperm of prize bulls is stored. The sperm were to be used to inseminate women whose husbands were considered inferior to those who were invited to store their sperm. Muller's suggestion received considerable support from those men who considered themselves superior, and such a bank has been formed in California according to the press reports.

Muller's eugenic plan is what we call *positive eugenics*. Its aim is to further the increase in frequency of genes that produce desirable phenotypes. It differs from *negative eugenics*, which is essentially designed to be preventive and lower the frequency of undesirable genes that produce undesirable phenotypes such as mental retardation and the like. The sterilization programs we have discussed are examples of negative eugenics.

Plant and animal breeders use both negative and positive procedures. The breeder betters his stock by culling out the undesirable and retaining the desirable

IT'S BEEN LIKE THIS SINCE THAT NOBEL PRIZE SPERM BANK OPENED...

(Reproduced with permission of Mike Peters, *Dayton Daily News*, Dayton, Ohio.)

to produce the future generations. This method has obviously been highly successful in agriculture, but it can be fraught with great dangers when applied to the human species. The problem centers around determining what is desirable in people. One has a strong feeling that H. J. Muller would consider his own sperm to be prime candidates for inclusion in a sperm bank. Perhaps, but this kind of hubris about one's self is relatively harmless and palls in comparison to the decision of some tyrannical politician like Hitler to commit genocide because he believes he *knows* what is desirable in a human being.

Negative eugenics attempts to prevent the increase in the numbers of persons with almost universally acknowledged undesirable phenotypes. These would include severe mental retardation of the type requiring institutional care, and such conditions as Lesch–Nyhan syndrome and Tay–Sachs disease. Genetic counseling and prenatal diagnosis are both in part applications of negative eugenics, since they can lead to the reduction in the number of defective offspring by nonreproduction on the part of heterozygous parents or by the abortion of an obviously defective fetus that might not abort naturally.

What may seem desirable to some in the application of negative eugenics may not to others. Antiabortionists in general are not in sympathy with prenatal diagnosis, and thorny legal problems are associated with both counseling and screening. Among these are the right to bear children as well as the civil rights of those born to live a "wrongful life."

Genes and Jurisprudence

The laws passed by the various states in the first part of this century pertaining to the rights of citizens to reproduce or not to reproduce through sterilization or contraception have presented the courts with some of their most difficult decisions. The subject is an extremely emotional one for all, including judges, and involves religious and ethical principles as well as rather technical biological issues not readily appreciated or understood by all laypeople.

A landmark decision dealing with the Habitual Criminal Sterilization Act passed by the Oklahoma Legislature was handed down by the United States Supreme Court in 1942. In the Court's decision (Skinner v. Oklahoma, 316 U.S. 535), written by Justice William O. Douglas, the Court declared that the right to have offspring is basic to the perpetuation of the race and questioned whether the state had the right to interfere with this individual right. More than 20 years later, this decision came to bear on other types of cases dealing with state laws banning the distribution of birth control information as well as contraceptive devices and their use. In two cases, Griswold v. Connecticut (381 U.S. 479, 1965) and Eisenstadt v. Baird (405 U.S. 438, 1972), the Court handed down decisions that essentially guaranteed individuals, either married or single, "to be free from unwarranted government intrusion into matters so fundamentally affecting a person as the decision whether to bear or begat a child." These decisions in turn led to the 1973 majority opinion of the Court in Roe v. Wade (410 U.S. 13) affirming the right to terminate pregnancy by induced abortion during the first trimester with no interference by the states. It is this decision, an outgrowth of decisions reaching back to the 1942 Skinner v. Oklahoma case, which is currently raising considerable controversy, especially from those who oppose abortion for any reason. We see then that all of the judicial opinions involved, including abortion, center around and are intertwined with eugenics.

Wrongful Life

Wrongful life is a concept formulated by the courts as an aspect of tort law. One principle of tort law is that a person who is injured by the negligence of another is legally entitled to recompensation. This principle has been used by children to bring action as plaintiffs against whoever is accused of being responsible for their condition. A child in New York was conceived after her mother, an inmate of a mental institution, was raped. The child sued the State of New York because it was negligent in its supervision of her mother. Another sued a physician because while *in utero* the mother contracted rubella and the physician assured her the virus would have no effect on her child. The child was born with birth defects nonetheless and demanded recompense from the physician. The Court decided in 1967 in the physician's favor. Several other similar cases followed with similar verdicts, but it is apparent that there is a growing tendency for the courts to become more lenient and sympathetic with the child plaintiffs and less with the physicians who gave bad advice to their parents or, in some cases, none at all.

Parents have also sued physicians and in some cases hospitals because they claimed that they were not sufficiently warned about the possibility of their children being defective. Professional genetic counselors also are subject to these same kinds of tort actions. It is important that physicians and counselors be completely frank with their patients. If a parent is forewarned about the possibility of a child being defective, there is no ground for legal action even if the child is defective.

Genetic Evidence and the Law

The courts in general are conservative, sometimes in the extreme, about the kind of evidence they will admit in civil and criminal cases. They have been especially conservative about evidence based on human genetics.

Disputed Paternity. Karl Landsteiner discovered the ABO system of blood groups and their pattern of inheritance in 1900, but it was not until 1936 that the first court admitted it as evidence in paternity suits in the United States. But even so, in 1946 in a paternity suit brought against Charlie Chaplin, the jury decided against Chaplin in spite of the fact that the ABO test excluded him as the father. The test cannot always determine paternity or maternity, but it can exclude *some* individuals as possible parents, as shown in Table 23.1. However, in the last 10–20 years, so many new blood types have been discovered that it is now possible nearly always to exclude a wrongfully accused defendant and approach certain identification of the man who in fact is the father.

Blood types have been used extensively to establish that a person born and living

**TABLE 23.1
Chart Showing Exclusion of Paternity for the ABO Blood Groups**

For each mother-child combination, the corresponding entries show which paternal blood groups are excluded (if any) and the frequency of the excluded groups in the U.S. white population. Thus the father of a type A child whose mother is type B must have contributed an A allele to the child. He would be either type A or AB. Men of type B or O, who comprise 55% of the population, would be excluded as the father.

The likelihood of a falsely accused man being excluded by genetic systems is a function of the gene combinations in the mother and the child and of the frequencies of the alleles in the population. Different loci vary in the efficiency with which they are useful in exclusions. The HLA locus, with its very large number of haplotypes, is highly efficient and will exclude well over 90% of falsely accused men. Indeed, if an accused man is compatible with the child at the HLA locus, some courts now consider that strong evidence that he is indeed the father, since so few men would be compatible.

		Blood Group of Mother				Frequencies of Blood Groups in U.S. Whites
		A	B	AB	O	
Blood Group of Child	A	N	B, O 55%	N	B, O 55%	A 41.3%
	B	A, O 87%	N	N	A, O 87%	B 9.9%
	AB	A, O 87%	B, O 55%	O 45%	Mat Exc	AB 3.5%
	O	AB 3.5%	AB 3.5%	Mat Exc	AB 3.5%	O 45.3%

N = none excluded; Mat exc = maternal exclusion.

in a foreign country has rights to citizenship in the United States because that person's father was a citizen at the time of birth. After the Chinese Revolution of 1948, thousands of Chinese applied for entrance on these grounds, and many were admitted. The reasons for this situation arising were related to the events and laws of the late nineteenth century. After the Civil War, railroads were being built in the West and gold was being intensively mined, especially in California, Nevada, and Montana. These activities required a great deal of cheap labor, which was to a large extent supplied by poor Chinese immigrants, 250,000 of whom entered the United States prior to 1882. In that year the Chinese Exclusion Acts were passed by Congress. These Acts severely restricted the entrance of Orientals to the United States. The Chinese men here could not find wives. They were prohibited by the miscegenation laws from marrying white women. They therefore returned to China to marry and sire children. Generally they also returned to the United States to make enough money to support their famiies in China.

After the repeal of the Exclusion Laws, it still was difficult to enter the United States as a permanent resident, whatever one's country of birth or race. But if one could prove that one's father was a citizen, entry became easy. Hence the rush to avoid the Communist takeover in China. By this time, in the 1950s, the courts were accepting blood types as evidence, and a great many of these children were able to provide evidence of having an American citizen as a father.

Blood Types and Criminal Law. When Sherlock Holmes first meets his future roommate, aide, and chronicler, *the* Dr. Watson, he is engaged in demonstrating that human blood can be identified chemically. This finding, he says, will make great changes in the ability of the authorities to apprehend and convict perpetrators of crimes that involve the spilling of blood. Actually that did not occur. At the beginning of this century, the significance of fingerprints was discovered and soon gained dominance in the identification of persons to the exclusion of most other methods.

More recently, however, the analysis of blood stains at the scene of a crime has been developed to the point that such stains, even if quite old, can be typed. Sometimes this can give important clues about the identity of the criminal. But it has definite limitations, just as in the case of determination of paternity. If the blood type is O in the ABO system, the information is of limited value. About half the population is type O. On the other hand, if a series of different loci can be determined, the chances of linking a suspect to the site of a crime and of eliminating innocent persons are considerably enhanced.

Although not yet of great importance in criminal investigations, forensic genetics may become immensely important in the future as techniques are developed to identify more and more immunological and other molecular differences that we all have. A criminal can wear gloves and avoid leaving fingerprints, but seldom can one avoid leaving other signs, such as hair, skin flakes, spittle, or sometimes blood, semen, or urine.

The Inheritance of Criminal Tendencies. The question of inheritance of criminality has been with us since at least the time of Plato, and there is still no good answer to it. Like most questions involving the inheritance of behavior patterns in humans, there are three schools of thought. One school says yes, it is inherited; another says genes have nothing to do with it, it's all a matter of nurture; and the third is in the middle. The question is far from being purely academic. The important problems of penal policy are significantly related. Are there criminals who can never become law-abiding citizens? Questions such as this are of extreme importance in our society.

Another problem that arises is a judicial one. If a man has genes that drive him to commit crimes, can he be held responsible for his criminal acts? This kind of question comes up in judicial proceedings in different guises, usually involving insanity. The most publicized cases have involved men who have committed violent criminal acts and have the karyotype 47,XYY. In the general population, about 0.11% of white males have a 47,XYY karyotype. The major phenotypic effects of the extra Y chromosome are a height above average and a strong tendency to develop acne. Neither of these attributes, separate or combined, mean that a male is XYY. Only a karyotype analysis can establish that.

The question of whether the 47,XYY constitution tends to promote aggressive behavior and criminality has been raised because 2% of the men in mental–penal institutions (frequently called institutions or hospitals for the criminally insane) are XYY. However, it is dangerous to jump to the conclusion from this twenty-fold difference between males in the general population and incarcerated males that an extra Y causes criminal behavior or mental problems in every case. Noncriminal men in mental institutions and sane men in penal institutions have no higher incidence of the 47,XYY karyotype than those in the general population.

The present consensus is that although there is a significant association between having an extra Y and being incarcerated for violent acts, the reasons are not understood. This conclusion obviously is equivocable, but one must be aware that many XYY men are behaviorally quite normal and law abiding. It may be, of course, that the extra Y together with certain other alleles at other loci promotes abnormally aggressive behavior. The absence of these alleles would insure that the XYY would be otherwise quite normal.

During the 1960s, four highly publicized murder trials, in England, France, Germany, and Australia, involved XYY defendants who pleaded not guilty by reason of insanity induced by an extra Y chromosome. The trial in France involved the strangling of an elderly prostitute. The accused was convicted but given a light sentence, perhaps because of his extra Y. In Germany and England, the two murderers were convicted and incarcerated despite the pleas of their attorneys that they were not responsible. But in Australia, an XYY defendant was declared not guilty by reason of an extra Y. The courts in the United States have generally not been impressed with the XYY plea. The general consensus among jurists seems to be that it is premature to conclude that an extra Y entitles a criminal to a plea of genetic predisposition toward commiting a crime.

The arguments over genetic predisposition to antisocial behavior will continue for a long time to come. Meanwhile, the wisest course to follow is to avoid the simplistic conclusions that many people seem inclined to make. Certainly biological determinism is important, but grave injustices under the law can be inflicted on society and on individuals by the blind, emotionally dominated use of the concept that all our acts are genetically determined. In that case, none of us is ever responsible for his or her own actions.

Some social scientists get around this dilemma by making the nurture plea the overriding factor determining our behavior. Hence they can argue that we are not responsible for our own actions because of our upbringing. Here we see the strains of Lamarckian thinking, which introduces the topic of genetics and politics, to be discussed shortly.

Genes and Jobs

One very controversial area of genetics and public policy has been the screening of persons for employment. On the face of it, this sounds like a most reprehensible practice. Its justification has been the protection of persons from exposure to industrial agents or experiences to which they are especially sensitive. In the case of pregnant women, the concern has often been for the fetus. Opponents of genetic screening have charged that it is an excuse for discrimination against women and minorities and that it is an excuse for not cleaning up the industrial environment in which people work.

Genetic effects are different from many other toxic effects in that there probably is no threshold. By this we mean that no matter how little the exposure to a mutagen, there will be a corresponding increase in some effect such as mutation or cancer. Other toxic effects such as immediate illness may disappear without any permanent damage. This difference is responsible for much confusion. The fact that people exposed to a mutagen may recover promptly from any immediate effects or may show no ill effects at all does not mean that mutations in their offspring or cancer 20 years later will not occur.

One frequent response to questions of occupational exposure to mutagens and other toxic substances is to reduce the exposure to zero or to a level that is not harmful. No one will argue that avoidance of unnecessary exposure should be a standard practice. The implementation is not so simple, however. In many situations, zero exposure cannot be achieved. One must then decide whether the risk from the exposure is justified by the value of the product. If the product has any value at all, what is the risk in its manufacture compared to the competing risk of doing without it? Another situation occurs with exposure that can be reduced but only at great expense. If reducing exposure to the minimum level increases the cost of an essential product tenfold, does that cost increase simply shift the burden of ill health to the consumer who can no longer afford to buy it?

The ethics and economics of occupational health are indeed complex. In this section we will consider only some examples that involve genetic considerations.

Men, Women, and Work

The most obvious human genetic polymorphism—so obvious that it is often overlooked—is the XX/XY chromosome variation that determines sex. Males and females are not biologically equal. But how many of the commonly perceived differences are biological and how many are cultural? And where there are real biological differences, how often are these germane to the question of employment?

From the point of view of genetics, we will ask the question in more specific terms: Are men and women equally sensitive to mutagens and carcinogens? No clear answer is available. For radiation, female mice are more resistant to mutation than are exposed male mice. Whether this is true for all loci and for other agents is not known. Various types of cancer occur with unequal frequencies in males and females, an obvious example being breast cancer, which is much more frequent in females. Most cancers do not show such marked differences. Among survivors of the atomic bombs in Japan, the increased frequency of leukemia was observed equally in males and females.

The principal reason offered for limiting women access to certain industrial jobs involving exposure to potentially detrimental agents is the increased sensitivity of embryos and fetuses to teratogens and probably to mutagens and carcinogens also. In several much-publicized instances, women were denied specific jobs unless they were sterilized. An interesting legal issue is that a woman cannot sign a release exonerating the employer from damages to a child that may be born. Should a child be born with defects, he or she can sue the employer because no one can sign away the rights of another person, in this case a person who was an embryo or fetus at the time of the possible damage. For this reason, companies are understandably reluctant to expose themselves to legal action by assigning women to jobs where a potential risk exists.

It has been asserted that men are also at risk for causing birth defects when they work around certain agents. Men and women clearly are not equal in this respect. The only mechanisms known by which men may transmit defect to an offspring are genetic—mutations, structural changes in chromosomes, and nondisjunction. These are matters of concern to be sure. An embryo, through direct exposure to environmental agents, may develop defects through interference with the various stages of differentiation and development. The exposure that occurs after the zygote is formed involves the mother and not the father.

Sickle Cell Anemia: Much Ado About Nothing

A classic study of how not to use genetics is provided by occurrences related to sickle cell hemoglobin. The problem has been largely the mistaken idea about sickle cell trait, the designation given to persons heterozygous for Hb S. Such persons are entirely normal. The only substantial question that has ever been raised is the effect of very low oxygen pressure on their red cells, pressure such as might be encountered at high altitudes—in an aircraft or in mountainous areas.

Suggestions that heterozygotes might occasionally have difficulty in flying have not been supported by later observations. African athletes at the Mexico City Olympic Games, some of whom were known to be heterozygous for Hb S, showed no problem with sickling after strenuous exercise at that altitude. There is no evidence that heterozygotes perform differently from homozygous Hb A persons under these extreme conditions. Yet, heterozygotes have been denied insurance or have had to pay higher insurance premiums, have been denied jobs in industry, and have been denied admission to the United States Air Force Academy. Most of these actions are now reversed, but they remind us of the dangers of inadequate or wrong information.

Genes That Do Increase Sensitivity to External Agents

Although it is easy to be upset by misuse of genetic information, what should be the policy when there are real differences in sensitivity to environmental agents? The best documented example is glucose-6-phosphate dehydrogenase (G6PD) deficiency. Persons who have this X-linked recessive trait may have an acute hemolytic reaction when exposed to a variety of agents, mostly drugs used in therapy but also such common chemicals as naphthalene (mothballs). Some 15% of U.S. black males and 3% of black females are G6PD deficient. Interestingly, there appear to be no publicized instances in which G6PD deficiency has been an issue in employment. Affected persons can withstand very low levels of exposure to agents to which they are sensitive without obvious ill effects, and few are tested routinely for their G6PD status.

Other examples of inherited sensitivity include ataxia telangiectasia, an autosomal recessive trait in which homozygotes are very sensitive to ionizing radiation, and xeroderma pigmentosum (Chapter 9), in which homozygotes cannot repair the damage from ultraviolet radiation.

The Right to Work

Many of the issues introduced above can be boiled down to the question of whether every person has an equal right to any job, regardless of his biological condition. One's first impulse is to say yes. But there are exceptions that most would endorse. No one wants to fly with a pilot who has heart trouble. We expect certain professions to be filled by persons who are both intelligent and well trained, in spite of the support given one candidate for the Supreme Court several years ago by a Senator who thought that people who are mediocre should be represented by a Justice of mediocre abilities. We accept the proposition that jobs should be filled by persons who can discharge the required duties and that not every person may be able to discharge those duties.

These principles are not based on whether a person's qualifications are inherited or acquired. We only insist (or should) that the qualifications be relevant to the job. But many persons unfortunately inherit traits that disqualify them. Persons with mental retardation, inherited or not, are not serious candidates for faculty positions at universities. People are not created equal in their biological makeup and talents. Rather, they are equal under the law, quite a different issue.

We must be prepared to face situations in which persons of different genotypes are not equally suitable for employment in a specific job. We often can do a great deal to minimize these differences by modifying the environment of the workplace. But the competing costs, including risks, will not always permit us to provide equal access to specific jobs. It is hoped that the biological diversity of the human population can be matched by a diversity of employment opportunities so that no one group will be excluded from participation in the economy.

Genes and Politics

Karl Marx (1818–1883) and Friedrich Engels (1820–1895), the founders of modern communist socialism, were admirers of Darwin and his theory, and like him they also tended to hold Lamarckian doctrine as a reasonable explanation for the origin of variation. The Lamarckian concept that environmental influences that mold the phenotype of an individual can be passed on to his or her offspring was a popular one in the nineteenth century, and Marx and Engels were definitely nineteenth century men. Out of this way of thinking about a purely biological matter came support for a political theory that a good society makes for good men.

Certainly this is not a bad idea, and certainly it is in part true. But it can be pushed to an absurd conclusion, namely, that genes don't really matter, it's all a matter of environment. Nurture, not nature, is the controlling factor in what a person is to be. This point of view had a powerful influence on the thinking of not only the adherents of Marxist socialism, but also those whose political ties lay elsewhere. Psychologists of a previous generation were particularly enamored with it. Many believed that an infant is a *tabula rasa* and can be molded into just about anything—musician, artist, mathematician, or what have you—simply by the kind of environmental influences brought to bear on it. Indeed, books on the subject early in this century advised pregnant mothers to play the piano every day if they wanted their child to be musically inclined.

This kind of outlook on the relative values of nature and nurture carried over into Russian Marxist–Leninist political philosophy and resulted in a celebrated confrontation between orthodox genetics and communist orthodoxy as represented by Josef Stalin.

Lysenkoism

Genetics of the kind that we have been expounding in this book was accepted and research in it encouraged and supported for a decade after the Russian Revolution of 1917. This brand of genetics excludes Lamarckianism as a viable explanation for the origin of heritable variations. No one has ever succeeded in providing experimental evidence that acquired characteristics are inherited, even though many have tried.

However, despite all good scientific evidence to the contrary, the stage for the emergence of Lamarckian concepts was set in Russia because of Marx and Engels, whose ideas were accepted as gospel, and because a horticulturist by the name of I. V. Michurin (1855–1935), an anti-Mendelian with pro-Lamarckian ideas, exerted great influence on the practical plant breeders. There entered then in the mid 1930s a practical plant breeder, T. D. Lysenko, who denounced what he called "Mendelian–Weismannian–Morganian" genetics as metaphysical and idealistic in contrast to his own materialistic theory of heredity. This theory was essentially Lamarckian in nature, and he claimed that by applying it rapid strides would be made in developing new strains of crop plants, such as wheat and potatoes, that would quickly satisfy the needs of the Russian population. Though he was a wretched geneticist, he was a superb politician, and in 1948 his views were adopted as official doctrine by the Soviet Communist Party. Most important, his views received the complete support of Stalin.

Lysenko took complete control of all plant and animal breeding in the Soviet Union. Those geneticists who did not accept his views, at least outwardly, were banished to Siberia to work on projects not considered important to agriculture. The result was that all the good geneticists who could contribute their expertise and talents to the development of Soviet agriculture were removed from the scene. This was a tragedy because some of the world's best geneticists were Russian. Now they were put on ice so to speak, and a number of them, including I. V. Vavilov, one of the most eminent of the exiles, died in labor camps.

After the death of Stalin and the coming of N. Khrushchev to power, Lysenko began to lose power. By the end of the 1960s, his influence had become essentially nil. However, Russia lost a whole generation of geneticists and as a result her agriculture suffered enormously.

This is a prime example of political meddling in science, but it is not the only incident. Soviet adherence to the dialectic of Marxism is a brand of fundamentalism in principle, not unlike that found in other countries, including the United States (see Box 23.1). Politicians such as Lysenko can be found in every country. And they will always take the opportunity to enhance their standing by attacking

Box **23.1**

Creationism and Evolution in America

The conflict between those who draw conclusions without scientific evidence and those who base their conclusions on scientific method is nowhere better delineated than in the Opinion written by U.S. District Court Judge William R. Overton in McLean v. Arkansas Board of Education (reprinted in its entirety in *Science* 215: 934–943, 1982). In 1981, the Arkansas Legislature passed Act 590 requiring "balanced treatment for creation-science and evolution-science" in the public schools. The Act was challenged in the courts by a number of plaintiffs, and it was voided by Judge Overton. The following are excepts from his Opinion. (References to footnotes are omitted.)

The religious movement known as Fundamentalism began in nineteenth century America as part of evangelical Protestantism's response to social changes, new religious thought and Darwinism. Fundamentalists viewed these developments as attacks on the Bible and as responsible for a decline in traditional values.

The various manifestations of Fundamentalism have had a number of common characteristics, but a central premise has always been a literal interpretation of the Bible and a belief in the inerrancy of the Scriptures. Following World War I, there was again a perceived decline in traditional morality, and Fundamentalism focused on evolution as responsible for the decline. One aspect of their efforts, particularly in the South, was the promotion of statutes prohibiting the teaching of evolution in public schools. . . .

In the early 1960's, there was again a resurgence of concern among Fundamentalists about the loss of traditional values and a fear of growing secularism in society. The Fundamentalist movement became more active and has steadily grown in numbers and political influence. There is an emphasis among current Fundamentalists on the literal interpretation of the Bible and the Book of Genesis as the sole source of knowledge about origins. . . .

In the 1960's and early 1970's, several Fundamentalist organizations were formed to promote the idea that the Book of Genesis was supported by scientific data. The terms "creation science" and "scientific creationism" have been adopted by these Fundamentalists as descriptive of their study of creation and the origins of man. . . .

Section 4 of the Act provides:

"Definitions, as used in this Act:

(a) "Creation-science" means the scientific evidences for creation and inferences from those scientific evidences. Creation-science includes the scientific evidences and related inferences that indicate: (1) Sudden creation of the universe, energy, and life from nothing; (2) The insufficiency of mutation and natural selection in bringing about development of all living kinds from a single organism; (3) Changes only within fixed limits of originally created kinds of plants and animals; (4) Separate ancestry for man and apes; (5) Explanation of the earth's geology by catastrophism, including the occurrence of a worldwide flood; and (6) A relatively recent inception of the earth and living kinds.

(b) "Evolution-science" means the scientific evidences for evolution and inferences from those scientific evidences. Evolution-science includes the scientific evidences and related inferences that indicate: (1) Emergence by naturalistic processes of the universe from disordered matter and emergence of life from nonlife; (2) The sufficiency of mutation and natural selection in bringing about development of present living kinds from simple earlier kinds; (3) Emergence by mutation and natural selection of present living kinds from simple earlier kinds; (4) Emergence of man from a common ancestor with apes; (5) Explanation of the earth's geology and the evolutionary sequence by uniformitarianism; and (6) An inception several billion years ago of the earth and somewhat later of life. . . ."

Box 23.1

(*continued*)

Creationists have adopted the view of Fundamentalists generally that there are only two positions with respect to the origins of the earth and life: belief in the inerrancy of the Genesis story of creation and of a worldwide flood as fact, or belief in what they call evolution. . . .

The emphasis on origins as an aspect of the theory of evolution is peculiar to creationist literature. Although the subject of origins of life is within the province of biology, the scientific community does not consider origins of life a part of evolutionary theory. The theory of evolution assumes the existence of life and is directed to an explanation of *how* life evolved. Evolution does not presuppose the absence of a creator or God and the plain inference conveyed by Section 4 is erroneous.

As a statement of the theory of evolution, Section 4(b) is simply a hodgepodge of limited assertions, many of which are factually inaccurate.

For example, although 4(b)(2) asserts, as a tenet of evolutionary theory, "the sufficiency of mutation and natural selection in bringing about the existence of present living kinds from simple earlier kinds," Drs. Ayala and Gould both stated that biologists know that these two processes do not account for all significant evolutionary change. They testified to such phenomena as recombination, the founder effect, genetic drift and the theory of punctuated equilibrium, which are believed to play important evolutionary roles. Section 4(b) omits any reference to these. . . .

More precisely, the essential characteristics of science are:

(1) It is guided by natural law;

(2) It has to be explanatory by reference to natural law;

(3) It is testable against the empirical world;

(4) Its conclusions are tentative, i.e., are not necessarily the final word; and

(5) It is falsifiable.

Creation science as described in Section 4(a) fails to meet these essential characteristics. First, the section revolves around 4(a)(1) which asserts a sudden creation "from nothing." Such a concept is not science because it depends upon a supernatural intervention which is not guided by natural law. It is not explanatory by reference to natural law, is not testable and is not falsifiable. . . .

The methodology employed by creationists is another factor which is indicative that their work is not science. A scientific theory must be tentative and always subject to revision or abandonment in light of facts that are inconsistent with, or falsify, the theory. A theory that is by its own terms dogmatic, absolutist and never subject to revision is not a scientific theory. . . .

. . . For example, the defendants established that the mathematical probability of a chance chemical combination resulting in life from non-life is so remote that such an occurrence is almost beyond imagination. Those mathematical facts, the defendants argue, are scientific evidences that life was the product of a creator. While the statistical figures may be impressive evidence against the theory of chance chemical combinations as an explanation of origins, it requires a leap of faith to interpret those figures so as to support a complex doctrine which includes a sudden creation from nothing, a worldwide flood, separate ancestry of man and apes, and a young earth. . . .

Implementation of Act 590 will have serious and untoward consequences for students, particularly those planning to attend college. Evolution is the cornerstone of modern biology, and many courses in public schools contain subject matter relating to such varied topics as the age of the earth, geology and relationships among living things. Any student who is deprived of instruction as to the prevailing scientific thought on these topics will be denied a significant part of science education. Such a deprivation through the high school level would undoubtedly have an impact upon the quality of education in the State's colleges and universities, especially including the pre-professional and professional programs in the health sciences. . . .

those with unpopular views that conflict with a current though false view of nature. So did Copernicus and Galileo have their problems in their time with the authorities.

Conclusions

Genetic factors have been of concern to us humans since at least the dawn of civilization, especially if we reckon with the breeding of animals and plants. We have written evidence of human concerns with inheritance in our species going back to ancient Egypt, the earliest writings of the Hindus, and the Old Testament writings of the ancient Hebrews. Many of the Greek philosophers dealt with the matter of reproduction and inheritance at great length. Most of this was passed on to the Romans, who added little to it but at least preserved some of the Greek thought in their writings and incorporated some into their laws.

Continued interest in the subject that we now call genetics was expressed in writings through the Middle Ages, the Renaissance, and the Modern Age in which we now live. All through this long period of thousands of years, genetics has influenced human society in many ways, including public policy as expressed in the law, government, the philosophies of politics, economics, and even religion.

We started a new period in the human epic in 1900 with the beginnings of modern genetics. Now we are entering a second important period, the result primarily of the application of molecular genetics in genetic engineering. Where this will take us is not possible to foresee clearly. But one thing is certain: Genetics will continue to become more and more important in human affairs. To put it in a few words, we have it in our power now to direct the course of evolution of life on this planet, including our own species. For this reason, a knowledge of genetics is almost mandatory for all thinking persons.

One of the most important things we can do as responsible citizens is to recognize the great genetic heterogeneity of the human population. And, while with our laws we should guarantee all persons equal rights under the law, we must also recognize, in the words of J. B. S. Haldane, "that any satisfactory political and economic system must be based on the recognition of human (genetic) inequality." There will always be a tension between equal civil rights and genetic inequality. We must help our legal systems to reduce these tensions as intelligently as possible. Since we have chosen to be paragons, we must try to act as intelligent ones.

Review Questions

1. Why did Darwin's theory of natural selection lead to the founding of the eugenics movement?
2. What eugenic practices have been applied in the human population in the past? What practices are now being used?
3. How do genetic principles influence political philosophy?
4. What is Darwinian Socialism?

5. Distinguish between negative and positive eugenics. Can you think of examples of or proposals for positive eugenics?

6. What is the concept of "wrongful life"?

7. A man is accused in court of being the father of a child. He denies this. His blood group is AB. The mother's blood group is A, and the child is type O. Could he have been the father?

8. What does the term "tabula rasa" mean in connection with genetics?

9. The Lysenko controversy in Russia centered around what biological concept?

10. Equality of persons before the law and genetic equality are two quite different things. As concepts, they can be clearly separated, but in practice they often are in conflict. How and why?

References and Further Reading

Baer, A. S. (ed.). *Heredity and Society*, 2nd ed. Macmillan, New York. This is a series of readings in social genetics written by a number of authors. It deals with concepts and problems considered in both this and the previous chapter in some depth.

Hilton B., D. Callahan, M. Harris, P. Condliffe, and B. Berkeley (eds.). 1973. *Ethical Issues in Human Genetics.* Plenum Press, New York. 455 pp. Articles by some of the leading geneticists, lawyers, and ethicists.

Lappe, M., and R. S. Morison. 1976. Ethical and scientific issues posed by human uses of molecular genetics. *Ann. N.Y. Acad. Sci.* 265:1–208. A collection of papers presented at a conference attended by some of the outstanding scientists and others interested in applications of genetics.

Milunsky, A., and G. J. Annas (eds.). 1976. *Genetics and the Law.* Plenum Press, New York. 532 pp.

Milunsky, A., and G. J. Annas (eds.). 1980. *Genetics and the Law II.* Plenum Press, New York. 480 pp. These two volumes contain papers presented at conferences that examined the many legal problems raised by new genetic technologies.

Oosthuizen, G. C., H. A. Shapiro, and S. A. Strauss. 1980. *Genetics and Society.* Oxford Univ. Press, Capetown. 200 pp. Reports from a conference on scientific, legal, ethical, and religious aspects of genetics. Includes views from non-Western religions.

Reilly, P. 1977. *Genetics, Law and Social Policy.* Harvard Univ. Press, Cambridge, Mass. 275 pp. A book easily read by the layman by an author who is both a lawyer and a geneticist.

acid   a chemical compound that readily donates or releases hydrogen ions, for example, $HA \rightleftharpoons A^- + H^+$, where HA is the acid and H^+ the hydrogen ion. *See also* carboxyl group.

acrocentric   a term referring to a chromosome in which the centromere is near one end so that the chromosome has one long arm and one very short arm.

active site of a protein   the part of the protein in its tertiary or quaternary configuration that reacts directly with the substrate if the protein is an enzyme or with whatever the protein is supposed to react with, as, for example, O_2 in the case of hemoglobin.

aerobic   describes an organism that uses free O_2 to respire. True of virtually all eukaryotes (very few exceptions).

allele   one of two or more alternate forms of a gene. Alleles occupy homologous positions on a particular type chromosome.

amino acid   an organic acid with an amino group ($-NH_2$). Hundreds of different kinds exist in nature, but only 20 are involved in gene-directed synthesis of proteins.

amino group   a group consisting of nitrogen and hydrogen, generally written $-NH_2$; found as part of amino acids and many other organic compounds in living organisms.

amniocentesis   withdrawal of amniotic fluid, usually for prenatal diagnosis of disease in the fetus.

amnion   a structure associated with some vertebrate embryos. It is a membrane containing a liquid-filled cavity within which the embryo develops.

Glossary

amniotic cavity   the liquid-filled cavity within which the vertebrate embryo develops.

anabolism   that part of the metabolic processes in which the body chemical constituents are synthesized from smaller molecules; for example, the synthesis of proteins from amino acids.

anaerobic   describes an organism that can metabolize and survive without the presence of free O_2.

anaphase   that phase of mitosis and meiosis in which the chromosomes or chromatids separate and go to opposite poles of the dividing cell.

androgen   a hormone produced primarily in the testis with male-determining properties; any hormone that stimulates male secondary sexual characteristics.

aneuploid   having other than the basic complement of chromosomes.

anion   an ion that bears a negative charge and hence moves to the anode in an electric field.

anthropocentrism   the assumption that humans are the center of all things.

antibody   an immunoglobulin with specificity for a particular antigen.

anticodon   a stretch of three contiguous nucleotides in a transfer RNA molecule that is complementary to the codon in mRNA for the amino acid specific for that tRNA.

antigen   a chemical compound that elicits the formation of a specific antibody or immunoglobulin.

atom   the smallest particle of an element with the chemical properties of that element.

atomic nucleus   the central core of an atom containing protons and neutrons.

atomic number   the number of protons in the nucleus of an atom. Each chemical element has a different atomic number.

atomic weight   the relative mass of an element measured against the mass of the common isotope of carbon, ^{12}C. The atomic weight of an element depends primarily on the number of protons and neutrons in its nucleus.

autoimmune   a condition in which an individual develops antibodies to his or her own body constituents.

autosome   a chromosome that is not a sex chromosome.

auxotroph   a mutant organism (or cells in culture) that has a requirement for a nutritional component (e.g., an amino acid or vitamin) not required by the normal or wild type member of that species or type.

bacteriophage   virus that infects bacteria. Generally referred to simply as *phage*.

Barr body   a densely staining body found in the interphase nuclei of somatic cells of female (XX) mammals; also known as *sex chromatin*. Two and three may be found in aberrant XXX and XXXX individuals. The Barr body represents an X chromosome that remains inactive, presumably because of its condensed state.

base   a chemical compound that accepts protons (hydrogen ions), such as those released by acids.

binary fission   the term applied to the asexual reproduction of single-celled organisms in which one cell divides to form two equivalent daughter cells.

blastocyst   the very early mammalian embryo several days after the fertilization of the egg.

blending theory of inheritance   the theory, popular in the nineteenth century but no longer accepted, that the male and female contributions to the fertilized egg blend so as to give offspring intermediate between the parents.

blue-green algae   a primitive type of photosynthesizing algae that, like bacteria, have no nuclei. They are therefore prokaryotes and fundamentally different from the higher algae, which are eukaryotes.

B-lymphocytes   the lymphocytes (white blood cells) that produce the immuno-globulin antibodies circulating in the blood plasma and responsible for humoral immunity.

carbohydrate   a chemical compound consisting of C, H, and O, with many C—OH groups; includes the simple sugars, which may form polymers such as starch, glycogen, and cellulose.

carboxyl group   the active group of organic acids, which ionizes to release

$$\text{hydrogen ions: } -\overset{\displaystyle O}{\overset{\|}{C}}-OH \rightleftharpoons -\overset{\displaystyle O}{\overset{\|}{C}}-O^- + H^+.$$

carcinoma   the type of cancer that arises in epithelial tissues such as skin and the lining of the digestive tract.

catabolism   the breakdown of compounds in metabolism to smaller molecules. Energy is often released and stored as ATP.

cation   an ion that bears a positive charge and hence moves to the cathode in an electric field.

catalysis   the process by which a catalyst increases the rate of a chemical reaction.

cell hybrid   the fusion of two somatic cells in culture and the subsequent fusion of their nuclei.

cell theory   the theory, now accepted as fact, that cells are the units of life.

cell wall   the outer inanimate covering of a cell, such as the cellulose cell wall of plant cells. The cell wall is quite distinct from the plasma membrane, which is an integral part of the cytoplasm. Animal cells usually do not have cell walls.

cellular immune system   that part of the immune system dependent on T cells and responsible for tissue histocompatibility.

centimorgan   the percent of crossing over as measured by the recombination between two loci on the same chromosome.

centriole   a tiny organelle near the nuclear membrane of animal cells. It divides at the beginning of mitosis, and the two daughter centrioles participate in the mitotic process by serving as centers for spindle formation.

centromere   a specialized region of a chromosome to which spindle fibers attach during cell division.

chemical bond   the force that holds one atom to another and thus leads to the formation of molecules.

chemical compound   a molecule that consists of two or more atoms, usually involving different elements.

chemical element   a form of matter that cannot be decomposed, transformed into, or composed from other elements; one of the ultimate building blocks of which matter is composed.

chromatid   one of the two duplicated forms of a chromosome prior to division of the centromere.

chromatin   the substance of which chromosomes are composed, so named because it is easily stained in microscope preparations.

chromosome   a structure in the cell nucleus that stores genetic information.

chromosome aberration   an inherited modification of the structure of a chromosome.

chromosome arm   one of the two main segments of a chromosome separated by the centromere.

chromosome mutation   generally used synonymously with chromosome aberration.

cis-trans test   a genetic test in which the effects of two different gene mutations are tested when they are on the same chromosome (cis) as compared to when they are on different members of a pair of homologous chromosomes (trans) in a heterozygote.

clone   a population of cells or organisms derived from a single cell by binary fission or by mitosis. This term is such misused in the popular media. Only the above definition is accepted in scientific use.

codon   the nucleic acid coding unit for an amino acid in a polypeptide. A codon consists of three nucleotides.

coenzyme   a low molecular weight substance that, in combination with a protein (*apoenzyme*), acts as a catalyst.

coevolution   evolving together.

complementarity   When used with reference to DNA, complementarity means that one chain complements the other because adenine pairs with thymine and guanine with cytosine. It also refers to the similar relation between the transcribed DNA chain and its RNA transcript.

complementation   When two mutant genes are tested together in the trans position in heterozygotes and give a phenotype similar to wild type, they are said to be complementary.

conception   the union of the sperm and egg nuclei following fertilization.

congenital defect   a defect in a newborn child. It may or may not be inherited.

conjugation   a general term referring to the connection, coupling, pairing, or uniting of cells or organelles.

covalent bond   a chemical bond formed by the sharing of electrons between atoms.

crossing over   the process of exchange between homologous chromatids that leads to recombination of linked genes.

cytoplasm   the part of the eukaryotic cell that is outside the nucleus.

cytoplasmic inheritance   inheritance through cytoplasmic DNA (mitochondrial and chloroplast) rather than nuclear DNA.

dalton   a unit of molecular weight equal approximately to the weight of one hydrogen atom.

deficiency   *See* deletion.

deoxyribonucleic acid (DNA)   the chemical substance in which genetic information is stored. DNA is a polymer of deoxyribonucleotides.

deletion   absence of a section of a chromosome.

differentiation   the process by which an embryonic cell changes phenotypically to become a specific kind of cell with respect to function and structure.

dihybrid cross   a cross between two double heterozygotes, for example, *AaBb* × *AaBb*.

diploid   having twice the haploid or basic set of chromosomes.

discontinuous trait   an inherited trait for which the alternatives do not overlap phenotypically, as opposed to a continuous trait such as height.

disulfide bond   a covalent bond between two sulfur atoms. Disulfide bonds formed between cysteine side chains of proteins help stabilize the proteins.

dizygotic twins   twins each of which came from a different zygote and hence have the same genetic relationship as any two siblings.

DNA polymerase   an enzyme that catalyzes bond formation between nucleotides to make polynucleotides or nucleic acids. Several kinds occur in cells, some of which are involved in repair of DNA and some in replication.

dominance   An allele is *dominant* if it affects the phenotype in heterozygous combination with other alleles.

doubling dose   the amount of radiation or other mutagen needed to increase its biological effect twofold.

ectoderm   one of the three primordial tissues of the embryo. The ectodermal cells differentiate into the nervous system (including the brain), the skin, and most parts of the sense organs.

electron   a unit negatively charged particle, such as those outside the nucleus of an atom.

electron transport system   the metabolic system in the mitochondria that generates ATP from the breakdown of pyruvic acid in the tricarboxylic acid cycle. The electrons reduce O_2 to form H_2O.

electrophoresis   the movement of charged molecules in an electric field.

endoderm   one of the three primordial tissues of the embryo. The endodermal cells provide the linings of the gut, other parts of the digestive tract, and the lungs.

endometrium   the lining of the uterine wall.

endoplasmic reticulum   a system of membranes in the cytoplasm of cells involved in the synthesis of proteins.

entelechy   a philosophical concept derived from the Aristotelian idea of an essence or soul that directs the development of an egg into a specific kind of organism.

enzyme   proteinaceous substance that catalyzes metabolic chemical reactions.

epigenesis   the hypothesis that embryos are derived from undifferentiated living material in the eggs of their mothers. The implication is that the egg contains directions rather than a preformed organism. It is thus in contradiction with preformation.

episome   a plasmid or phage that has the capacity to become an integral part of the host chromosome DNA, replicate with it, and finally to leave the host chromosome and replicate on its own in the cytoplasm.

epistasis   the dominance of one gene over a nonallelic gene, or the interaction of genes at different loci in general.

estrogen   a class of female steroid hormones produced primarily in the ovaries. Estrogens stimulate the development of female secondary sex characteristics and the rate of cell division in various parts of the body.

eukaryote   an organism having cells with true nuclei that contain chromosomes.

exon   a sequence of DNA that is translated into protein.

extraembryonic membranes   the membranes derived from the embryo that become the placenta and amnion.

F_1   the first filial generation in a genetic cross.

F_2   the second filial generation; the offspring of crosses involving the F_1 generation.

fertilization   the union of the sperm and egg followed by the union of their nuclei and the formation of the zygote nucleus.

fetus   the human embryo after about 8 weeks of development becomes a fetus by definition.

frameshift mutation   a type of gene mutation in which one or more nucleotide base pairs are either deleted or added.

gamete   a germ cell, either female (ovum) or male (spermatozoon).

gene   the unit of inheritance identified by its function and passage from generation to generation unchanged except by mutation. Physically, it is a segment of DNA.

gene mutation   a change either by substitution, deletion, or addition of nucleotides within the DNA of a single gene. Also called *point mutation*.

gene regulation   the process of regulating the quantitative and temporal aspects of gene activity.

genetic code   the code constituted of 64 triplets of nucleotides in DNA or RNA. Each of 61 triplets specifies a particular amino acid; the three exceptions cause termination of translation.

genetic segregation   the separation of alleles in a heterozygote during meiosis.

genome   a haploid set of chromosomes.

genotype   the genetic endowment or capacity of an individual.

genotypic ratio   the ratio of the different genotypes resulting from a cross.

germ line   the cells in the gonads that are capable of becoming gametes. The primordial germ cells are called oogonia in females and spermatogonia in males.

glycolysis   the breakdown of glucose to three-carbon compounds such as pyruvate. Oxygen is not required in glycolysis.

Golgi apparatus   a membranous structure in the cytoplasm of cells. The Golgi apparatus is especially involved in synthesis of proteins that are to be secreted.

gonad   a general term for ovary or testis.

haploid   having a single genome or set of nonhomologous chromosomes.

hematopoietic center   a tissue in which red blood cells and other blood corpuscles are made.

hemizygous   the state of having only one copy of a locus, as, for example, loci on the X chromosome in XY males.

heritability   the proportion of the total phenotypic variance for which genotypic differences are responsible in a population. It is a statistic.

hermaphrodite   an individual with both ovarian and testicular tissues.

heterochromatin   chromatin that stains differently from other chromatin (euchromatin), particularly in the interphase of the cell cycle.

heterokaryon   a cell in which two or more nuclei of different genotypes exist.

heteromeric   consisting of two or more different kinds of subunits. Proteins are often heteromeric because they consist of more than one kind of polypeptide subunit, each with a different amino acid sequence.

heteromorphic   existing in two or more different forms.

heterozygote   a cell or individual that has two different alleles of a particular gene.

histocompatibility gene   a gene that regulates the immunological properties of cellular antigens. Such antigens are highly variable among individuals and are responsible for the rejection of tissue transplants.

histone   a kind of basic protein associated with the DNA in chromosomes.

hnRNA   the acronym for heterogeneous nuclear RNA. This RNA is the initial product of transcription of structural gene DNA. It is processed to form the final, mature, functional mRNA.

homologous chromosomes   chromosomes that bear similar or identical genes and that pair in the prophase of meiosis.

homomeric   consisting of only one kind of subunit. Many proteins consist entirely of one kind of polypeptide chain, even though there may be several copies in the functional protein.

homozygote   a cell or individual that is diploid or of higher ploidy and that bears identical alleles at the particular locus of interest.

homunculus   an imaginary miniature human in a sperm or egg.

hormone   a substance produced by an endocrine gland that regulates physiological processes.

humoral immune system   the system responsible for the production of circulating antibodies.

hybrid   a cross between two distinct races or subspecies. The term is also used to describe crosses between species when such is possible as well as the products of fusion of cells of different origin.

hydrophilic   water-loving. Applied to chemical compounds that are readily soluble in water.

hydrophobic   water-hating. Applied to chemical compounds not readily soluble in water but that are soluble in fat solvents.

hydroxyl group   —OH. A hydroxyl group is the functional group characteristic of alcohols.

idiogram   a diagrammatic representation of a karyotype based on microscopic examination of stained chromosomes.

immune system   the system in animals responsible for the production of antibodies.

immunoglobulin   an antibody molecule.

implantation   in mammalian embryology, the interdigitation of the blastocyst into the uterine wall of the mother.

inbreeding   matings between genetically related individuals.

independent assortment   the random assortment of homologous chromosomes during meiosis.

interphase   the period in the cell cycle consisting of the G_1, S, and G_2 phases during which the cell is not undergoing division.

intron   a sequence of DNA in the interior of a structural gene that is not translated into protein. Also called *intervening sequence*.

inversion   an intrachromosomal structural change in which a segment of the chromosome is released by two breaks and then reinserted in inverted sequence.

ion   a molecule or atom with an electrical charge.

ionic bond   an electrostatic bond between an ion with a positive charge and one with a negative charge.

isotope   one of two or more forms of a chemical element with the same atomic number, hence, the same numbers of protons and electrons, but with different numbers of neutrons in the nucleus.

karyotype   the chromosome complement of an individual or species as exhibited in the number and morphology of the chromosomes, usually at mitotic metaphase.

law of genetic continuity   Life is a continuous stream, originating only from pre-existing life by cell division.

lethal mutation   a mutation that results in death at an early stage, that is, from the zygote to the late embryo or fetus. May be dominant or recessive.

linkage   as used by geneticists, designates two or more genes that do not segregate independently and hence presumably are on the same chromosome.

lipid   a general term for fats.

locus   the site or location of a gene on a chromosome.

Lyonization   the inactivation of one of the X chromosomes, reflected in the formation of Barr bodies in XX female mammals. Named after Mary Lyon, who first proposed an explanation for this phenomenon.

lysosome   a cell organelle that contains many of the enzymes that break down cell constituents

macromolecule   a very large molecule made up of smaller molecules. Proteins and nucleic acids are macromolecules.

major histocompatibility complex (MHC)   a complex of closely linked genes that determine the most important antigens involved in tissue transplant compatibility.

map distance   the distance between two gene loci on the same chromosome as determined from recombination data and crossover estimates.

map unit   the unit of measurement of distance between linked genes, as determined from recombination data. One unit represents one percent recombination.

maternal inheritance   inheritance through other than nuclear chromosomes. Cytoplasmic inheritance and extranuclear inheritance are more or less synonymous terms.

mating type   the mating property or capacity of an organism, usually genetically determined. Individuals with the same mating type do not mate with one another but only with those of other mating types.

meiosis   a process in eukaryotic cells that consists of two cell divisions but only one replication of chromosomal DNA. The result is the formation of four daughter nuclei, each with half the number of chromosomes of the mother cell. Each daughter cell also has only one chromosome of each homologous pair of chromosomes present in the mother cell.

Mendelism   the type of inheritance in eukaryotes in which characteristics are transmitted by chromosomes and the genetic determinants (genes) segregate in germ cell formation. To be distinguished from cytoplasmic inheritance.

mesoderm   one of the three primordial germ layers or tissues of the embryo. The mesoderm differentiates into muscle, bone, blood, and much of the substance of the other body organs.

messenger RNA   processed RNA that has been transcribed from structural gene DNA and that is ready for translation into protein.

metabolism   the sum total of the chemical processes that occur in the living organism.

metacentric   a term referring to a chromosome with the centromere at or near the center.

metaphase   the stage in eukaryotic cell division at which the chromosomes are maximally condensed and ready to be distributed to the daughter cells.

missense mutation   a gene mutation in which a nucleotide substitution results in a codon change that, in turn, results in an amino acid substitution in a polypeptide chain.

mitochondrion   cell organelle in eukaryotes. Mitochondria carry out essential metabolic functions and are the centers of respiration. They are often called the powerhouses by virtue of their production of ATP.

mitosis   cell division in eukaryotes in which a mother cell gives rise to two daughter cells, each with the same number and kind of chromosomes found in the mother cell.

molecule   a chemical entity ordinarily consisting of two or more atoms held together by chemical bonds.

monogenic trait   a trait whose alternate forms are inherited as a single gene difference.

monohybrid cross   a cross involving only a single pair of alleles.

monosomy   having only one copy of a particular chromosome rather than the normal diploid two.

monozygotic twins   twins that have developed from a single egg and hence are identical genetically, at least for nuclear genes.

morula   in vertebrate embryogenesis, the ball of cells formed by the first series of cleavages of the egg.

mosaic   an individual with cells of different genotypes derived from the same zygote. The different genotypes arise during the individual's lifetime by gene or chromosomal mutation, nondisjunction, or changes in ploidy.

mRNA   the acronym for messenger RNA.

Mullerian inhibitor   a product of the testes that inhibits further development of the Mullerian ducts in the XY embryo. These ducts, if allowed to develop as in the XX embryo, will form the uterus, Fallopian tubes, and part of the vagina.

multiple alleles   more than two alleles at a single genetic locus.

mutagen   an agent that causes mutations.

mutation   an inherited change in the genetic material.

neutron   an electrically neutral particle found in the nuclei of most atoms. A neutron has approximately the same mass as a proton.

nondisjunction   the failure of sister chromatids to go to opposite poles during mitosis or for homologues to disjoin or segregate properly during meiosis.

nonsister chromatids   the chromatids of different homologous chromosomes of a pair, as differentiated from sister chromatids, the latter being derived by replication from one chromosome.

nuclear membrane   the double membrane surrounding the nuclear material in eukaryotes.

nucleic acid   a polymer of nucleotides. The two major types are ribonucleic acid (RNA) and deoxyribonucleic acid (DNA).

nucleolus   an organelle in the eukaryotic nucleus that is the site of ribosomal RNA synthesis.

nucleosome   a body associated with eukaryotic chromosomes. Each of the many thousands of nucleosomes associated with a chromosome is composed of DNA and histones.

nucleotide   the unit of structure or basic building block of nucleic acids. Each nucleotide consists of a pentose sugar, a phosphate group, and a purine or pyrimidine base.

nucleus   the eukaryotic organelle that contains the chromosomes.

oncogene a gene that may include the genome of a tumor virus such as a retrovirus. Oncogenes are associated with development of cancer.

oöcyte The egg cell derived by mitosis from the oögonium is called a primary oöcyte. It starts meiosis and forms the secondary oöcyte and a polar body. The secondary oöcyte undergoes the second meiotic division to form the ovum.

oögenesis meiosis and development of the female primary oöcyte to the stage when it becomes a fully developed haploid egg.

oögonium a female primordial sex cell in the ovary. Oögonia become primary oöcytes at the initiation of meiosis.

organelle an identifiable structural entity in a cell. Examples are mitochondria, nucleoli, and lysosomes.

oxidation in the broadest sense, the removal of an electron from the outer shell of an atom. In many systems, including those of aerobic organisms, oxygen is the ultimate acceptor of the electron.

paragon a type of excellence or perfection.

parthenogenesis development of an embryo from an unfertilized egg.

particulate inheritance the kind of inheritance in which inherited characteristics are determined by entities (genes) that pass unaltered from generation to generation.

peptide bond the bond between the two amino acids in which the amino group nitrogen of one forms a covalent bond with the carboxyl carbon of the other.

periodic table of elements the table in which the elements are arranged in a fashion logical to chemists.

pH a measure of the hydrogen ion concentration of a solution. pH $= -\log\,(\mathrm{H}^+$ concentration). A pH of 7 is neutral; values lower than 7 are acid, and values greater than 7 are basic (alkaline).

phage short for *bacteriophage;* a bacterial virus.

phenotype the result of the action of the genotype in a particular environment; the characteristics of an organism.

phenotypic ratio the ratio of the different phenotypes resulting from a cross. It may or may not be the same as the genotypic ratio.

photosynthesis the process by which green plants use light energy to synthesize high-energy carbon compounds such as sugars.

pinocytosis taking into the cell by a process of engulfing.

placenta the extraembryonic membranous organ that provides the connection between the developing embryo and fetus and the uterine tissue of the mother. The placenta is the major part of the afterbirth.

plasma membrane the membrane surrounding the cytoplasm of a cell; also called the cell membrane.

plasmid an element in bacterial cells that consists of a circle of DNA that replicates in the cytoplasm independently of the host chromosome. Plasmids often can also become an integral part of the host chromosomal DNA.

polar body a cell consisting mostly of nuclear material produced during oögenesis of most animals. Polar bodies do not participate ordinarily in fertilization but degenerate.

polymer   a chemical compound made up of repeating subunits of smaller molecules. Examples are polysaccharides (glycogen, starch), proteins, and nucleic acids.

polymorphic   existing in multiple forms. In genetics, polymorphic genes are those with two or more common alleles.

polypeptide   a chain of amino acids bound together by peptide bonds.

preformation   the hypothesis that one or the other of the animal germ cells contains a preformed adult in miniature.

primary structure of a protein   the amino acid sequence of the polypeptide.

proband   the affected person with whom study of a particular inherited characteristic begins.

progestin   a hormone produced by the ovary; important in maintaining pregnancy.

prokaryote   an organism whose cells lack nuclei, mitochondria, and the other kinds of organelles found in eukaryotic cells. This group includes all bacteria and the blue-green algae (Cyanophyta).

prophase   the first phase of cell division; ends the G_2 phase.

propositus   same as proband.

protein   a polymer of amino acids joined by peptide bonds.

proton   a positively charged atomic particle located in the atomic nucleus.

prototroph   descriptive of "wild-type" cells that do not require the nutrients necessary for mutant, *auxotrophic* cells.

purebred   generally, an individual whose pedigree is known for several generations without evidence of outbreeding to other races or subspecies. Many purebred domestic animals are also highly inbred, but this need not be so.

pure line   a descendant of a highly inbred line of plants or animals.

purine   a class of organic compounds, generally described as bases. Two purines, adenine and guanine, are always present in DNA and RNA.

pyrimidine   a class of organic compounds generally described as bases. Two pyrimidines, cytosine and thymine, are always present in DNA. Cytosine and uracil are always present in RNA.

quaternary structure of proteins   structure at the multimeric level, in which two or more polypeptide subunits with specific tertiary structure join to form a single functional molecule.

race   a population that differs genetically from other populations of the same species. Equivalent to *subspecies*.

radioactive isotope   an isotope of a chemical element that is unstable and that emits radiation when an atom disintegrates.

radioactivity   the spontaneous emission of subatomic particles (neutrons, protons, electrons, alpha particles) and high-frequency electromagnetic radiation (gamma rays, X rays) by the unstable isotopes of certain elements.

recessive   an allele that is expressed phenotypically only when an allele dominant to it is absent.

recombination   the reassortment of genes in meiosis to give rise to cells or individuals with new combinations of genes. Reassortment at a particular locus can only be detected if the parents differ genetically at that locus or if both are heterozygous.

recombination frequency   the frequency of recombinant types among the offspring of a cross.

regulatory gene   a sequence of DNA that in some way regulates the functioning of a structural or other gene either directly or through its products.

repetitive DNA   DNA nucleotide sequences that repeat from several to hundreds of thousands of times.

repulsion   a term used to describe the situation when the mutant alleles of linked genes in heterozygotes are on different members of a pair of homologous chromosomes.

respiration   the metabolic process in which oxygen is consumed and ATP, H_2O, and CO_2 produced.

restriction endonuclease   an enzyme that breaks the phosphodiester bonds between certain nucleotides in DNA. The enzyme "cuts" only at specific sites that it recognizes by the adjacent nucleotide sequences.

restriction enzyme   *See* restriction endonuclease.

retrovirus   a generic name for RNA viruses that infect animals and in some cases result in tumor formation.

reverse transcriptase   a polymerase that catalyzes synthesis of a DNA strand complementary to an RNA strand. Its action is opposite to that of RNA polymerase.

ribosomal RNA   the RNA of ribosomes.

ribosome   an organelle composed of rRNA and proteins. Ribosomes have an essential role in protein synthesis in the cytoplasm.

RNA polymerase   an enzyme that catalyzes synthesis of RNA complementary to a DNA template.

rRNA   the acronym for ribosomal RNA.

sarcoma   a cancer of mesodermal origin.

secondary structure of a protein   the alpha-helical structure of a polypeptide.

segregation   *See* genetic segregation.

sex   a biological mechanism for insuring recombination of genetic material.

sex chromatin   *See* Barr body.

sex chromosome   a chromosome whose presence or absence plays an important role in the determination of sex.

sex determination   the determination of the sex of an individual by either genetic or environmental mechanisms or by some combination of these.

sex hormone   one of the hormones secreted especially by the gonads and that affects the development of the secondary sex characteristics as well as the functioning of nearly all cells of the body.

sex-linked   a gene on an X chromosome; also called *X-linked.*

sex ratio   the ratio of males to females in a population.

sexual dimorphism   the differences between male and female of a species. Vive la différence!

sib (sibling)   a brother or sister. A set of offspring from a particular mating pair is called a *sibship.*

sister chromatids   the chromatids derived from an original mother chromosome. They are therefore identical in genetic constitution.

soma   the totality of the somatic cells and tissues of an individual, as distinguished from the germinal tissue.

somatic cell   any cell not in the germ cell line.

somatogenetics   the genetics of somatic cells.

spermatogenesis   the meiotic process leading to the production of sperm from spermatogonia via spermatocytes.

spermatogonia   the primordial male sex cells in the testes.

sperm midpiece   the part of the sperm between the tail and the head. This part consists primarily of mitochondria and provides the power to drive the tail.

spindle fibers   microtubules of protein, RNA, lipid, and carbohydrate that form fibers beginning in early metaphase. Some of these attach to the centromeres of chromosomes, others do not. The whole set of fibers forms the spindle, which functions in the mitotic and meiotic anaphases.

spontaneous generation   the hypothesis that life arises from formless, lifeless ooze or matter.

stable isotope   an isotope that does not emit radiation.

stem cell   a cell that is only partly differentiated. On dividing, it produces one daughter cell that becomes differentiated (e.g., a blood cell) and one that remains undifferentiated like the mother cell.

symbiosis   a relationship between two organisms, usually of quite different kinds, that is mutually beneficial to both.

synapsis   the pairing of homologous chromosomes, as in the prophase of meiosis.

synteny   Genes that are on the same chromosome are *syntenic.*

tabula rasa   literally, a "clean table" [L.]. As used in psychology and genetics, it denotes a mind that begins at birth with no innate (inherited) ideas or abilities, hence, a "clean slate."

telocentric   a term referring to a chromosome in which the centromere is at one end so that the chromosome has only one arm.

telophase   the last stage of mitosis or meiosis in which the chromosomes become highly extended and are re-enveloped by new nuclear membranes.

teratogen   an agent that causes congenital defects.

tertiary structure of a protein   the structural stage that involves folding of the alpha-helix.

tetrad of chromatids   the bundle of four chromatids of two paired (synapsed) homologous chromosomes during the prophase of the first meiotic division.

T lymphocyte   a lymphocyte involved in the production of antibodies in the cellular immune system, in distinction to the humoral immune system.

toxoid   the altered toxin of a disease-producing bacterium used as an antigen to induce antibodies to and prevent the effect of the toxin.

transcription   the process of synthesizing a nucleic acid strand complementary to an existing nucleic acid strand, which serves as a template. Transcription is catalyzed by polymerases.

transduction the process by which genetic material from one cell is transmitted and incorporated into the genetic material of another by a virus or phage.

transfer RNA the RNA that becomes "charged" with a specific amino acid and functions in translation by placing this amino acid properly in the polypeptide being synthesized.

transformation the transfer of genetic information from one cell to another by means of naked extracellular DNA.

translation the process of synthesizing a polypeptide using the code in a messenger RNA.

translocation the unilateral or reciprocal exchange of segments of one chromosome with another nonhomologous chromosome.

tricarboxylic acid cycle the metabolic cycle that is part of the breakdown of glucose to CO_2, H_2O, and energy (ATP).

triploid having three haploid sets of chromosomes.

trisomy having three copies of a chromosome rather than the usual two in a diploid cell.

tRNA the acronym for transfer RNA.

umbilical cord a stalk connecting the embryo with the placenta. Through it pass the embryonic and fetal blood into the placenta to be purified and to obtain nutrients and then be returned to the fetus.

unit character the term used by the early geneticists to describe inherited characteristics that segregate according to the Mendelian laws.

valence a measure of the combining ratios of atoms of elements in forming chemical bonds.

virus a particle consisting of nucleic acid and protein that is capable of reproducing itself in a suitable host cell.

X chromosome one of the two kinds of sex chromosomes.

X-chromosome inactivation the inactivation of X chromosomes in excess of one in mammalian cells. *See also* Lyonization.

Y chromosome one of the two kinds of sex chromosomes. Presence of a Y chromosome in mammals causes a zygote to differentiate into a male.

zygote a fertilized egg.

Chapter 2

1. Preformation stated that living things were all created at the beginning in their present form, one generation inside another. Spontaneous generation stated that there is essentially no physical link between generations. Each individual arises from a formless mass of matter, and the directions for development are assumed to be spiritual.

2. Life comes only from pre-existing life. Life is a continuous stream, a consequence of the transmission of genetic blueprints from one generation to the next.

3. Yes. It contains elements of preformation in the sense that we inherit a set of preformed chromosomes containing a set of directions for our formation from the unorganized matter that we call nutrients.

4. The important life processes, such as reproduction and metabolism, only occur in cells.

5. Mendel's technique of analysis employed two important approaches. (1) He chose easy-to-identify, simple characteristics that had clear alternative states (long versus short, yellow versus green). (2) He inbred each kind of alternative until it produced offspring of only that one type and then used these inbred plants as parents to obtain the hybrids. These F_1 hybrids then showed a segregation of the alternative characteristics in the F_2 offspring.

6. His characteristics maintained a constancy from generation to generation. They

Answers to
Review Questions

could disappear in hybrid crosses and then reappear in subsequent generations. No other explanation based on blending inheritance accounted for the
persistence of the recessive traits.

7. The answer to this is given in part by the answer to question 6. In blending
inheritance, one would not expect a characteristic to maintain its identity from
generation to generation. Rather, one would expect it to be diluted.

8. Mendel's ratios were explained by the segregation of the chromosomes in
meiosis. Cell biology and Mendelism were united to form the beginnings of
modern genetics.

9. The group of geneticists at Columbia University who, during the period from
about 1910 to 1916, demonstrated that the genes of the fruit fly *Drosophila*
resided on the chromosomes. The members of the group were T. H. Morgan and
his students, C. Bridges, H. J. Muller, and A. H. Sturtevant.

Chapter 3

1. When an atom is broken apart, it no longer has the identity of the element it
formerly represented. It is the smallest bit of matter with the properties of that
element. A cell is the smallest bit of matter capable of reproducing itself and
possessing the properties of life. They are different insofar as one can reproduce itself and the other can't.

2. Not really. It is sometimes said that DNA reproduces itself, but that is an
exaggeration. It can only be reproduced as part of a cell or in the laboratory in
the presence of enzymes. It is synthesized in its own image, but it does not self-
reproduce.

3. About 90 occur naturally. Altogether, 103 are known to occur. Thirteen are
produced artificially by physicists, and most of these have a very short
existence.

4. Twenty-seven, of which only twenty are universally present.

5. Matter that consists of a single kind of atom.

6. A substance that is composed of atoms of one or more elements bound
together in specific ratios by means of chemical bonds.

7. It becomes ionized and has a positive charge.

8. Protons have a positive charge, whereas electrons are negative. A proton has a
mass about that of a hydrogen atom. A neutron is like a proton in mass but has
no charge. The electron has a negative charge and a mass less than 1/1800 that
of a proton.

9. The number of protons in an atomic nucleus.

10. A dalton is a unit of molecular weight and is approximately equal to the mass
of one hydrogen atom.

11. A stable isotope is very long lived and does not emit radioactivity. An unstable
isotope decomposes into other atoms and subatomic particles with the
release of radioactivity.

12. The splitting or fission of atomic nuclei to produce atoms with smaller nuclei.

13. An element is radioactive when its atomic nuclei are unstable and fly apart,
forming smaller nuclei and releasing energy in the form of radiation and
subatomic particles.

14. ^{40}K (potassium-40).
15. It means that it has a charge and is formed by the loss of an electron from its shell.
16. It measures the concentration of hydrogen ions (H^+), which is a measure of the acidity of the solution.
17. An organic molecule includes carbon atoms bound by covalent bonds. All other molecules are inorganic.
18. Covalent bonds are about 10 to 20 times stronger than hydrogen bonds, which are so weak that they may break at temperatures above 50° C.
19. A hydrophilic group such as —OH or —COOH is readily soluble in water, whereas a hydrophobic group such as —CH$_3$ is not.
20. The amino group (—NH$_2$), the carboxyl group (—COOH), and the side chain (the remainder of the molecule).
21. The amino and carboxyl groups.
22. The sequence of its amino acids, i.e., its primary structure.
23. Forming an α-helix and then folding.
24. Lipids, proteins.
25. Nucleotides.
26. Adenine, guanine, cytosine, thymine, uracil.
27. Uracil and thymine.
28. Adenine in one chain forms hydrogen bonds with thymine in the other, and cytosine forms hydrogen bonds with guanine.
29. The structure of the ribose sugar which forms the phosphodiester bonds with the phosphate backbone.
30. Oxygen. Photosynthetic organisms, such as green plants.

Chapter 4

1. The prokaryotes differ from eukaryotes primarily in their cell structure. Some of the more significant of these differences are

prokaryotes	eukaryotes
a. no nucleus	nucleated cells
b. a single small chromosome	large chromosomes; more than one kind per cell
c. no mitosis in cell division	mitosis
d. no meiosis	meiosis characteristic
e. cells small, ca. 1.5 μm	cells large enough to contain organelles the size of bacteria
f. no mitochrondria or chloroplasts	all have mitochondria; some also have chloroplasts

2. The genotype is the genetic endowment or program received from our parents. Our phenotype is the result of the expression of that genotype or program.
3. DNA, proteins, and some RNA.
4. The DNA runs as a single strand through the length of the chromosome. It is

intimately involved with four kinds of histone proteins in structures called nucleosomes, and these in turn are associated with a fifth type of histone and a set of other proteins frequently called nonhistone or acidic proteins to distinguish them from the histones, which are basic. The histones are not enzymes, whereas most of the nonhistones appear to be enzymes or to carry out specific roles in regulation. The RNA occurs in part as hnRNA, rRNA, and tRNA, all in the process of being transcribed.

5. The two principal are the nucleus and the cytoplasm. The nucleus contains the chromosomes, which have the bulk of the cell's genetic program in the form of DNA. The cytoplasm contains mitochondria, endoplasmic reticulum, Golgi apparatus, lysosomes, and centrioles. The mitochondria generate energy in the form of ATP, the endoplasmic reticulum is the site of protein synthesis, and the Golgi apparatus is involved in the production of certain kinds of macromolecules that are composites of proteins and carbohydrate. Lysosomes contain enzymes that break down fats and other large molecules.

6. In the process of synthesizing carbohydrates and other carbon compounds with high energy content, green plants emit free oxygen while using CO_2:

$$CO_2 + H_2O + \text{light energy} \longrightarrow C_xH_yO_z + O_2$$

Animals emit CO_2 while "burning" $C_xH_yO_z$:

$$ADP + Pi + C_xH_yO_z + O_2 \longrightarrow CO_2 + H_2O + ATP$$

The CO_2 generated by animals is used by the green plants to produce more O_2 and $C_xH_yO_z$. Thus one is dependent on the other.

7. The processes of glycolysis and oxidative phosphorylation. The latter process is by far the more important. It occurs in the mitochondria and results in the formation of CO_2, H_2O, and ATP while using O_2.

8. The changes that occur to cells during development of a eukaryote. The changes are phenotypic rather than genotypic for the most part.

9. Replication, transcription, and regulation.

10. The aerobic organism requires O_2, the anaerobic does not. Current theory proposes that life arose when no O_2 was present free in the atmosphere. Hence, it is probable that the first organisms were anaerobic.

Chapter 5

1. The four phases are (1) G_1, in which the chromosomes are extended and probably active in transcription; (2) S, in which replication of the DNA occurs; (3) G_2, in which each of the chromosomes consists of a pair of chromatids; and (4) M, in which mitosis and cell division occur.

2. The DNA double helix unwinds. The DNA polymerase begins synthesis of two new complementary strands. The new double helices are reformed.

3. Mitosis starts with a prophase in which G_2 chromosomes begin to become visible as they coil and shorten. By the end of prophase, each chromosome can be seen to consist of two chromatids. In metaphase, the chromosomes line up

in a plane across the cell, and in anaphase the sister chromatids separate and go to opposite poles. Division of cytoplasm starts. In telophase, the new chromosomes begin to extend, a new nuclear membrane is formed, and the daughter cells separate and enter the G_1 phase.

Meiosis begins with prophase I, in which the homologous chromosomes pair to form tetrads of four chromatids. Crossing over occurs during this phase. The homologous pairs begin to separate with the onset of metaphase I and finally do separate and go to opposite poles in anaphase I. Telophase I may be brief or prolonged. Prophase II, metaphase II, and anaphase II resemble mitosis, except that only a haploid complement of chromosomes is present. The end result of meiosis is four cells, each with a haploid set of chromosomes.

4. Mitosis consists of one cell division to give two daughter cells, each with a complete chromosome complement and roughly the same amount of cytoplasm. Meiosis consists of two cell divisions, but only one round of replication of chromosomes precedes it. Therefore the chromosome number is reduced by one-half.

5. The statement is a half truth. Besides reducing the number of chromosomes, meiosis must also occur in such a way that each meiotic product receives a full haploid complement of chromosomes, or, to put it another way, each must receive a representative of each pair of homologous chromosomes.

6. Homologous chromosomes have the same or allelic genes and pair during prophase I of meiosis.

7. Ninety-two in mitotic prophase and meiotic prophase I.

8. Twenty-three, since there are 23 pairs of homologs.

9. 24.

10. In the G_1 phase, chromosomes but not chromatids exist by definition. In the G_2 phase, $2 \times 23 = 46$ chromatids exist.

11. In oögenesis, thee polar bodies are formed. They are essentially haploid nuclei that do not function in fertilization. Only one functional gamete, the egg or ovum, is formed. In spermatogenesis, four functional sperm are formed from a single spermatogonium.

Chapter 6

1. The term gene has always been difficult to define. It has been, and still is, a concept with many different facets. Perhaps the closest we can come today to a meaningful definition is to say that it is a segment of DNA with an identifiable function in forming the phenotype. The kinds of genes that are recognizable can be grouped according to their function(s): (1) structural, (2) regulatory, (3) rRNA, (4) tRNA. Structural, rRNA, and tRNA genes have readily identifiable specific functions. Regulatory genes generally are not well understood, especially in eukaryotes. The label is often used to cover ignorance.

2. Transcription involves the synthesis of an RNA chain using a DNA chain as a template, whereas translation involves transferring the information represented in the RNA as a triplet code into a sequence of amino acids. In both cases, information is being transferred, but in the case of transcription one

chain in DNA is used to form a complementary chain in RNA. In translation, an entirely new kind of molecule is formed. The "language" of nucleic acids is translated into the "language" of proteins.

3. hnRNA is the primary product of structural gene transcription. It is much longer than the mRNA derived from it. It is formed by transcription of regions of DNA that code for polypeptides as well as surrounding regions that do not. Much of the transcript which is hnRNA is then eliminated to form the mature mRNA.

4. Exons are the segments of a structural gene that code for polypeptide sequences. Introns are transcribed but are removed during processing of the hnRNA to form mRNA.

5. More than one codon encodes for one amino acid. All amino acids except methionine and tryptophan have more than one codon.

6. The anticodon of the tRNA is complementary to the codon in mRNA specific for the amino acid carried by that tRNA.

7. Two α-globins and two β-globins.

8. A partial list would be (1) enzymatic, (2) regulatory (hormones, for example), (3) structural (collagen, for example), and (4) transport (hemoglobin, for example).

9. General proteins are the kinds found in all types of cells in our bodies. They are those that carry out the necessary housekeeping functions required in all cells, such as respiration, protein synthesis, nucleic acid synthesis, etc. Differentiated proteins are the specialists that give each type of differentiated cell its particular characteristics.

10. tRNA carries the amino acid to be inserted into the polypeptide. The ribosome "holds" the mRNA that contains the codon that the tRNA recognizes. It is at the point at which the mRNA is in contact with the ribosome that the tRNA attaches and "deposits" its amino acid in the growing polypeptide chain.

Chapter 7

1. (a) (1) Since the woman's mother was albino and therefore had the genotype *aa*, the woman must be heterozygous *Aa*.

    (a) (2) Half of their children should be albino and half should be normal. This is the statistical expectation, since the cross is *aa* × *Aa*, and the mother will produce two kinds of gametes, *A* and *a*, in equal numbers.

    (b) (1) If the woman had normally pigmented parents, she is most probably *AA* in genotype. This is supported by the fact that all six children are pigmented and therefore must have received *A* alleles from her.

    (b) (2) If one child is albino, the mother must be *Aa* despite the large deviation from the 1:1 ratio expected.

2. We will designate the man's genotype as gal^-/gal^-, h^+Y and the woman's as gal^+/gal^+, h^+/h^-. The woman has to be heterozygous for hemophilia because her father has hemophilia and could only transmit to her his h^- X chromosome. The cross is written then as

$$gal^+/gal^+, h^+/h^- \times gal^-/gal^-, h^+Y$$

The best procedure with a problem of this sort is to list the possible gametes. For the mother they will be gal^+h^- and gal^+h^+ in equal numbers. For the father they will be gal^-h^+ and gal^-Y in equal numbers. The results of gamete combinations will be

mother	father	offspring
$0.5\ gal^+h^+$	$0.5\ gal^-h^+$	$= 0.25\ gal^+/gal^-h^+/h^+$
	$0.5\ gal^-$Y	$= 0.25\ gal^+/gal^-h^+$/Y
$0.5\ gal^+h^-$	$0.5\ gal^-h^+$	$= 0.25\ gal^+/gal^-h^+/h^-$
	$0.5\ gal^-$Y	$= 0.25\ gal^+/gal^-h^-$/Y

From this we see that their daughters ($gal^+/gal^-h^+/h^+$ and $gal^+/gal^-h^+/h^-$) will be phenotypically normal. Half of their sons will have hemophilia (gal^+/gal^-h^-/Y) and the other half will be phenotypically normal (gal^+/gal^-h^+/Y). All daughters will be heterozygous for galactosemia, as will be all sons. Half of the daughters will be heterozygous for both hemophilia and galactosemia.

3. Sons carry their mother's X chromosome and their father's Y. The Y chromosome bears very little genic material compared to the X. In addition, the evidence is strong that sperm do not contribute to the mitochondrial DNA of the zygote. Therefore, we all carry only our mother's mitochondrial DNA.

4. The Y chromosome is transmitted only via the male line. Therefore, a male child must receive his Y chromosome from his father's paternal grandfather and never from his father's maternal grandfather.

5. (a) The probability is ½.
(b) The probability that the grandchild's parent will bear the gene is ½ as in (a). The probability that the child will receive a particular allele of a pair is ½. Since the chance that the parent carries the dominant allele is ½, the chance that the grandchild will get it is ½ × ½ = ¼. Of course, if it becomes known that the grandchild's parent has the disease, the chances increase to ½.

6. This is a recessive sex-linked characteristic. The colorblind man is cY and his wife is cc. All of their X-chromosome-bearing gametes will carry the c allele. Therefore their children should be either cY or cc and colorblind.

7. For this problem it is best to draw a pedigree:

grandparents	$C/-$ $\top$ cY	cY $\top$ $C/-$
parents	Cc ——————	cY
woman in question	Cc or CC $\top$	cY
	?	

The woman has a 50% chance of carrying the c allele. If she does carry it, the mating is $Cc \times c$Y. Each son will have a 50% chance of being colorblind, and each daughter will also have a 50% chance of being colorblind. If she does not carry the c allele, the mating is $CC \times c$Y, and neither the sons nor the daughter will be colorblind.

8. The man is heterozygous for Tay-Sachs disease and is presumably homozygous for the normal allele for galactosemia. We designate his genotype as t^+/t^-,

g^+/g^+ and his wife's genotype as $t^+/t^+, g^+/g^-$. The gametes will be

sperm

eggs		$t^+ \; g^+$	$t^- \; g^+$
	$t^+ \; g^+$	$t^+t^+ \; g^+g^+$	$t^+t^- \; g^+g^+$
	$t^+ \; g^-$	$t^+t^+ \; g^+g^-$	$t^+t^- \; g^+g^-$

From this diagram, it is apparent that (a) all children should be normal; (b) ¼ of the children should be homozygous normal at both loci; (c) ¼ of the children should be heterozygous for both diseases.

9. Crossing over is a physical exchange between chromatids of homologs. It may or may not lead to recognizable recombination. A recombination is a reassortment of genetic material that may be the result either of crossing over or independent assortment of chromosomes.

10. This is an example of maternal inheritance, probably due to cytoplasmic factors. The trait might be caused by a variant form of mitochondrial DNA, for example.

11. It means that they are different forms of the same gene. The term gene is used here in a generic sense. Its alleles are the specifics.

12. When the phenotype of a heterozygote is different from either possible homozygote, the phenotypic and genotypic ratios will be the same. If one allele is completely dominant to the other, the ratios will be different.

13. Epistasis is dominance of the expression of one gene over the expression of another nonallelic gene. It is not related to the dominance of one allele over another. Neither has anything to do with crossing over, which is the physical exchange of homologous chromatid segments in meiotic prophase.

Chapter 8

1. By crossing over between the loci in question on nonsister chromatids.

2. Ten percent of the time.

3. In many instances, there is no real difference. Strictly speaking, a syntenic group includes all genes located on the same chromosome. A linkage group refers to loci that are thought to be on the same chromosome because of reduced recombination between them. If two loci are sufficiently far apart, they may recombine freely but still be syntenic.

4. Because the farther apart genes are, the more likely they will have single of even double crossovers between them. This will result in 50% recombination or random recombination characteristic of genes on different chromosomes.

5. The woman in question will have the genotype $Gd^- \; c/Gd^+ \; C$. If no crossovers occur between the two genes, the chances of a son having the X with both Gd^- and c will be 50%. However, since the map distance between the two is approximately 5 centimorgans (5% recombination), we expect 5% of the woman's gametes to be either $Gd^- \; C$ or $Gd^+ \; c$. Thus only 47.5% (½ × 95%) of her gametes will be $Gd^- \; c$. Her son's chance of being both G6PD deficient and

colorblind is 47.5%. Since her daughters also receive an X chromosome from their father, they will have neither trait unless the father is affected.

6. This kind of hybrid generally loses human chromosomes rather than mouse chromosomes. The human chromosomes are distinguishable from the mouse after staining. Thus, when one is lost, the observer can identify what is lost. By comparing the presence or absence of human gene products with the presence of specific human chromosomes, one can often show that a particular human gene product is made only when a specific human chromosome is present. From this one concludes that the gene for that product is on that chromosome.

7. On the X chromosome, because the X chromosomes of mammals generally bear the same kinds of genes.

8. Linkage of two traits identifies their genes as being on the same chromosome. The association of two traits is more often due to different expressions of the same gene, as in the case of blond hair and light skin.

9. Genes that are very close together on the same chromosome do not recombine very often. Therefore, one can predict the alleles present at one locus if one knows the alleles at the other locus and knows how the alleles are coupled in the particular family. In the case of a hereditary disease, the person transmitting the allele for the disease must be heterozygous at both loci. If the coupled allele is not transmitted, the disease allele could only be transmitted if crossing over occurred between them. This approach is used only rarely now because too few loci have been mapped on human chromosomes.

10. 23, since there are 23 pairs of chromosomes.

Chapter 9

1. A mutation is any inherited change in DNA.

2. Gene mutations can't be detected by examining chromosomes under the microscope. Chromosomal mutations can be detected by this technique.

3. A missense mutation may alter a codon by base substitution to code for a different amino acid, or it may form a stop or termination codon, either UAA, UGA, or UAG. In the latter case, translation will stop because tRNAs do not recognize these codons.

4. A frameshift mutation is a deletion or addition of a base or group of bases while a missense mutation is a substitution of one base for another.

5. A gene locus with two or more common alleles.

6. In general, a metabolic block is the result of a mutation in a structural gene that alters its protein product to the extent that it can no longer act as an enzyme to catalyze a specific metabolic reaction. Other causes may be a null type mutation in which no protein at all is produced. This may be the result of a deletion of the structural gene or a mutation of a regulator controlling its function.

7. A lethal mutation is a mutation of a gene that causes death at an early stage, usually in the embryonic or fetal stage in humans. There is no difference at the DNA level from other kinds of gene or chromosomal mutations.

8. By gene mutation, followed by an increase in frequency of the mutant allele.

9. An allele is a form of a gene that differs from other forms of the gene at that locus.

10. Sickle cell hemoglobin is a result of a missense mutation in the β-chain of Hb A. (a) At the molecular level, the β-globin chain has a substitution of the amino acid valine for glutamic acid at position 6. This causes the deoxygenated Hb S to be less soluble. (b) Sickled red cells block blood capillaries, causing oxygen deficiency in tissues in which the block occurs. The sickled red cells are also easily destroyed, leading to anemia.

11. Since he has most of a chromosome 21 fused to chromosome 15, he has a good chance of passing the translocated 21,15 chromosome along with his untranslocated 21 to a sperm that, when it fertilizes an egg from his wife containing a chromosome 21, will produce a zygote with three chromosomes 21 or trisomy 21. He also has the potential of producing sperm with only one chromosome 21. (See Fig. 9.11).

12. A genetic block in metabolism due to a defective or missing enzyme.

13. Euploidy is the condition in which the cell contains a complete set of the proper chromosomes. This means it contains a complete maternal and complete paternal set. Aneuploidy is the condition in which one or more of the chromosomes is either missing in whole or part or is present in excess.

14. Heterozygosity for an inversion can lead to the production of gametes with an aberrant genetic constitution (aneuploidy) because of crossovers within the inversion (see Fig. 9.8). If, however, the inversion is homozygous, crossovers do not generate genic imbalance.

15. Damage to DNA in somatic cells can lead to serious difficulties, especially if the cells reproduce after the damage. If the cells do not reproduce, the damage may be less but could still be phenotypically deleterious. For example, many mutations in our nerve cells could conceivably cause death of the cell, even though our nerve cells do not reproduce.

Chapter 10

1. Germ cells are those set aside during development to differentiate into oögonia or spermatogonia. Somatic cells are those in our body that are not germ cells.

2. Cells that are derived from an original single cell by mitotic divisions and thus presumably have an identical genotype. All our body cells are a clone derived from the fertilized egg.

3. Yes.

4. A mosaic. If an aneuploid cell arises by loss or gain of a chromosome, this could be detected by examining the karyotypes of various tissues. Culture of cells from the body to establish clones may make it possible to detect gene mutations that cause genetic blocks in metabolism. Genetic mosaicism for skin pigmentation might also be easy to recognize, although there are also nongenetic explanations for the "patchiness" of skin pigmentation.

5. The cell changes its growth characteristics and becomes "immortalized" in the sense that it gives rise to a clone that grows indefinitely in culture so long as someone is around to care for it.

6. The gene for APRT is most probably on human chromosome 16.

7. The two different types of XP cells have different mutations affecting repair. Thus they make up for each others' deficiencies. This is called complementation.

Chapter 11

1. (1) Height in humans is continuously distributed. (2) Although in general children resemble their parents in height, the trait does not segregate into simple alternatives.
2. The fact that height in general is not inherited in a discontinuous fashion does not mean that genes segregating in the population cannot mutate to alleles that severely restrict growth. These alleles can be either dominant or recessive, and they may produce a phenotype that is discontinuous with the distribution of heights in normal persons.
3. Heritability is the portion of the variation in the population attributable to genetic variation. Since everyone is blond, there is no variation and, hence, there can be no genetic variation that affects the phenotype. Study of another population might show blond hair to be highly inherited.
4. Fraternal twins are derived from different zygotes. Identical twins are clones from the same zygote.
5. Of all twin pairs, ⅓ are MZ, of whom ½ are both girls; ⅔ of all pairs are DZ, of whom ¼ are both girls. For any pair of twin girls, the probability that they are DZ = (prob DZ)/(prob DZ + prob MZ) = $(⅔)(¼)/[(⅔)(¼) + (⅓)(½)]$ = ½.
6. Monozygotic twins are genetically identical. Therefore, any differences between members of a MZ twin pair are due to environment or to random events. The definition of environment in this instance is very broad and includes intrauterine as well as postnatal influences.
7. When the egg underwent its first division, it was a normal division, and each daughter cell gave rise to a fetus. However, early in the following cell divisions a nondisjunction occurred in one of the embryos, resulting in trisomy 21. These trisomic cells became the dominant population of cells in that individual who was then born with Down syndrome. One could test this hypothesis by demonstrating that the Down twin is actually a mosaic for normal diploid cells and for trisomy 21 cells.
8. Monozygotic twins reared together have much the same environment, but those reared apart have somewhat different environments. Therefore, such differences that show up later in their lives can logically be considered to be brought about by environmental differences.

Chapter 12

1. It is a biological state, condition, or phenomenon (choose one) designed to assure genetic recombination.
2. Sexuality is the state of having a sex of one type or another, male versus female, for example. Sex is the process by which sexuality occurs.
3. Sexual heteromorphism is a state in which the different sexes within a species

are morphologically different, as in male and female humans. A mating type is a physiological state that assures that one type of cell fuses with another of a different type to produce a zygote.

4. The mechanism by which it is determined whether an individual is to be either male or female.

5. In humans, the presence of a Y chromosome determines that a zygote will develop into a male. In *Drosophila* the determining factor is the ratio of X chromosomes to autosomes.

6. He produces two kinds of sperm with respect to the sex chromosomes: the Y-bearing and the X-bearing sperm.

7. A male-determining factor whose presence depends on the short arm of the Y chromosome. The H-Y factor possibly is the male-determining factor.

8. They have a Y chromosome even though they have two X chromosomes and, as a result, are sex-chromatin positive.

9. Persons with Turner syndrome have only one X chromosome (XO) and do not have sex chromatin (Barr bodies). All X's are functional.

10. Nondisjunction of X and Y chromosomes during meiosis when eggs and sperm are formed or during mitosis early in development.

11. A true hermaphrodite has both male and female gonads. They are believed to arise by nondisjunction early in development of an individual who started out as a male (XY). Nondisjunction of a Y would start a line of XO cells in addition to the normal XY.

12. Testicular feminization is the result of the mutation of a gene on the X chromosome that normally produces the androgen receptors in cells. Failure to produce this receptor means that the androgen testosterone does not bind, and the Wolffian ducts do not differentiate into male structures. The XY individuals develop primarily female characteristics and are pseudohermaphrodites.

   The Mullerian duct syndrome is the result of a mutation of another gene whose product normally inhibits development of certain internal female organs, such as the uterus and Fallopian tubes. Affected males have normal male genitalia but, in addition, have these female organs internally.

13. X (XO) is female; XXX is female; XYY is male; XXY is male; XXXX is female.

14. She is a functional mosaic because in her cells only one X chromosome is fully functional. The other X chromosome is largely inactive. Which one functions is a matter of chance; in some cells one of the two is functional, in other cells the other X is functional. Since she will always be heterozygous for some genes on the X chromosome, she will be mosaic for those loci.

Chapter 13

1. Structural genes determine the primary structure of polypeptides and, hence, their function is qualitative. Regulatory genes, on the other hand, have a quantitative and temporal role. They determine how much of a polypeptide is to be made and when.

2. The differences among eukaryotes are basically rather superficial. If we consider animals, for example, the differences at the biochemical and

metabolic levels are not substantial and in many cases they are trivial. Our proteins are mostly of the same kind, whether we be human, ape, monkey, cow, or mouse. What makes us different is the degree to which our regulatory genes differ. Thus, during our development, they control what we will finally be by controlling the synthesis of our various kinds of proteins and, therefore, the metabolism of our bodies.

3. The sperm meets the egg in the oviduct or Fallopian tube.

4. The implantation of the blastocyst in the wall of the uterus.

5. They cover the embryo and fetus with a liquid layer in the amnion, and they maintain the contact between the fetus or embryo and the mother through her uterine wall.

6. It is of the utmost importance, since the embryo and fetus cannot develop and grow without nutrients, and the mother provides all of these. Hence, she must maintain a balanced diet containing all the nutrients in the form of amino acids (proteins), carbohydrates, vitamins, etc., required by both her and the developing creature within her.

7. When the blood first begins to form in the embryo, it has a form of hemoglobin not found in the adult human. As the embryo and fetus continue to develop, several other forms of hemoglobins appear and disappear. This illustrates the principle that, during development, structural gene actions are regulated.

8. A congenital defect is one that is present at birth. The study of congenital defects is called teratology.

9. Most congenital defects are the result of an alteration of development from the normal pathways.

10. Thalassemias are the result of a mutation that changes the quantity of a hemoglobin subunit, such as the α- or β-globin chains, so that only a very small amount is produced or perhaps none at all. The sickle cell mutation involves a change in the structure of the β-globin chain, but the amount produced is not the main problem, although the amount is less than normal. Rather, the solubility of Hb S is greatly reduced in the absence of oxygen.

11. The great majority of embryos or fetuses that have microscopically visible abnormal chromosome complements are aborted. Some 50% of early abortions are chromosomally abnormal.

12. In experimental animals, and presumably in humans, the two are quite closely intertwined. Some congenital defects will develop only under certain environmental circumstances, such as the mother being given a certain drug and the embryo having a certain genotype. Also the mother's genotype may be quite important.

Chapter 14

1. Asexual reproduction can result in the rapid reproduction of many individuals of the same genotype. This can be a great advantage to the organism if this genotype is well adapted to the ambient environment. However, it can be disastrous if the environment suddenly changes to a state where that genotype becomes unsuitable or unadapted. Sexual reproduction has the advantage that it produces many different genotypes. When the environment changes, it is likely that one or more of them will be adaptable and the species will survive. Its disadvantage is that a considerable number of individuals will be formed that are not adapted to the current environment. In that sense, it is a wasteful process compared to asexual reproduction.

2. Conjugation and transduction.

3. Phage are bacterial viruses. Viruses are particles containing DNA or RNA with a proteinaceous coat. They can only reproduce inside a living cell. Plasmids are rings of naked DNA that can reproduce inside living cells. They differ from viruses insofar as they do not have protein coats and cannot therefore exist as infectious agents outside the cell. Episomes are plasmids or viruses that possess DNA capable of integrating into the host's DNA and reproducing with it.

4. On being transferred from one bacterium to another, R factors can carry genes for resistance to an antibiotic. Hence, they can spread such resistance throughout that bacterial population.

5. DNA transformation involves the entrance of a piece of naked DNA into a cell, where it becomes integrated into the host's genome. Transduction is essentially the same process but the DNA is carried by a virus and enters the host cell via viral infection.

6. They reproduce by first forming a DNA chain complementary to their RNA via a reverse polymerase. The DNA chain forms another DNA chain complementary to itself, and the double helical strand then reproduces as DNA in the usual fashion. More viral RNA can form from the DNA by transcription.

7. By the DNA forming from them as described in 6 and becoming integrated into our genomes and replicating with our chromosomes.

8. It is a virus that has a number of forms that are associated in divers ways with a number of infectious diseases and at least one type of cancer in humans.

9. They "cut" DNA strands by hydrolyzing phosphodiester bonds, but they do so only at specific sites. They recognize these sites by the sequences of bases adjoining them. They are useful in genetic engineering because they allow the breaking of a circle of DNA at a specific site. This allows the insertion of a piece of DNA with the desired sequence into a circle that can then be replicated.

10. They are the circular forms of DNA that can replicate within a bacterium. As they replicate, they can also replicate foreign DNA from whatever source that has been inserted within them as described in 9. Under proper circumstances, they will also allow for the transcription of the foreign DNA, and the RNA so formed may act as a messenger that the bacterial host cell may translate into a protein.

11. The problem may be the result of the presence of R factors that confer resistance to antibiotics. Another possibility is a mutation in the genome of the bacterium that confers resistance to antibiotics.

Chapter 15

1. The cells in the skin graft from person A elicit an immune response in the recipient B. B forms T cells that specifically destroy A cells, and the graft is rejected.

2. The immune response to a foreign protein is the formation of antibodies specific for the protein by B lymphocytes.

3. Vaccination is the injection of an attenuated or killed virus or bacterium that causes an immune response specific to that virus or bacterium. The presence of antibodies then gives protection against infection by the virus or bacterium.

4. The child possesses many of the antibodies formed by the mother to disease organisms to which she has been subjected. The child loses the maternal antibodies over a period of time, thus losing its resistance to the diseases against which the antibodies provided protection.

5. Immunoglobulins are antibodies, and their function is to combine with specific antigens. The function is highly specific. Any change in the antigen structure will generally diminish the reaction with antibody. The specificity of the antibody resides in the primary amino acid sequence of the "variable regions."

6. The immunoglobulin genes have segments that are cut out and discarded. The remaining ends are rejoined and then transcribed. Many variants may be formed in the somatic cell lines in which this occurs. This type of rejoining of variable to constant regions does not occur in structural genes in general.

7. Joyce and William; Grace and John; Hope and Greg.

8. This is the designation for the human major histocompatibility complex (MHC). This complex of genes produces a large number of different antigens in the human population that account for the differences among us that are important for transplantation.

9. The ABO group produces the A and B antigens. When the antigens are not formed, as in blood group O, the antibodies to them *are* formed. Type A persons form B antibodies, type B form A antibodies, and type AB form neither of the antibodies. This occurs even though the persons have never been transfused with cells that have these "foreign" antigens. Antibodies to M and N antigens are almost never formed, even when a blood donor and recipient are not matched for MN type.

10. When a mother who is Rh− bears a child who is Rh+, she will produce antibodies in response to the child's Rh+ antigens. The child's antigens react with the mother's antibodies, and hemolytic disease is initiated. Generally this does not occur with the first Rh+ child. It takes two or more for the mother to build up sufficient antibody levels.

11. The Rh− mother of an Rh+ child can be injected with antibodies to Rh+ cells shortly after birth of the child. This helps remove fetal cells with Rh+ antigen that may have gotten into her blood stream. She would then be less likely to make antibodies against the Rh+ cells, and any subsequent Rh+ child would be less likely to have hemolytic disease of the newborn.

Chapter 16

1. Dominant mutations require the presence of only one mutant allele in order to be expressed. The mutation can thus be recognized in the first generation in which the mutant gene appears.

2. $41/(2 \times 4,664,799) = 4.4 \times 10^{-6}$ or 4.4 mutations per million gametes.

3. Fairly well. Mutation rates range from 10^{-5} to 10^{-6} per generation for most organisms.
4. 0.4 to 0.04.
5. Roentgens measure the extent of ionization in air produced by a radiation source. Rads measure the amount of radiation energy absorbed by a given mass of medium. Rems are based on the amount of radiation absorbed and the biological effects of that radiation in man. For X rays, the three units are very similar, but for other kinds of radiation, they may differ.
6. The nucleus of a stable isotope does not disintegrate at a discernible rate. A radioactive element has unstable nuclei that constantly break apart and emit subatomic particles and ionizing radiations.
7. ^{40}K.
8. A dose of 100 rem is the amount necessary to double the mutation rate over the spontaneous rate.
9. All life on earth has been exposed to ionizing radiation since the beginning. Consequently, during its evolution, it has developed repair systems against the types of damage caused by ionizing radiation. Chemicals made by humans, however, may cause other types of damage for which there are no repair systems.
10. They have repair systems of different degrees of efficiency.

Chapter 17

1. For class discussion.
2. Those who produce the most offspring are the fittest. The most fecund shall inherit the earth.
3. Characteristics acquired by an individual during his or her life can be passed on to the children.
4. The population would have to be large enough to prevent chance drift in gene frequencies.
5. The frequency of the *pk* allele is the square root of 1/15,000 = 0.008. The frequency of the *pk*$^+$ allele will therefore be $1 - 0.008 = 0.992$. The frequency of heterozygotes will be $2 \times 0.008 \times 0.992 = 0.016$. There are 0.016 $\times$ 220,000,000 = 3,520,000 persons in the U.S. population who are heterozygous *pk*$^+$/*pk*.
6. The gene is on the X chromosome and is recessive. Therefore, a colorblind woman must carry two genes for colorblindness whereas a colorblind man need carry only one.
7. New ones are constantly arising in the population by mutation.
8. (a) 9,626/11,565 = 0.9123 for *A*. *B* = $1 - 0.9123 = 0.0877$. (b) Type B persons = $(0.088)^2(11,565) = 89$. Type AB = $2(0.912)(0.088)(11,565) = 1851$.
9. The population will arrive at an equilibrium called a balanced polymorphism. In this population, the deleterious allele will be preserved at a relatively high level because those carrying it in heterozygous combination have a selective advantage, i.e., they are more fit. This would be true even though the fitness of one of the homozygotes is zero.
10. For class discussion.

11. It should not be important, since most cancers take their toll after a person has passed the child-bearing period.

Chapter 18

1. The enzyme phenylalanine hydroxylase is defective in persons with phenylketonuria. The result is that phenylalanine is not converted to tyrosine and phenylalanine accumulates in the blood. This causes brain damage and mental retardation. The knowledge that phenylalanine accumulates led to a successful treatment based on reduction of the phenylalanine in the diet.
2. Galactosemia can be effectively treated by eliminating galactose (from milk) from the diet.
3. Males more often because they are hemizygous for genes on the X chromosome, and X-linked recessive alleles are fully expressed. This is in agreement with the experience that males in institutions tend to express more sharply defined inherited defects.
4. By the appearance of the X chromosome in cultured lymphocytes.
5. The Lesch-Nyhan syndrome.
6. In studies comparing adopted children with their biological and social parents, the children resemble the biological parents more.
7. Schizophrenia is sometimes difficult to diagnose clearly, and it is probably due to a number of different causes.
8. Most of the time trisomy 21 is the result of nondisjunction in the oöcyte that gives rise to the affected individual. Hence, only one person is affected. It is not transmitted as a Mendelian trait.
9. The mother's age could not have been a factor since the nondisjunction must have occurred in the father. Most XYY males develop normally; hence, no special advice or warning to the parents is indicated. They should be informed that many scientists believe that there is a small increased risk of a behavioral problem in XYY males.
10. Sociobiology is the study of the biological basis of societal behavior. The remainder of the question is for class discussion.

Chapter 19

1. The binomial system of nomenclature is used to provide all organisms with their official scientific name that is recognized worldwide. The name is latinized and consists of two parts: the genus or generic name, which is capitalized, and the species or specific name, which is not capitalized unless it applies to a plant named after a person. Both names are generally italicized (underlined) by convention under the rules of the International System of Nomenclature.
2. A taxon is a taxonomic category such as a species, genus, family, class, etc.
3. Typological thinking is the view that variation within a species represents deviation from an ideal type for that species. Population thinking treats all species as populations, with each member of the population contributing equally to the definition of the species.

4. A species is a population of individuals that can interbreed with one another. Like most generalizations in biology, this definition has exceptions.

5. Mutation of genes that alter fertility, food preferences, or attraction of one sex toward another, or cause sterility in interspecific crosses or in hybrids if they are produced.

6. A coadapted group of genes is one that is well balanced with respect to members of the group and gives high fitness to the individuals that carry the group.

7. A eutherian is a mammal with a true placenta. The advantage is that a fetus can develop in a protected place for a long period to reach an advanced stage before leaving the uterus. One disadvantage is that there is not much room in a single uterus, and eutheria therefore cannot bear many young at one time.

8. Aside from the obvious morphological similarities and behavioral characteristics that we seem to have in common with the apes, there are fundamental similarities in our chromosome structures and most importantly in the structure of our proteins and therefore our structural genes. We are virtually identical to the gorilla and chimpanzee at the molecular level.

9. By the fusion of two ape acrocentrics (2p and 2q in the ape) to form the human submetacentric chromosome 2.

10. The apes and, among the apes, the gorilla and chimpanzee.

11. *Ramapithecus.*

12. *Homo erectus* differed from earlier hominids by having an upright posture, a larger brain, and the ability to fashion tools and use fire. But they were smaller than *Homo sapiens*, with smaller brains and limited culture.

13. The position of the foramen magnum and the size of surfaces where neck muscles are attached.

14. For class discussion.

15. *H. sapiens* can occupy many different environmental niches and is adaptable to new environments.

16. For class discussion.

Chapter 20

1. Subspecies and race are virtually synonymous in their meaning. Both are subcategories of species. They are generally morphologically distinguishable groups that are completely fertile with members of other subspecies and races within that same species. In general they differ mainly in the frequencies of a number of alleles.

2. To say that racial differences are nonadaptive means that the differences arose by chance, not because they conferred greater fitness to the various groups in their particular environments.

3. This terminology is applied to alleles that have the same fitness. Hence, there is no selective advantage of one over the other.

4. Genes that shorten the span of fertility by several years would have little impact in a society in which few if any lived to the end of the child-bearing years. But with increased life span, such genes confer a disadvantage.

5. One of the probable reasons may be that persons heterozygous for this disease may be fitter than those who are not. Hence, the deleterious cystic fibrosis allele, like the sickle cell allele, is maintained at a high frequency as a balanced polymorphism.

6. The volume of an object is calculated as a cube, x^3, and hence increases with the cube of the dimensions. On the other hand, surface area is calculated as the square, x^2. Therefore, as an animal increases in size, the ratio of its surface area to its volume decreases: $x^2/x^3 = 1/x$. If the animal is warm blooded and needs to maintain a body temperature above the ambient temperature of its surroundings, it will be more able to do so if its surface through which it loses heat is small compared to its volume which generates heat.

7. A polymorphism is the coexistence of two or more alleles in a population. Since races are different populations, the concept of polymorphism does not apply to genetic differences among races.

8. Malaria, especially falciparum malaria, appears to have imposed a heavy selective burden on human populations for many centuries. Genes that are thought to confer some degree of resistance to malaria include the polymorphic hemoglobin variants, thalassemia, glucose-6-phosphate dehydrogenase deficiency, and the Duffy-negative blood group.

9. Lactase breaks down lactose, the principal sugar in milk. Virtually all infants have lactase, but many persons lose the enzyme as they get older. Persistence of lactase into adulthood is due to a dominant gene L. Since the frequency of L is 0.83, the frequency of l must be $1 - 0.83 = 0.17$. Only the homozygous ll individuals will get sick from drinking milk. If the Swedish population is in equilibrium for these allele frequencies, then $(0.17)^2 = 0.03$ of the population should get sick. The population of Sweden is approximately 8,000,000. Therefore, $0.03 \times 8,000,000 = 231,200$ Swedes should avoid milk.

10. Since l is recessive, 23% of the U.S. white population is ll. To calculate the frequency of l, we take the square root of $0.23 = 0.48$.

11. Possibly the B allele originated in Asia and has spread to other populations. If this were true, the B allele must have some selective advantage not yet identified.

Chapter 21

1. Select those individuals with desirable characteristics and use them for breeding. Cull out the undesirable and do not allow them to breed.

2. This biblical version of genetic engineering is more Lamarckian than Mendelian, but it is not pure Lamarckian either. It doesn't seem to work anymore.

3. The pool of genes in the original wolf population, supplemented by a few mutations, has produced all the varieties of present dogs, ranging from tiny hairless Chihuahuas to giant Irish Elk Hounds. The genetic principle that natural populations are generally highly heterozygous is proved along with the related fact that selective breeding can produce most remarkable results.

4 *Triticum aestivum* is a hexaploid that probably arose from a cross of a tetraploid with a wild diploid wheat.

5. An amphidiploid acts like a diploid during meiosis. During the meiotic prophase I, homologs pair to form bivalents. Each bivalent is a tetrad of four

chromatids. In this sense, an amphidiploid is like an ordinary diploid. But they are actually tetraploid hybrids formed by crossing two different species whose chromosome do not pair in meiosis. The hybrids are sterile until their chromosome numbers are doubled to form the amphidiploid, also called allotetraploids.

6. The haploid number of the cabbage and radish is 9. Therefore, the hybrid will have 18 chromosomes. When doubled it will have 36 and be an amphidiploid.

7. These constitute an irreplaceable gene pool that is useful for "synthesizing" new strains of crop plants.

8. Corn was derived from a single ancestral species by a process of intensive selection over a period of many years. The present cultivated wheat was derived by crossing a number of different species of wild wheat to produce the present allohexaploid (a hexaploid that acts like a diploid).

9. By crossing homozygous strains to produce a highly heterozygous hybrid that is superior to the parental strains.

10. Heterosis, also known as hybrid vigor, is the result of heterozygosity. The heterozygotes are more fit and have more desirable characteristics than the homozygotes.

11. Inbreeding increases homozygosity, which may result in low fertility.

12. A disease may arise that wipes out that particular variety.

13. Plant and animal species that won't cross can nonetheless have their DNA introduced into one another by genetic engineering.

14. Superovulation is the induction of ovulation of more than one or two eggs per ovary. The embryos that develop after superovulation are numerous. They can be transplanted to the uteri of a number of surrogate mothers, giving each as many as she is capable of nurturing. In this way, embryos from superior parents can be raised to term, producing many more superior offspring than would be possible otherwise.

15. For class discussion.

Chapter 22

1. Since both are heterozygous, ¼ of the zygotes they produce should statistically be homozygous for the lethal and the embryos from them aborted. Their six normal children should therefore represent ¾ of the total zygotes. Two embryos may have been aborted. In the small families characteristic of humans, chance may cause substantial deviation from these expectations.

2. The cost to society of the detrimental alleles compared to normal alleles.

3. For class discussion.

4. It provides prospective parents with information on risks of genetic defect.

5. The woman has a probability of $2 \times 0.01 \times 0.99 = 0.02$ of carrying the albino gene. All her husband's sperm will carry the albino allele. Hence, the chances that their child will be albino is $(1/2)(0.02)(1.0) = 0.01$

6. Since the frequency of babies born with galactosemia is 1/65,000, the gene frequency is the square root of the population frequency = 0.004. The family background of the wife shows a case of galactosemia, but the relationship of the galactosemic to the wife is not specified. This makes it difficult to be more specific about her carrying the gene for galactosemia than to say that the chances are better than 0.004. The sperm from her husband has only a probability of 0.004 of carrying the galactosemia gene. Hence, the risk to a child is very low.

7. When the fetus is not known to be at increased risk for a disease. High risk pregnancies are those in which the parental genotypes make it possible to have an affected child. Advanced maternal age is also an indication for Down syndrome and other disorders due to meiotic nondisjunction.

8. Prenatal screening refers to screening before birth to detect an affected fetus. Genetic screening ordinarily refers to screening of infants or adults to test for genetic disease or the possibility of transmitting genetic disease.

9. For class discussion.

10. For class discussion.

11. The removal of a sample of amniotic fluid from the amniotic cavity during the prenatal period.

Chapter 23

1. Because the emphasis on selection impressed upon many socially minded individuals the possibility of bettering the human race by culling out the unfit.

2. The principal one was infanticide, which was probably more common than most of us realize. It wasn't something that one talked about. At present, sterilization is used, but in the U.S. it is used only with court permission. An effective eugenic practice, usually done for other reasons, is the institutionalization of persons so that they have less opportunity to reproduce.

3. For class discussion.

4. Darwinian socialism or social Darwinism is simply the practice of eugenics with the aim of reducing the incidence of the "unfit" in the population.

5. Negative eugenics is limitation on reproduction of persons with undesirable traits; positive eugenics is the encouragement of reproduction of persons with desirable traits. In both instances, the assumption is that the traits are in part inherited.

6. The concept that someone is responsible for the birth of a defective child.

7. No.

8. A child is born with a "clean slate" with no inborn behavior patterns. What it becomes or develops in the way of abilities, talent, etc., are all the result of experiences after birth. It is one way of saying that nurture is important, nature is not.

9. Lamarckianism.

10. For class discussion.

Index